The Minerals, Metals & Materials Series

Jonghyun Lee · Samuel Wagstaff ·
Guillaume Lambotte · Antoine Allanore ·
Fiseha Tesfaye
Editors

Materials Processing Fundamentals 2020

TMS

Springer

Editors
Jonghyun Lee
Iowa State University
Ames, IA, USA

Samuel Wagstaff
Novelis
Kennesaw, GA, USA

Guillaume Lambotte
Boston Metal
Woburn, MA, USA

Antoine Allanore
Massachusetts Institute of Technology
Cambridge, MA, USA

Fiseha Tesfaye
Åbo Akademi University
Turku, Finland

ISSN 2367-1181 ISSN 2367-1696 (electronic)
The Minerals, Metals & Materials Series
ISBN 978-3-030-36555-4 ISBN 978-3-030-36556-1 (eBook)
https://doi.org/10.1007/978-3-030-36556-1

Cover illustration: Left and top right: From Chapter “Numerical Simulation on the Multiphase Flow During the KR Process Using the Eulerian–Eulerian Modeling”, Yanyu Zhao et al., Figure 1: Calculation domain geometry (a) and mesh distribution (b) and Figure 5: Free surface distribution under different conditions, (a) without gas injection, (b) with gas injection, (c) with baffle, and (d) with slope bottom. https://doi.org/10.1007/978-3-030-36556-1_15. Bottom right and bottom left: From Chapter “Control Center Segregation in Continuously Cast GCr15 Bloom by Optimization of Solidification Structure”, Hanghang An et al., Figure 6: Simulated results of (a) fraction solid distributions and (b) morphologies of bloom samples with steel nails and Figure 16: Simulated solidification structure of central equiaxed grain zone: (a) without M-EMS; (b) with M-EMS. https://doi.org/10.1007/978-3-030-36556-1_18.

This Springer imprint is published by the registered company Springer Nature Switzerland AG
The registered company address is: Gewerbestrasse 11, 6330 Cham, Switzerland

Contents

About the Editors

Jonghyun Lee is Assistant Professor in the Department of Mechanical Engineering at Iowa State University. He has been conducting multiple industry and government-funded projects in the field of materials processing as PI and Co-PI. He is the recipient of the Young Leaders Professional Development Award in 2013 from The Minerals, Metals & Materials Society where he has been serving as a Co-organizer and Co-editor of the Materials Processing Fundamentals Symposium since 2014 and as Vice-Chair of the Process Modeling and Technology Committee since 2017. Prior to joining his current institution, he was a Research Assistant Professor at the University of Massachusetts, Amherst. He also had nearly 5 years of industry experience and worked as a Post-doctoral Associate for Tufts University, Medford, Massachusetts. He earned his M.S. and Ph.D. in Mechanical Engineering from the University of Massachusetts Amherst and his B.S. in the same discipline from Inha University in Incheon, South Korea.

Samuel Wagstaff began working in the aluminum industry at age 14 with Novelis in Spokane, Washington and now works for the same company in Kennesaw, Georgia as a Process Scientist. In 2013, he received his Bachelor of Science from Cornell University in Mechanical and Aerospace Engineering. He continued his education at the Massachusetts Institute of Technology in the Department of Materials Science and Engineering. His Ph.D. on the minimization of macrosegregation through jet erosion of a continuously cast ingot uses a turbulent jet to reduce the uneven distribution in aluminum alloy ingots by over 70%. He finished his masters and doctorate at MIT in September 2016 after just 3 years. He has published more than a dozen articles on DC casting and macrosegregation and holds 12 patents.

Guillaume Lambotte is R&D Director at Boston Metal, a Massachusetts Institute of Technology (MIT) spin-off startup focusing on the development of an environmentally friendly and energetically efficient primary metal extraction process. Dr. Lambotte primarily focuses on computational thermodynamic modeling, electrochemistry, and high-temperature equilibrium. Prior to joining Boston Metal, he conducted research as a post-doctoral associate at the University of Massachusetts Amherst and MIT. Before his graduate studies, Dr. Lambotte worked as a production assistant manager at Alcan Extruded Products (Crailsheim, Germany).

Dr. Lambotte obtained his bachelor's degree from the European Engineer School for Materials Science (Nancy, France). He received an M.Sc. and a Ph.D. in metallurgical engineering from Ecole Polytechnique of Montreal (Montreal, Canada).

Dr. Lambotte was the recipient of the 2015 TMS EPD Young Leaders Professional Development Award. The same year he was one of the TMS representatives at the Emerging Leaders Alliance Conference.

Antoine Allanore is Associate Professor of Metallurgy in the Department of Materials Science & Engineering at MIT. He received his higher education in Nancy (France) where he earned a chemical process engineer diploma from Ecole Nationale Supérieure des Industries Chimiques and an M.Sc. and Ph.D. from Lorraine University. Dr. Allanore joined MIT in 2012 as a faculty member, leading a research group that develops sustainable materials extraction and manufacturing processes. He has developed numerous alternative approaches for metals and minerals extraction and processing. With an emphasis on electrochemical methods for both analytical and processing purposes, his group combines experimental and modeling approaches to promptly investigate the ultimate state of condensed matter, the molten state. He teaches thermodynamics and sustainable chemical metallurgy at both the undergraduate and graduate levels. He received the Vittorio de Nora Award from TMS in 2012 and the TMS Early Career Faculty Fellow Award in 2015.

Fiseha Tesfaye is Senior Researcher and Adjunct Professor in metallurgical thermodynamics at the Johan Gadolin Process Chemistry Centre (PCC) of Åbo Akademi University, Finland. He received his master's degree in materials processing technology in 2009 from Helsinki University of Technology and his Ph.D. degree in metallurgy in 2014 from Aalto University, Finland. During his Ph.D. period, he focused his research on the electrochemical investigation of the thermodynamic properties of sulfide and intermetallic materials.

After a post-doctoral position in the Laboratory of Inorganic Chemistry at Åbo Akademi University from 2015 to 2017, which was focused on the sulfosalts and sulfates characterizations, Dr. Tesfaye attracted a large research project related to the thermodynamic investigation of complex inorganic material systems in renewable energy and metal production processes. From September 2017 onward, his research activities have been focused mainly on the sulfate-oxide systems database development with the FactSage software package, as well as rigorous theoretical and experimental investigations for promoting improved recovery of values from waste streams. Dr. Tesfaye was also appointed as a Visiting Research Scientist at Seoul National University, South Korea, for 6 months between March and August 2018. His current research interests are also within the scope of metallurgical engineering and circular economy of metals.

Dr. Tesfaye is a regular contributor and member of TMS and is the 2018 TMS Young Leaders Professional Development Award winner. He has served on TMS committees including Recycling and Environmental Technologies, Energy, and Professional Development. He has edited scientific research books and has served as a guest editor for *JOM*. His personal achievements include significant improvement of experimental research applying the solid-state EMF technique for thermodynamic studies, as well as contribution of new experimental thermodynamic data of several chalcogenide materials. He has published more than 40 peer-reviewed articles.

Part I
Nucleation, Crystallization, and Solidification

Demineralization of a High Ash Coal in Acidic Salt Solution

A. A. Adeleke, L. O. Jimoh and S. A. Ibitoye

Abstract The research investigated the reduction of the ash and sulphur contents of the high ash and high sulphur Nigerian Akunsa coal. The coal as received contained 38 and 1.98% of ash and sulphur, respectively. The coal samples were leached using sodium carbonate, nitric acid, and sodium carbonate acidified in nitric acid at varying combinations of concentrations, contact time, and at a constant temperature of 90 °C. The results obtained showed that sulphur and ash were effectively removed by nitric acid and sodium carbonate, respectively. It was further observed that the 1.5 M sodium carbonate acidified in 1 M nitric acid gave the highest leaching efficiency for the same solid-to-liquid ratio as the ash was reduced by 76.8–8.8% and the sulphur reduced by 82.3–0.35%. The coal concentrate obtained contains permissible ash and sulphur that render it suitable for blending in cokemaking.

Keywords Coal · High ash · High sulphur · Leaching · Concentrate

Introduction

Coal is a combustible fossil fuel and a sedimentary organic rock which is composed mainly of carbon, hydrogen, and oxygen. Coals with high ash content requires higher transport cost and also increases erosion in plant parts, which makes coal grinding difficult, lowers flame temperature, and thus increases energy requirement. The use of coals with lower ash has been found to reduce plant parts erosion rates by more than 50% and maintenance costs by 35%, an increase in thermal efficiency by 4–5% in pulverized coal boilers, and reduction in carbon dioxide emissions by 15%. Fluidized bed combustors have also been found to operate more efficiently with higher grade coals [1].

The chemical leaching methods to reduce ash and sulphur contents of coals include molten caustic leaching and agitation caustic leaching, mineral acid leaching and have been successfully employed to reduce the ash and sulphur contents of coals [2, 3]. Nigeria is endowed with coal deposits such as those of Lafia-Obi and Akunsa

A. A. Adeleke (✉) · L. O. Jimoh · S. A. Ibitoye
Department of Materials Science and Engineering, Obafemi Awolowo University, Ile-Ife, Nigeria
e-mail: abrahamadeleke@gmail.com

J. Lee et al. (eds.), *Materials Processing Fundamentals 2020*, The Minerals, Metals & Materials Series, https://doi.org/10.1007/978-3-030-36556-1_1

in Nassarawa state in the north central region of the country. The coal deposits are found in the middle region of the Benue trough, a sedimentary basin stretching about 1000 km long from the Gulf of Guinea to the Lake Chad basin. While the Lafia-Obi deposit had been extensively explored and found to hold about 22 million tons of coal, the Akunsa deposit reserve estimate is not available [4].

In this research, the high ash and high sulphur Akunsa coal was effectively demineralized in sodium carbonate acidified with nitric acid.

Materials and Methods

Materials

Sample Preparation

About 3.5 kg of Akunsa coal lumps was received for the laboratory-scale research. The coal lumps were crushed, air dried, and properly ground manually and later homogenized in a mortar by pounding with a pestle.

Methods

Particle Size Analysis Using Sieving Method

The sizing ground coal sample was carried out by employing sieves and an automated vibrating sieve shaker. The sieves were selected based on the square root of two rules. The weights of the sieves—355, 250, 150, 125, and 90 μm were taken empty and stacked in that decreasing order of mesh size with a collecting pan at the bottom. 200 g of the crushed and ground coal sample was poured into the uppermost coarsest sieve, and the stack sieves was shaken in a vertical plane for 60 min. After the required time, the stack of sieves was taken apart, and the mass of coal retained on each sieve was weighed.

Proximate Analysis

Proximate analysis was carried out on both the representative sample and the leached Akunsa coal to determine their weight per cent moisture content, ash, volatile, and fixed carbon content using modified ASTM methods.

Preparation of Leach Solutions

1.5 M solution of sodium carbonate was prepared. The 1 and 0.5 M solutions were then produced from the 1.5 M solution by serial dilution. 1.5 M of nitric acid solution was also produced, and 1 and 0.5 M solutions were produced from it by serial dilution.

Leaching Demineralization Tests

Atmospheric leaching in oven

The coal sample was subjected to leaching at atmospheric pressure in the oven and on the magnetic stirrer hot plate using 3^2 factorial design as shown in Table 1, with reagent concentration and contact time varied at three levels while the temperature was held constant at 90 °C. The reagent concentrations of $C_1 = 1.5$, $C_2 = 1$, and $C_3 = 0.5$ M and contact times of $t_1 = 30$, $t_2 = 60$, and $t_3 = 90$ min were used.

One gram of coal sample was weighed on a clean watch glass and transferred into a clean conical flask containing 25 ml of the prepared solution. The slurry in the conical flask was then homogenized by stirring for 5 min, covered with Al foil and transferred into the oven for leaching in combinations shown in Table 1. The resulting concentrates were skimmed off, air dried for 48 h, oven dried at 105 °C for 60 min, and weighed. The leaching was further repeated in the oven with the t_3C_3 leaching combination that gave the highest weight loss but using 0.25, 0.5, 0.75, 1.0, and 1.5 M sodium carbonate solutions acidified with 1 M nitric acid.

The leaching on magnetic stirrer hot plate was carried out by selecting the t_3C_3 combination and that of acidic mixture that gives the best reduction in mineral matter. One gram of coal sample was weighed and transferred into a clean conical flask containing 25 ml of prepared solution of t_3C_3 combination and 1.5 M sodium carbonate acidified with 1 M nitric acid. The slurry in the conical flask was then homogenized

Table 1 Factorial combination of time and concentration for atmospheric oven leaching

S/N	Time (min)	Concentration (M)	Combinations
1	t_1	C_1	t_1C_1
2	t_1	C_2	t_1C_2
3	t_1	C_3	t_1C_3
4	t_2	C_1	t_2C_1
5	t_2	C_2	t_2C_2
6	t_2	C_3	t_2C_3
7	t_3	C_1	t_3C_1
8	t_3	C_2	t_3C_2
9	t_3	C_3	t_3C_3

by stirring for 5 min, covered with Al foil and heated for leaching. Leaching experiments were conducted on magnetic stirrer at constant temperature of 90 °C and for 30 min contact time. The temperature was monitored with a thermometer. The resulting concentrate was skimmed off, air dried, oven dried, and weighed.

Determination of Total Sulphur by Eschka Method

The determination was carried out as described by Francis and Peters [5].

Results and Discussion

The results obtained showed that 30.40% of the coal was retained on the sieve size 355 μm after crushing and grinding. This suggests that the coal is relatively hard and thus respond slowly to crushing and grinding as less than 70% pass the relatively coarse 355 μm sieve size. The results obtained for moisture, volatile matter, ash, and fixed carbon of the coal sample are presented in Table 2. The high ash content of 38% for the coal as received showed that Akunsa coal is a high ash coal whose ash content far exceeds the maximum of 10% allowed in coals for use in metallurgical cokemaking [6]. A low-quality, high ash coal creates problems in both ironmaking and power generation. For instance, in India, it has been found that the supply of high ash coals with 30–50% ash causes increased transport cost due to the carriage of large amounts of ash forming minerals and creates shortages of rail cars and port facilities.

A high ash coal aggravates erosion in plant parts, makes coal pulverization difficult, lowers flame temperature and radiative heat transfer, and leads to production of excessive amount of fly ash containing large amounts of un-burned carbons. The use of beneficiated coal has been found to cause a reduction in erosion rates by 50–60% and maintenance costs by 35%, an increase in thermal efficiency by 4–5% in pulverized coal boilers, and reduction in carbon dioxide emissions by 15%. Even fluidized bed combustors have been found to operate more efficiently with higher grade coals [1].

Table 2 Proximate analysis of Akunsa coal

Parameter	% Content
Moisture	4.4
Volatile matter	10.2
Ash	38
Sulphur	1.98
Fixed carbon	47.4

Coal beneficiation has been found to be a low-cost solution for high ash Indian coals and a means of reducing imports of high grade foreign coals. Raw coals for power generation are cleaned to contain below 34% ash. The Indian Dadri power plant uses 34–35% ash coal from Indian Central Coalfield Limited. Typical landing cost of Run of Mine (ROM) high ash Indian coal is \$38.80/ton and transport over 400 km cost \$26.10/ton [1]. Considering the detrimental effects of high ash in coals for coal transport, cokemaking, and combustion for power generation, it is necessary to reduce the ash content of the Akunsa coal.

The sulphur content was determined to be 1.98% for the coal as received. The results obtained showed that the sulphur content of the Akunsa coal is much higher than the maximum of 0.9% required for coals for cokemaking [6]. High sulphur in coke on combustion forms sulphur dioxide, a very strong oxidizing agent. Additionally, an appreciable amount of sulphur in coke passes into the molten iron. This reduces the quality of the iron in terms of mechanical strength, and the oxidizing nature of the sulphur dioxide complicates the operations of subsequent conversion of iron to steel [5, 7].

The results obtained on the leaching of the Akunsa coal samples with solutions of sodium carbonate, nitric acid, and their mixture are presented in Figs. 1, 2, and 3, respectively. The results obtained showed that for the same leaching duration of 30, 60, or 90 min, the percentage weight loss in the samples was highest at the highest sodium carbonate molar concentration of 1.5 M with about 15% decrease in weight. Similarly, the results for 1.5 M nitric acid gave the highest weight loss of 6% at 90 min compared to the 15% for sodium carbonate, while the sodium carbonate acidified with nitric gave a much higher weight loss of 19% for the same leaching conditions at 1.5 M sodium carbonate solution. Thus, the weight loss obtained generally increased with increasing contact time and increasing concentration of the two leachants.

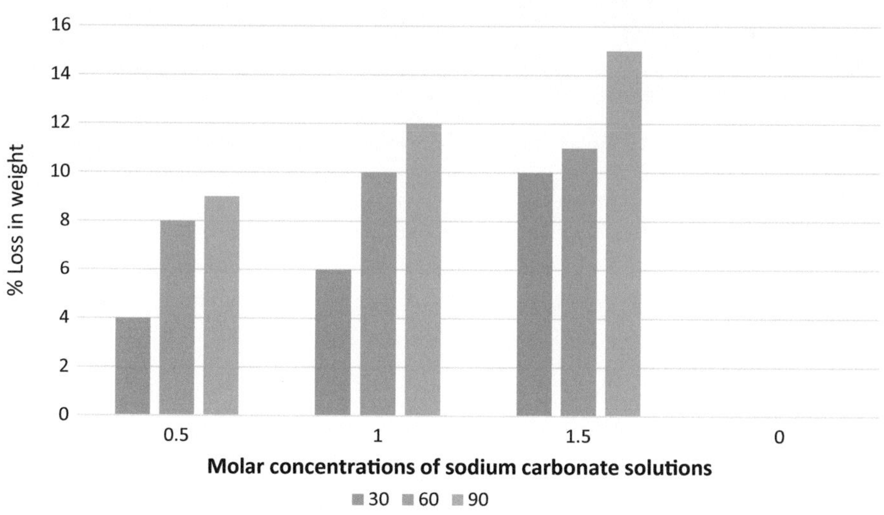

Fig. 1 Percentage weight loss in coal after sodium carbonate atmospheric oven leaching

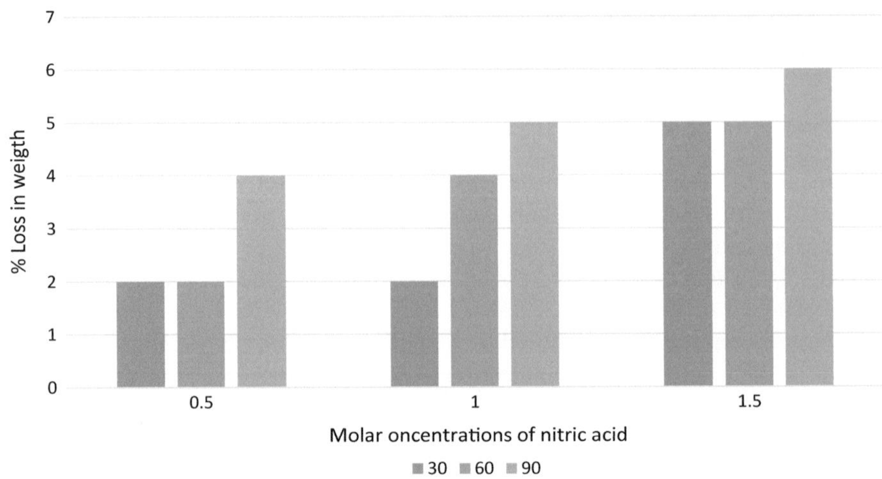

Fig. 2 Percentage weight loss in coal in nitric acid atmospheric oven leaching

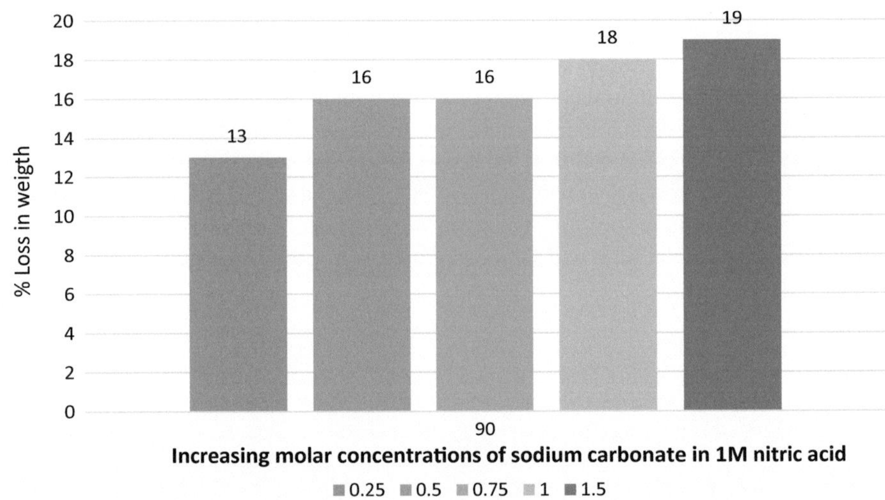

Fig. 3 Percentage weight loss in coal in acidified sodium carbonate atmospheric oven leaching

The results further showed that nitric acid is a weaker leachant for Akunsa coal than sodium carbonate as it gave a much lower 6% weight loss for the coal as against 15% for sodium carbonate at the same 1.5 M molar concentration and 90 min contact time. However, sodium carbonate in acidified solution of nitric is much stronger than nitric acid alone and gave 19% weight loss at 1.5 M sodium carbonate and 90 min contact time. The reduction in weight of the sample showed that the leaching was effective as part of the coal material was successfully converted to solution to remove them from the coal.

The higher weight loss in the acidified sodium carbonate may due to the increase in the reacting potency of sodium carbonate in the presence of nitric acid. The increasing weight loss at higher molar concentrations may be due to increasing collision between coal particles and the sodium carbonate ionic species at higher concentration. It has been established that leaching rate increases with increasing leachant concentration [8].

The results obtained for leaching on the magnetic stirrer hot plate with its agitating follower and a temperature of about 90 °C showed that weight losses of 26, 37, and 51% were obtained for the agitating leaching of Akunsa coal with nitric acid, sodium carbonate, and the mixture, respectively (Fig. 4). The weight losses were much higher than for oven leaching for all the reagent options used on the hot plate. The results showed that the effect of pulp agitation was very high on the leaching rate. Pulp stirring has been established as an important parameter for effective leaching [8].

The ash contents determined for the coal as received, the coal as leached by sodium carbonate solution, the coal as leached by nitric acid, and the coal as leached by sodium carbonate acidified in nitric acid by atmospheric oven and on magnetic stirrer hot plate leaching are shown in Tables 3 and 4, respectively. The results obtained gave 38% for the Akunsa coal as received and 19.7, 29.3, and 14.9%, respectively, for the coal oven leached by sodium carbonate solution, nitric acid solution, and sodium carbonate acidified in nitric acid. The agitation leaching on the magnetic stirrer yielded coal concentrates with 11.2, 19.2, and 8.8% ash contents, respectively. The results obtained translate to ash reduction per cents of 22.9, 48.2, and 60.8; and 49.5, 70.5, and 76.8% for oven and magnetic stirrer leaching, respectively.

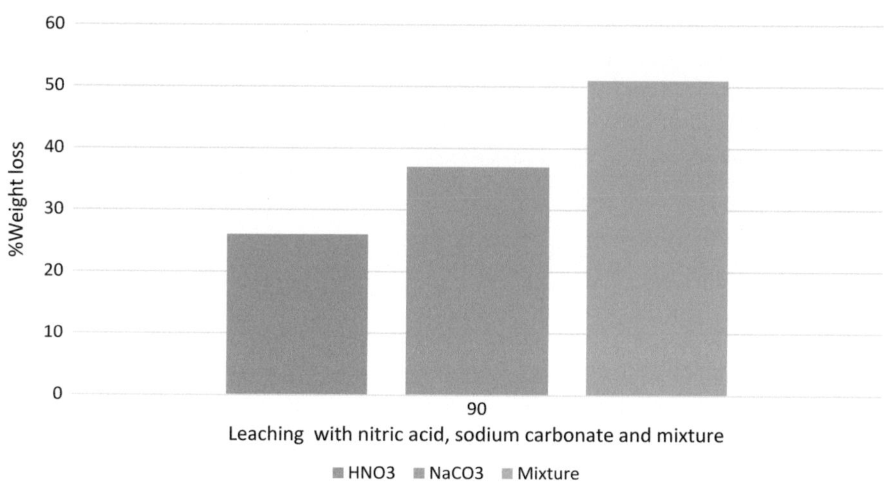

Fig. 4 Percentage weight loss in coal on magnetic stirrer for nitric, sodium carbonate, and acidified sodium carbonate after 30 min

Table 3 Ash content in leached coal in atmospheric oven and magnetic stirrer leaching

Leachant	% Ash after oven leaching	% Ash reduction	% Ash after agitation leaching	% Ash reduction
Nitric acid (1.5 M, 90 min)	29.3	22.9	19.2	49.5
Sodium carbonate (1.5 M, 90 min)	19.7	48.2	11.2	70.5
Mixture (1 M nitric acid, 1.5 M SC, 90 min)	14.9	60.8	8.8	76.8

Table 4 Effect of sodium carbonate, nitric acid, and acidic mixture on Akunsa coal sulphur in atmospheric oven and agitation leaching

Leachant	% *S* after oven leaching	% *S* reduction	% *S* after agitation leaching	% *S* reduction
Nitric acid (1.5 M, 90 min)	0.99	50	0.82	58.6
Sodium carbonate (1.5 M, 90 min)	1.38	30.3	1.14	42.4
Mixture (1 M nitric acid, 1.5 M SC, 90 min)	0.93	53	0.35	82.3

Similarly, the sulphur contents determined for the Akunsa coal as received were 1.98% and 0.99, 1.38 and 0.94%, respectively, for the coal oven leached by nitric acid solution, sodium carbonate solution, and sodium carbonate solution acidified in nitric acid, respectively. The agitation leaching on the magnetic stirrer yielded coal concentrates with 1.15, 0.82, and 0.36% sulphur contents, respectively. The results translate to sulphur reduction per cents of 50.0, 30.3 and 58.60 and 42.40 and 82.30%, respectively. The results obtained thus showed that sodium carbonate was more efficient than nitric acid in ash reduction, while nitric acid was more effective in sulphur removal. The theory of the leaching is that sodium carbonate reacts with ash components such as silica, iron oxide, and alumina to form water-soluble complex silicates like sodium silicate, sodium aluminium silicate, and others. It also forms soluble compounds of sulphur to reduce the sulphur content. Nitric acid also reacts with ash components and sulphur to form water-soluble compounds thus reducing the ash and sulphur contents.

Conclusion

The high ash and high sulphur Akunsa coal was successfully demineralized of its ash and sulphur contents using sodium carbonate, nitric acid, and sodium carbonate acidified with nitric acid in the oven and on the magnetic stirrer hot plate. However, the results obtained showed that sodium carbonate acidified with nitric acid gave the best concentrate during agitation leaching and reduced the ash content by 76.80% and sulphur by 82.30%.

References

1. Zamuda CD, Sharp MA (2007) A case for enhanced use of clean coal in India. https://fossil.energy.gov›Publications›coal_beneficiation_paper_zamuda. Accessed 5th Sept 2019
2. Adeleke AA, Ibitoye SA, Afonja AA (2013) Multistage caustic leaching de-sulphurization of a high sulphur coal. P&C 55(2):544–551
3. Adeleke AO, Makan RS, Adahama AB, Makan RS, Ibitoye SA (2007) An evaluation of the coking characteristics of Polish coking coals for cokemaking with non-coking Nigerian coals. P&C 49(1):1–6
4. Federal Ministry of Mines and Steel Development (2005) Coal exploration and power generation opportunities in Nigeria. http://www.mmsd.gov.ng/Downloads/Coal.pdf. 27th Nov 2010
5. Francis W, Peters MC (1980) Fuels and fuel technology. Pergamon Press, New York
6. Raw Materials and products specifications (1994) For Federal Government Steel Companies, Nigeria
7. Krivandin V, Markov B (1980) Metallurgical furnaces. Mir Publishers, Moscow
8. Ghosh A, Ray HS (1991) Principles of extractive metallurgy. Wiley, New York

Simulation for Solidification Structure of Continuous Casting Bloom Using Cellular Automaton-Finite Element Model

Yadong Wang, Dongbin Jiang, Sha Ji and Lifeng Zhang

Abstract A coupled cellular automaton-finite element (CAFE) model was used to simulate the solidification structure of continuous casting bloom. The simulated solidification structure and temperature field were validated by the industrial trial. The influence of superheat, secondary cooling water flow, and mold electromagnetic stirring (M-EMS) on the solidification structure was discussed. To simulate the intense flow field and uniform temperature with M-EMS, the thermal conductivity in liquid and the formation of crystals increased, separately. The ratio of equiaxed crystal zone decreases from 28.47 to 16.45% with the superheat increasing from 15 to 35 °C. With the secondary cooling water flow increasing 20% and decreasing 20%, the ratio of equiaxed crystal zone is 20.64% and 23.16%, respectively. The ratio of equiaxed crystal zone increases from 21.75 to 32.77% with the application of M-EMS.

Keywords Solidification structure · CAFE method · Mold electromagnetic stirring · Superheat · Secondary cooling water flow

Introduction

Solidification structure during continuous casting process has a significant effect on the internal quality of the final steel products especially for the internal cracks, macrosegregation, porosity, and so on. Homogenization heat treatment and rolling process are impossible to eliminate internal defects of bloom [1, 2]. Therefore, solidification structure should be properly controlled by optimizing the continuous casting parameters.

With the improvement of computational efficiency and solidification structure models, the simulation of solidification structure can be achieved. There are two main simulation methods including deterministic models and stochastic models [3,

Y. Wang · D. Jiang · S. Ji · L. Zhang (✉)
School of Ecological and Metallurgical Engineering, University of Science and Technology Beijing, Beijing 100083, China
e-mail: zhanglifeng@ustb.edu.cn

J. Lee et al. (eds.), *Materials Processing Fundamentals 2020*, The Minerals, Metals & Materials Series, https://doi.org/10.1007/978-3-030-36556-1_2

4]. The cellular automaton-finite element (CAFE) model is one of the stochastic models which was widely used in the last decades. Rappaz and Gandin [3] are the earliest researchers who simulated the microstructure of Al alloy using CAFE method. Thereafter, the CAFE method was gradually utilized in the continuous casting process [4–6]. Based upon a two-dimensional cellular automaton (CA) technique, Rappaz et al. [3] investigated the effect of the cooling rate upon the microstructure. The extent of the columnar region is reduced with the increasing of cooling rate. Gandin et al. [7] established a 3-D CAFE model for the prediction of dendritic grain structures formed during solidification. It was demonstrated that the 3-D CAFE model was feasible for the investment casting and continuous casting processes. Zhang et al. [8] studied the effects of undercooling, casting temperature, mold initial temperature, and filling rate on the solidification structure of the ingot based on a 3-D CA method. The undercooling is a critical parameter on the solidification; however, the initial temperature of the mold has little effect on the final solidification structure. Fang et al. [6] utilized the CAFE model to discuss the effect of secondary cooling conditions and superheat on the solidification structure of continuous casting bloom. The results shown that the secondary cooling conditions had little effect on the grain size and the percentage of center-equiaxed grains decreased with the superheat increasing.

In the current work, the CAFE model validated by trial results was used to simulate the solidification structure of continuous casting bloom. The effect of superheat, secondary cooling water flow, and M-EMS on the solidification structure was discussed.

Mathematical Formulation

The content of the 20CrMnTi bloom is listed in Table 1. The main parameters used in the current study are listed in Table 2.

In the current model, the nucleation and grain growth were calculated by the CA method in which the Gaussian distribution of nucleation was used. The continuous heterogeneous nucleation model was used in the CA method and was expressed by Eq. (1).

$$\frac{\mathrm{d}n}{\mathrm{d}(\Delta T)} = \frac{n_{\max}}{\sqrt{2\pi}\,\Delta T_\sigma}\exp\left[-\frac{(\Delta T - \Delta T_{\mathrm{ave}})^2}{2\Delta T_\sigma^2}\right] \tag{1}$$

where ΔT is undercooling in K; $n_{\max}$ is the maximum nucleation density in m^{-3}; ΔT_σ is the standard deviation in K; ΔT_{ave} is the mean undercooling in K. The growth

Table 1 Chemical compositions of 20CrMnTi bloom

Elements	C	Si	Mn	P	S	Cr	Ti	Al
Content (%)	0.20	0.24	0.89	0.02	0.01	1.09	0.06	0.02

Table 2 Parameters of continuous casting process

Parameters	Values
Bloom dimension (mm^2)	510 × 390
Effective mold length (mm)	680
Submerged depth of SEN (mm)	90
Casting speed (m/min)	0.42
Pouring temperature (K)	1811
Liquidus temperature of molten steel (K)	1786
Solidus temperature of molten steel (K)	1723

velocity of dendrite tip was calculated according to the simplified Kurz–Giovanola–Trivedi (KGT) model [8, 9].

$$V(\Delta T) = a_2 \Delta T^2 + a_3 \Delta T^3 \tag{2}$$

where a_2, a_3 are the fitting polynomial coefficients of the dendrite tip growth kinetic parameters. In current calculation, a_2, a_3 are 0.0 and 8.031×10^{-6}, respectively; $V(\Delta T)$ is the growth velocity of the dendrite tip.

The heat transfer during continuous casting was calculated by finite element (FE) model, and the governing equation was expressed by Eq. (3).

$$\frac{\partial T}{\partial t} = \frac{\lambda}{\rho C_p}\left(\frac{\partial^2 T}{\partial x^2} + \frac{\partial^2 T}{\partial y^2}\right) \tag{3}$$

where λ is the thermal conductivity in W/(m K), ρ is the density in kg/m^{-3}, and C_p is specific heat in J/(kg K). The procedure of CAFE model was as follows: The temperature field was calculated using coarse mesh, and then to combine the FE and CA calculation in the same model, the FE elements were subdivided into much smaller grids to ensure the calculating precision of the nucleation and growth. The latent heat released in the nucleation and growth progress was fed back to the thermal calculation, updating the temperature of the nodes in FE model. Finally, the CA and FE model were fully coupled [3, 10].

To improve the computational efficiency, a moving slice model was established [4]. From the meniscus to the cutting point, different cooling boundary conditions were applied to the model, shown in Table 3 [11, 12]. The parameters used in the CAFE model are shown in Table 4, where $\Delta T_{v,ave}$ is the mean undercooling of internal nucleation, $\Delta T_{v,\sigma}$ is the standard deviation of internal nucleation, $n_{v,max}$ is the maximum nucleation density of internal nucleation, $\Delta T_{s,ave}$ is the mean undercooling of surface nucleation, $\Delta T_{s,\sigma}$ is the standard deviation of surface nucleation, and $n_{s,max}$ is the maximum nucleation density of surface nucleation.

Table 3 Governing equations of cooling conditions

Cooling zone	Boundary conditions	Governing equations
Mold	q	$q = 2680000 - 315000\sqrt{t}$
Foot roller section	$q_s = h_s(T - T_w)$	$h = 5849.9 \times W^{0.451} \times (1 - 0.0075\theta)$
Section 2 in secondary cooling zone		
Section 3 in secondary cooling zone		
Section 4 in secondary cooling zone		
Section 5 in secondary cooling zone		
Air cooling zone	$q_a = \varepsilon\sigma(T^4 - T_a^4)$	$\varepsilon = 0.8$

Table 4 Parameters in the CAFE model

Parameters	$\Delta T_{v,ave}$ (K)	$\Delta T_{v,\sigma}$ (K)	$n_{v,max}$ (m^{-3})	$\Delta T_{s,ave}$ (K)	$\Delta T_{s,\sigma}$ (K)	$n_{s,max}$ (m^{-2})
Value	2.0	0.6	6.5×10^8	1.0	0.1	1.0×10^8

Model Validation

In the current study, the temperature field was calculated by the FE model. Obviously, the simulated temperature in the loose side center was agreement with the measured ones shown in Fig. 1. The comparison of the numerical and trial solidification structures without mold electromagnetic stirring (M-EMS) was shown in Fig. 2. The ratio of equiaxed crystal zone in calculation and industrial experiment

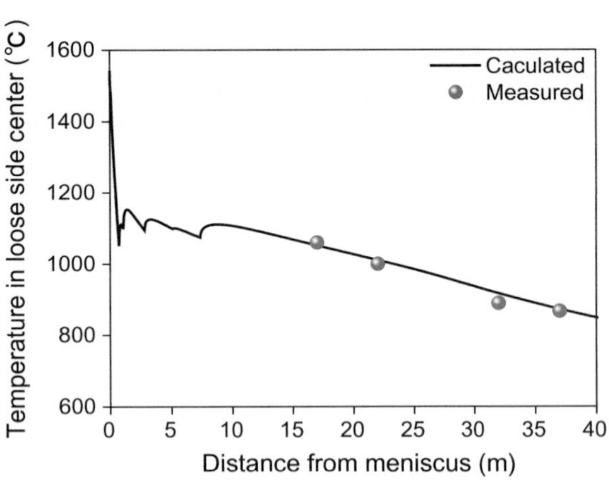

Fig. 1 Simulated and measured temperature in loose side center

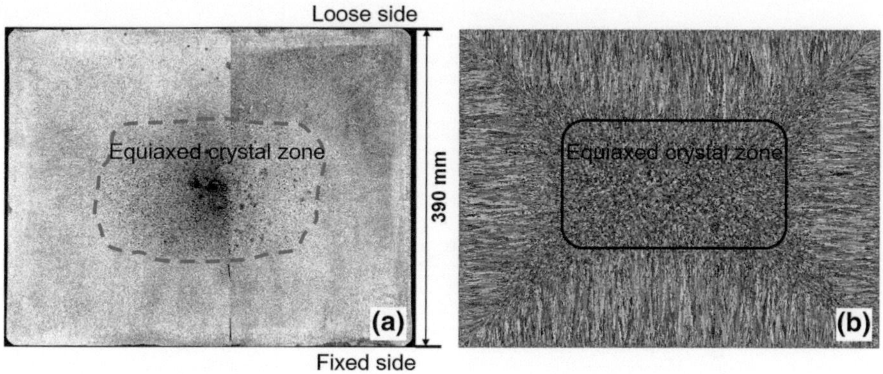

Fig. 2 Simulated and measured solidification structure without M-EMS

was 21.3% and 21.75%, respectively. Both the measured temperature field and the solidification structure agreed well with the numerical results, so the CAFE model was validated for further prediction.

Results and Discussion

To investigate the effect of superheat on the solidification structure, the microstructures of superheat from 15 to 35 °C were calculated, as shown in Fig. 3. The ratio of equiaxed crystal zone and the mean radius of grains under different superheats are shown in Fig. 4. With the superheats increasing from 15 to 35 °C, the ratio of equiaxed crystal zone decreases from 28.47 to 16.45%, and the mean radius of columnar crystal grains becomes larger. However, the mean radius of grains in equiaxed crystal zone is 1.06 mm approximately. So the mean radius of grains in whole section increases from 1.02 to 1.21 mm.

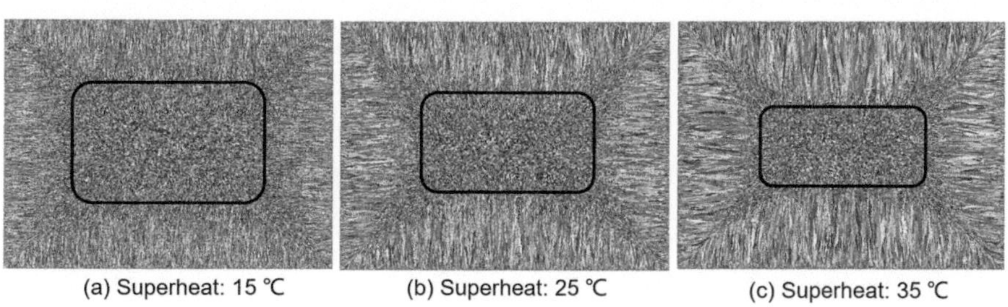

Fig. 3 Effect of superheats on the solidification structure

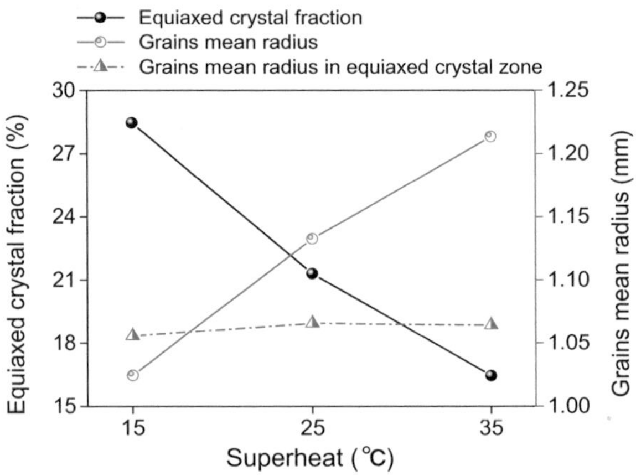

Fig. 4 Ratio of equiaxed crystal zone and mean radius of grains under different superheats

The simulated solidification structure, the ratio of equiaxed crystal zone, and mean radius of grains under different secondary cooling water flow are shown in Figs. 5 and 6. There are three cases including decreasing 20%, normal, and increasing 20% of the secondary cooling water flow. With the secondary cooling water flow increasing, the ratio of equiaxed crystal zone decreases from 23.16 to 20.64%, since enhanced cooling condition contributed to delaying columnar to equiaxed transition (CET). The mean radius of grains in equiaxed crystal zone decreases from 1.09 to 1.06 mm with the secondary cooling water flow increasing. The mean radius of grains in whole section is 1.13 mm approximately.

It has been reported that the mold electromagnetic stirring (M-EMS) plays an important role in improving the surface and inner quality of strand especially improving the ratio of equiaxed crystal zone [13, 14]. Electromagnetic stirring technology employs Lorentz force generated by multi-phase induction coils to stir the molten steel in the mold. The effects of rotating flow induced by electromagnetic force are as follows: increasing heat transfer between the solidification front and molten steel, breaking the dendrites by intense flow field, and increasing the formation of crystals [5, 15]. In current simulation, to simulate the intense flow field and increasing

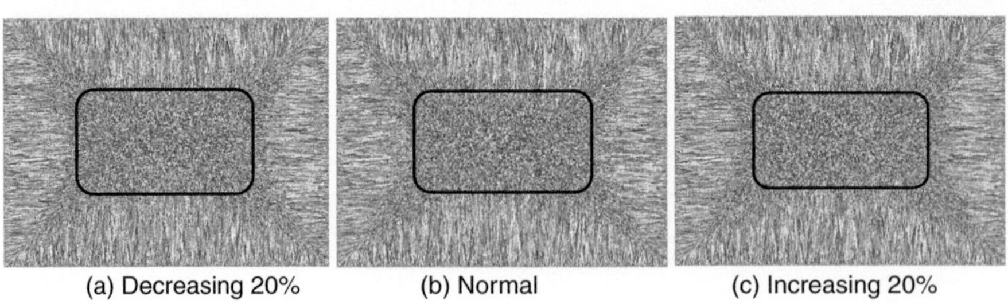

Fig. 5 Effect of secondary cooling water flow on the solidification structure

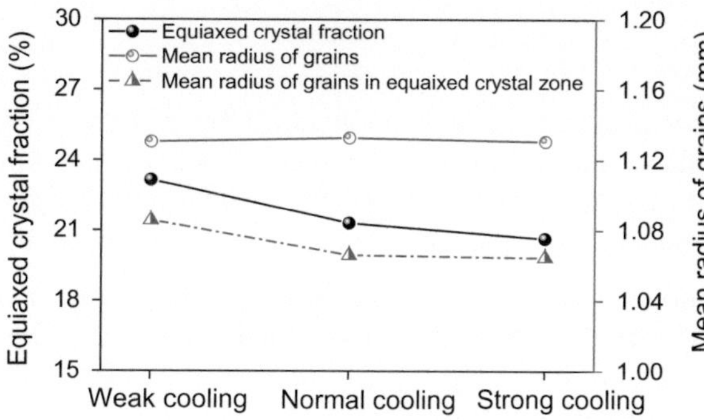

Fig. 6 Ratio of equiaxed crystal zone and mean radius of grains under different secondary cooling water flow

the formation of crystals, the maximum nucleation density of volume was changed from 6.5×10^8 to 7.0×10^8 m^{-3}. As for the increasing heat transfer between the solidification front and molten steel, a 1.2 times thermal conductivity was used with liquid fraction (f_l) larger than 0.7. In the solid zone ($f_l \leq 0.3$), the thermal conductivity remained original value. In the mushy zone with $0.3 \leq f_l \leq 0.7$, the thermal conductivity changed linearly.

The simulated and measured solidification structure with M-EMS is shown in Fig. 7. With the M-EMS considered, the ratio of equiaxed crystal zone increases from 21.75 to 32.77%, and the ratio of equiaxed crystal zone in calculation and industrial experiment was 31.7% and 32.77%, respectively. The results demonstrate that the present model simulating solidification structure formation during the continuous casting process of steel is reliable.

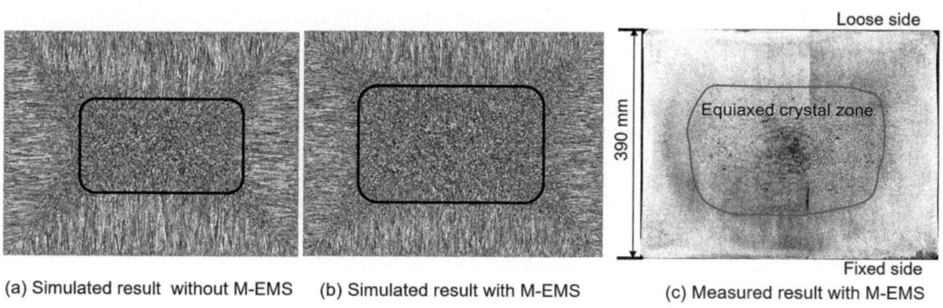

Fig. 7 Simulated solidification structure without M-EMS (**a**) and with M-EMS (**b**), measured solidification structure with M-EMS (**c**)

Conclusions

In the current paper, a coupled cellular automaton-finite element (CAFE) model was used to simulate the solidification structure of continuous casting bloom. The influence of superheat, secondary cooling water flow, and M-EMS on the solidification structure was discussed. The following conclusions can be obtained:

(1) The CAFE model was validated by the measured temperature and solidification structure, and then the CAFE model was reliable for further prediction.
(2) With the superheats increasing from 15 to 35 °C, the ratio of equiaxed crystal zone decreases from 28.47 to 16.45%, and the mean radius of grains increases from 1.02 to 1.21 mm. The mean radius of grains in equiaxed crystal zone is 1.06 mm approximately.
(3) With the secondary cooling water flow increasing, the ratio of equiaxed crystal zone decreases from 23.16 to 20.64%, and the mean radius of grains in equiaxed crystal zone decreases from 1.09 to 1.06 mm. The mean radius of grains in whole section is 1.13 mm approximately.
(4) With the M-EMS considered, the ratio of equiaxed crystal zone increases from 21.75 to 32.77%. The superheat and M-EMS are critical parameters on the solidification.

Acknowledgements The authors are grateful for support from the National Science Foundation China (Grant No. U1860206 and No. 51725402), Beijing International Center of Advanced and Intelligent Manufacturing of High Quality Steel Materials (ICSM), Beijing Key Laboratory of Green Recycling and Extraction of Metals (GREM), and the High Quality Steel Consortium (HQSC) at the School of Metallurgical and Ecological Engineering at University of Science and Technology Beijing (USTB), China.

References

1. Ludlow V, Normanton A, Anderson A (2005) Strategy to minimise central segregation in high carbon steel grades during billet casting. Ironmaking Steelmaking 32(1):68–74
2. Dong Q, Zhang J, Wang B (2016) Shrinkage porosity and its alleviation by heavy reduction in continuously cast strand. J Mater Process Technol 238:81–88
3. Rappaz M, Gandin CA (1993) Probabilistic modelling of microstructure formation in solidification processes. Acta Metall Mater 41(2):345–360
4. Hou Z, Jiang F, Cheng G (2012) Solidification structure and compactness degree of central equiaxed grain zone in continuous casting billet using cellular automaton-finite element method. ISIJ Int 52(7):1301–1309
5. Yamazaki M, Natsume Y, Harada H (2006) Numerical simulation of solidification structure formation during continuous casting in Fe-0.7 mass% C alloy using cellular automaton method. ISIJ Int 46(6):903–908
6. Fang Q, Ni H, Zhang H, Wang B (2017) Numerical study on solidification behavior and structure of continuously cast U71Mn steel. Metals 7(11):483
7. Gandin C-A, Desbiolles J-L, Rappaz M, Thevoz P (1999) A three-dimensional cellular automation-finite element model for the prediction of solidification grain structures. Metall Mater Trans A 30(12):3153–3165

8. Zhang H, Zhang L (2016) Modeling on the solidification structure of Fe-Ni-based Alloys using cellular automaton method. Metall Res Technol 113(4):410–413
9. Kurz W, Giovanola B, Trivedi R (1986) Theory of microstructural development during rapid solidification. Acta Metall 34(5):823–830
10. Gandin CA, Rappaz M (1994) A coupled finite element-cellular automaton model for the prediction of dendritic grain structures in solidification processes. Acta Metall Mater 42(7):2233–2246
11. Morales RD, Lopez AG, Olivares IM (1990) Heat transfer analysis during water spray cooling of steel rods. ISIJ Int 30(1):48–57
12. Savage J, Pritchard WH (1954) The problem of rupture of the billet in the continuous casting of steel. J Iron Steel Inst 178(3):269–277
13. Ayata K, Mori T, Fujimoto T (1984) Improvement of macrosegregation in continuously cast bloom and billet by electromagnetic stirring. Trans Iron Steel Inst Jpn 24(11):931–939
14. An H, Bao Y, Wang M (2018) Effects of electromagnetic stirring on fluid flow and temperature distribution in billet continuous casting mould and solidification structure of 55SiCr. Metall Res Technol 115(1):103
15. Takahashi T, Ohsasa K, Katayama N (1990) Simulation for progress of solid-liquid coexisting zone in continuous casting of carbon steels. Tetsu-to-Hagané 76(5):728–734

Effects of Al Substitution for Zn on the Non-equilibrium Solidification Behavior of Zn–3Mg Alloys

Yeqing Wang, Jianrong Gao and Ashwin J. Shahani

Abstract Zn-based alloys may find many applications because of their excellent corrosion resistance and biocompatibility. It is of fundamental and technical interest to investigate their solidification behavior in casting. In this work, solidification paths of Zn–3Mg–2Al alloys were investigated using in situ X-ray diffraction and thermodynamic calculations. Results showed that a stable and a metastable solidification path of the samples are not much different from those of binary Zn–3Mg alloys. In each path, the Laves compound $MgZn_2$ is the primary phase, and its growth is followed by growth of a metastable $MgZn_2$/Zn eutectic or by a stable Mg_2Zn_{11}/Zn eutectic. Such observations can be explained in terms of thermodynamic calculations of driving forces for nucleation of stable and metastable phases.

Keywords Zn alloys · Solidification path · Thermodynamic calculations · In situ X-ray diffraction · Competitive nucleation

Introduction

Zn-based alloys may find many applications because of their excellent corrosion resistance and biocompatibility [1–8]. Mechanical performance of Zn–Mg alloys depends on their microstructure and therefore on their solidification behavior. Previous studies reported that alloying with Al has significant influence on the solidification behavior of Zn–Mg alloys [9–11]. The influence of Al is dependent on the amount of Al addition. Recently, Zn–Mg alloys show a rare spiral eutectic morphology, which was attributed to screw-dislocation-assisted metastable eutectic growth of $MgZn_2$ with Zn in preference to the stable Mg_2Zn_{11} phase (Moniri S. et al., to be

Y. Wang · J. Gao (✉)
Key Laboratory of Electromagnetic Processing of Materials (Ministry of Education), Northeastern University, Shenyang 110819, China
e-mail: jgao@mail.neu.edu.cn

Y. Wang · A. J. Shahani (✉)
Department of Materials Science and Engineering, University of Michigan, Ann Arbor, MI 48109, USA
e-mail: shahani@umich.edu

J. Lee et al. (eds.), *Materials Processing Fundamentals 2020*, The Minerals, Metals & Materials Series, https://doi.org/10.1007/978-3-030-36556-1_3

published). For such reasons, it is of dual interest to investigate the influence of alloying additions on the solidification behavior of Zn–Mg alloys. Here, we investigated the effect of a minor Al addition on the solidification behavior and microstructure formation in Zn–3Mg eutectic composition. Results showed that the stable and a metastable eutectic appear simultaneously in solidification microstructure. Such a feature of the solidification microstructure can be understood in light of in situ synchrotron X-ray diffraction experiments and thermodynamic calculations of stable and metastable equilibria.

Experiments

Two ingots of bulk composition Zn–3wt% Mg–2wt% Al were prepared using pure elements of 99.999% purity. Each of them had a mass of about 1.0 g. The raw materials were sealed in a quartz tube with size of 6 × 30 mm and back-filled with pure argon after evacuation to a vacuum pressure of few Pa. They were heated and melted using a resistance furnace. The melts were held at 773 K for 15 min to get bulk homogeneity. After soaking, the quartz tube was moved out of the furnace. The melts were cooled and solidified in the quartz tube.

One ingot was taken out of the tube for differential scanning calorimetry (DSC) experiments. A small sample with a mass of 23.40 mg was cut and placed in an alumina pan. Then, it was loaded into the sample chamber of a TA Q100 DSC. The sample was heated and cooled in the temperature range 300–823 K at a rate of 15 K/min under the protection of a nitrogen atmosphere. The other ingot was kept in the sealed tube for in situ high-energy X-ray diffraction (HEXRD) experiments at the beamline 11-ID-C of the Advanced Photon Source (APS), Argonne National Laboratory. Before the HEXRD experiments, the sample was positioned in a cylinder-shaped slot of a specially designed resistance-heated copper block. The sample and the tube were heated to 723 K and held for 1 min to reach a thermal equilibrium with the copper block. The temperature of the copper block was controlled using a PID controller and monitored using a K-type thermocouple. After switching off the heating power, the sample was cooled and solidified. In heating and cooling, monochromatized X-rays with a wavelength of 0.1173 Å were incident on the sample through the quartz tube. The quartz tube was transparent to high-energy X-rays but provided protection of the sample from oxidation. X-rays were scattered by the sample, and those in the forward directions were detected and registered using a two-dimensional amorphous Si detector. An exposure time of 0.1 s was chosen for a single frame of HEXRD. More details of the HEXRD experiments and data analysis can be found elsewhere [12, 13]. After the HEXRD experiments, cross-sectional microstructure of the solidified sample was examined using an SS150 scanning electron microscope (SEM) in back-scattering electron imaging mode following standard procedures of metallographic studies. An X-ray energy-dispersive spectrometer (EDS) attached to the SEM was used to determine chemical composition of phase constituents.

Results

Phase formation sequence in solidification was predicted using thermodynamic calculations. With this in mind, phase equilibria in Zn–Mg–2Al composition were calculated using the CALPHAD method. As shown in Fig. 1, the calculated phase equilibria of Zn–Mg–2Al composition look similar to those of binary Zn-rich Zn–Mg composition [14]. Especially, the stable Mg_2Zn_{11} and a stable Mg_2Zn_{11}/Zn eutectic are predicted to form in the Zn-rich side of the phase diagram of Zn–Mg–2Al composition. Due to the addition of Al, the stable and the metastable eutectic reaction can occur in a range of composition and temperature during cooling. The stable eutectic can set in at a temperature of $T_E = 628$ K at composition Zn–2.6Mg–2Al, whereas the metastable eutectic can occur at a temperature of $T_{ME} = 620$ K at composition Zn–3Mg–2Al. Compared to the phase equilibria in the binary Zn–Mg systems, the minor Al addition shifts both eutectic reactions towards Mg-rich composition while it brings about a depression of the eutectic temperatures. For the Zn–3Mg–2Al composition, it now lies in a primary crystallization region of Mg_2Zn_{11}, and its equilibrium solidification ends with secondary crystallization of the stable Mg_2Zn_{11}/Zn eutectic. Such an equilibrium solidification path may change under non-equilibrium conditions which may involve a high cooling rate. A potential change is that the

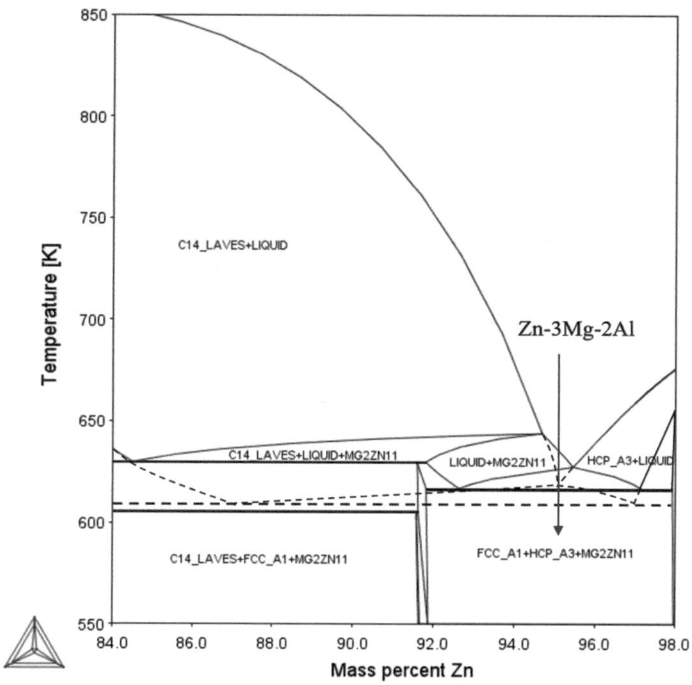

Fig. 1 Pseudo-binary section of Zn–Mg–2Al system. Dashed lines represent metastable extensions. Blue line is the investigated alloy composition

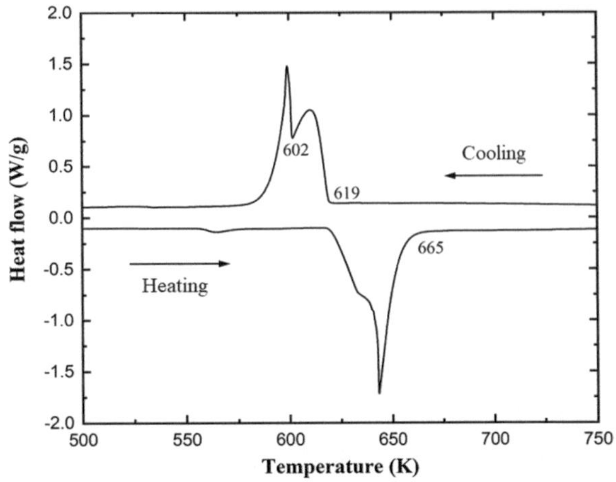

Fig. 2 DSC traces of a Zn–3Mg–2Al alloy in heating and cooling at a rate of 15 K/min

primary phase may be replaced by $MgZn_2$ as it has been shown that this phase has a higher nucleation rate than that of Mg_2Zn_{11} [15, 16]. In this case, Mg_2Zn_{11} may be crystallized by a metastable peritectic reaction of the primary $MgZn_2$ with liquid. Also, the stable Mg_2Zn_{11}/Zn eutectic may be replaced by the metastable $MgZn_2$/Zn eutectic [17]. Such predicted phase equilibria were checked by performing DSC and in situ HEXRD experiments on small- and bulk-sized samples.

Figure 2 illustrates DSC curves of the Zn–3Mg–2Al alloy sample in heating and cooling. Only the metastable solidification path was observed due to the high cooling rate. In heating, a small peak and a strong peak of heat flows were observed. The latter was close to the calculated liquidus temperature but had a shoulder, suggesting that two stages of melting are close to each other. In cooling, two strong and separated peaks were observed, suggesting a two-step solidification process. Attention was focused on the solidification process. The first peak commences at a temperature of 619 K and can be attributed to the metastable eutectic in accordance with the calculated phase diagram of Fig. 1. A second peak rises up at a temperature of 602 K. It lies well below the $MgZn_2$/Zn eutectic zone and may correspond to the formation of a $MgZn_2$/Zn/Al three-phase eutectic. No peaks of heat flow were observed around or below the predicted liquidus temperature (9 K higher than the first peak). The easy crystallization of primary $MgZn_2$ (due to its low solid-liquid interfacial energy [15]) needs to be justified by other experiments.

Figure 3 illustrates time-resolved HEXRD patterns of a bulk sample. The patterns were observed in two independent solidification processes of the sample showing different phase formation sequence. As shown in Fig. 3a, the first solidification path comprises primary $MgZn_2$, secondary Mg_2Zn_{11}, and sequential growth of Mg_2Zn_{11}, Zn and Al. The growth of $MgZn_2$ was indicated by a Bragg peak at $2\theta \approx 6.95°$. The growth of Mg_2Zn_{11} and Zn were observed at $2\theta \approx 4.82°$ and 5.02°, respectively. Later, diffraction peaks of Zn phase became stronger than that of Mg_2Zn_{11}, suggesting faster growth kinetics of this phase. In another solidification path (Fig. 3b),

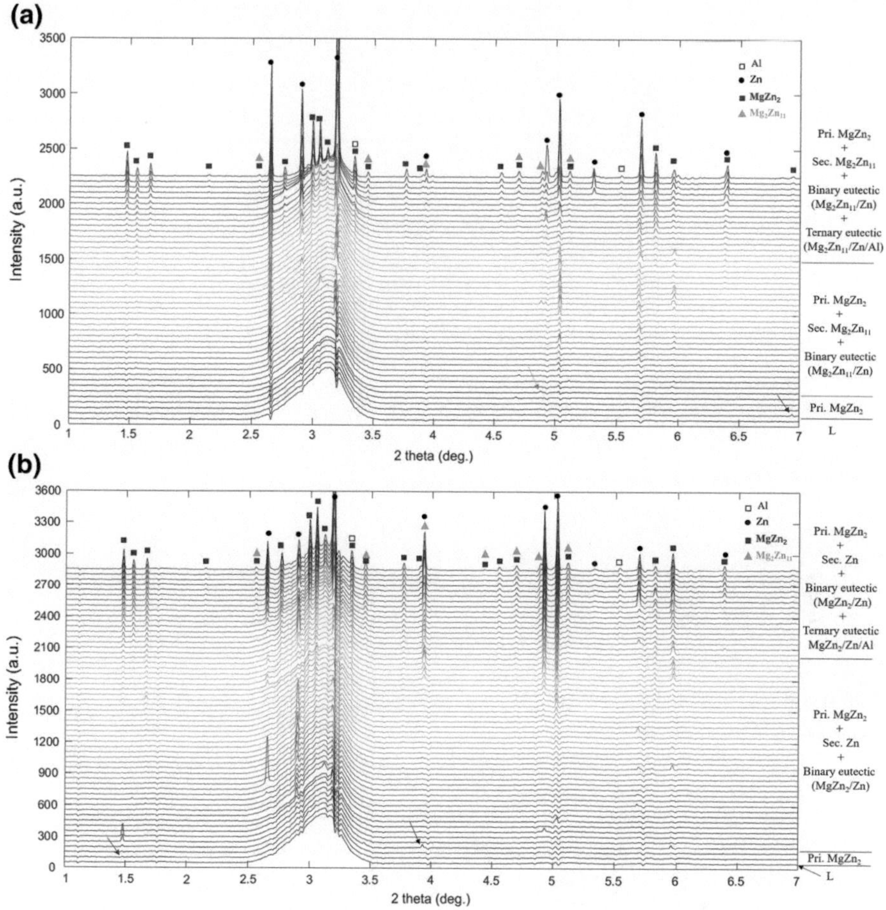

Fig. 3 **a** and **b** Illustration of integrated HEXRD patterns showing two different solidification paths of the Zn–3Mg–2Al sample. Patterns from the bottom to top are stacked with a time step of 0.1 s in each panel. Blue curves represent early times, and red curves represent later times. Note that the species of a secondary phase following primary $MgZn_2$ is different in the two paths

growth of primary $MgZn_2$ is followed by growth of secondary Zn, as evidenced by rapidly increasing intensities of diffraction peaks of Zn at $2\theta \approx 2.67°$ and $3.33°$. The solidification ends with sequential growth of $MgZn_2$, Zn, and Al. In this solidification process, faster growth of Zn occurs almost 1 s later than the simultaneous growth of $MgZn_2$ and Zn until it is terminated by the growth of Al as suggested by a peak observed at $2\theta \approx 5.55°$. This solidification path supports the observed two peaks corresponding to the metastable eutectic and ternary eutectic in the cooling state of Fig. 2.

Figure 4 illustrates solidification microstructure of the Zn–3Mg–2Al sample. In the micrographs of Fig. 4a with low magnification, the microstructure consists of

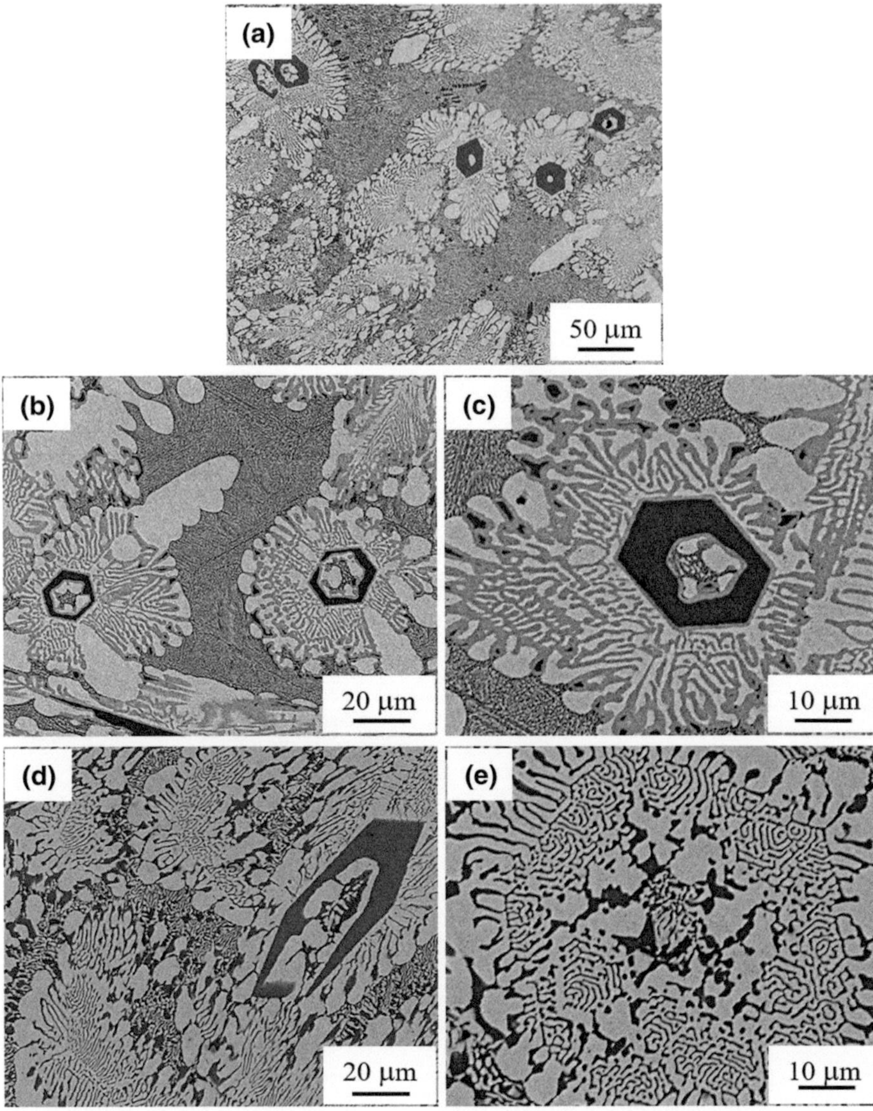

Fig. 4 Back-scattered SEM micrographs showing the solidification microstructure of the Zn–3Mg–2Al sample. Hollow hexagons with dark contrast are $MgZn_2$, lamellae and dendrites with bright contrast are Zn, eutectic phase with gray contrast is Mg_2Zn_{11}. Images (**b**, **c**) are magnified views of the right-hand side of (**a**), while (**d**, **e**) are magnified views of the left-hand side of (**a**)

hollow hexagons, coarse dendrites connected with two-phase eutectic clusters as well as fine three-phase eutectics. Magnified views of the microstructure in Fig. 4b, c showed that primary hexagonal $MgZn_2$ crystals are surrounded by a layer of Mg_2Zn_{11} and a few coarse irregular Mg_2Zn_{11}/Zn (stable) eutectics, wherein the Zn phase tends to grow into a dendritic morphology at the periphery of eutectic colonies. Regions between two-phase eutectic clusters are fine Mg_2Zn_{11}/Zn/Al three-phase eutectics. Such a type of microstructure is related to the first solidification path as determined through the HEXRD experiments (cf. Fig. 3a). Compared to these microstructure (Fig. 4b, c), magnified views of the microstructure in Fig. 4d, e showed that primary hexagonal $MgZn_2$ crystals are followed by subsequent irregular and lamellar metastable $MgZn_2$/Zn eutectic clusters. Regions between the metastable eutectic clusters are the $MgZn_2$/Zn/Al three-phase eutectics.

Discussion

The present work showed that a minor addition of Al to the binary Zn–3Mg eutectic composition promotes the competition between a stable and a metastable eutectic. Due to this competition, two solidification paths were observed through the HEXRD experiments. The competition may be governed by an advantage of the metastable eutectic phase $MgZn_2$ over that of a stable eutectic phase Mg_2Zn_{11}. According to the calculated phase equilibria in Fig. 1, the temperature-composition windows for the stable and the metastable eutectic overlap at relatively low temperatures. Thus, they can compete with each other at a slight liquid undercooling as in the binary Zn–Mg composition. It was explained elsewhere [15, 16] that the $MgZn_2$ Laves phase has a much lower solid/liquid interfacial energy than that of Mg_2Zn_{11} and Zn and therefore may attain a higher nucleation rate in solidification. In light of the present HEXRD experiments, the Al addition does not change the nucleation potency of primary $MgZn_2$ phase so much. Rather, it brings about a significant effect on the nucleation of a secondary phase following primary $MgZn_2$. The Zn–3Mg–2Al composition under study lies close to the metastable eutectic composition. As $MgZn_2$ nucleates and grows from the liquid, the liquid composition might follow the liquidus of Mg_2Zn_{11} until the stable Mg_2Zn_{11}/Zn eutectic composition is reached; then, stable eutectic crystallizes from the liquid. This process relies on a supersaturation driving force due to a liquidus composition difference between $MgZn_2$ and Mg_2Zn_{11}, which drives the peritectic reaction ($L + MgZn_2 \rightarrow Mg_2Zn_{11}$); the peritectic product Mg_2Zn_{11} can then nucleate the stable Mg_2Zn_{11}/Zn eutectic. This phase formation sequence is consistent with the HEXRD results in Fig. 3a and solidification microstructure of Fig. 4b, c. However, the growth of $MgZn_2$ may be extended to lower temperature following the liquidus of $MgZn_2$ until the liquid reaches the metastable $MgZn_2$/Zn eutectic composition. Then, the metastable $MgZn_2$/Zn eutectic will precipitate from the undercooled liquid. Since the faceted $MgZn_2$ Laves phase has higher interface temperature than that of Zn, it promotes the nucleation of Zn and the coupled growth with $MgZn_2$. This condition is met in the second solidification path in Fig. 3b and is in agreement with local microstructure shown in Fig. 4d, e. In each case, the three-phase eutectics can grow from the remnant liquid following either kind of two-phase eutectic. Compared to the effect of minor Ni on the solidification behavior and microstructures of Zn–3Mg alloys (Wang Y. et al., to be published), there is no such Mg_2Zn_{11}/Zn or $MgZn_2$/Mg_2Zn_{11}/Zn spiral structures observed. It might be due to the fact that the driving force for the peritectic reaction is smaller than that required to grow the metastable eutectic below the metastable eutectic temperature for the Zn–Mg–Al system.

Conclusions

The effects of a minor Al addition on solidification behavior and microstructure formation of Zn–3Mg alloys have been investigated. Two kinds of solidification paths have been determined for Zn–3Mg–2Al alloys. A major difference between the two paths is the species of the secondary phase. In the first pathway, primary $MgZn_2$ particles crystallize first, followed by secondary Mg_2Zn_{11} and Mg_2Zn_{11}/Zn eutectic, and finally the ternary Mg_2Zn_{11}/Zn/Al eutectic. In contrast, in the secondary pathway, the metastable $MgZn_2$/Zn eutectic grows following the nucleation of primary $MgZn_2$ particles. Then, the ternary $MgZn_2$/Zn/Al eutectic crystallizes as the last product of solidification. Selection of the paths has been explained by considering local undercooling and the driving force.

Acknowledgements YW and JG thank financial support from the National Natural Science Foundation of China (U1502272). YW also thanks the China Scholarship Council for a visiting Ph.D. studentship. All authors thank Yang Ren and Leighanne C. Gallington for their help in the HEXRD experiments and are grateful to the use of the resources of the beamline 11-ID-C of the Advanced Photon Source, a U.S. Department of Energy (DOE) Office of Science User Facility operated for the DOE Office of Science by Argonne National Laboratory under Contract No. DE-AC02-06CH11357.

References

1. de Bruycker E, Zermout Z, de Cooman BC (2007) Zn–Al–Mg coating: thermodynamic analysis and microstructure related properties. Mater Sci Forum 539–543:1276–1281
2. Hosking NC, Strön MA, Shipway PH, Rudd CD (2007) Corrosion resistance of zinc-magnesium coated steel. Corros Sci 49:3669–3695
3. Vojtěch D, Kubásek J, Novák P (2011) Mechanical and corrosion properties of newly developed biodegradable Zn-based alloys for bone fixation. Acta Biomater 7:3515–3522
4. Liu X, Sun J, Zhou F, Yang Y, Chang R, Qiu K, Pu Z, Li L, Zheng Y (2016) Micro-alloying with Mn in Zn–Mg alloy for future biodegradable metals application. Mater Des 94:95–104
5. Kang N, Na HS, Kim SJ, Kang CY (2009) Alloy design of Zn–Al–Cu solder for ultrahigh temperatures. J Alloys Compd 467:246–250
6. Galib RH, Hasan RA, Sharif A (2016) Study of off-eutectic Zn–xMg high temperature solder alloys. J Mater Sci Mater Electron 27:8734–8744

7. Prosek T, Nazarov A, Bexell U, Thierry D, Serak J (2008) Corrosion mechanism of model zinc-magnesium alloys in atmospheric conditions. Corros Sci 50:2216–2231
8. Yao C, Wang Z, Tay SL, Gao W (2014) Effects of Mg on morphologies and properties of hot dipped Zn–Mg coatings. Surf Coat Technol 260:39–45
9. Liang P, Tarfa T, Robinson JA, Wagner S, Ochin P, Harmelin MG, Seifert HJ, Lukas HL, Aldinger F (1998) Experimental investigation and thermodynamic calculation of the Al–Mg–Zn system. Thermochim Acta 314:87–110
10. Oh MS, Kim SH, Kim JS, Lee JW, Shon JH, Jin YS (2016) Surface and cut-edge corrosion behavior of Zn–Mg–Al alloy-coated steel sheets as a function of the alloy coating microstructure. Met Mater Int 22:26–33
11. Kubásek J, Pospíšilová I, Vojtěch D, Jablonská E, Ruml T (2014) Structural, mechanical and cytotoxicity characterization of as-cast biodegradable Zn–xMg (x = 0.8–8.3%) alloys. Mater Tech 48:623–629
12. Ren Y (2012) High-energy synchrotron X-ray diffraction and its application to in situ structural phase-transition studies in complex sample environments. JOM 64:140–149
13. Wang Y, Gao J, Kolbe M, Chuang A, Ren Y, Matson D (2018) Metastable solidification of hypereutectic Co_2Si-CoSi composition: microstructural studies and in-situ observations. Acta Mater 142:172–180
14. Okamoto H (1995) Comment on Mg–Zn. J Phase Equilib 16:474–475
15. Fransaer J, Wagner AV, Spaepen F (2000) Solidification of Ga–Mg–Zn in a gas-filled drop tube: experiments and modeling. J Appl Phys 87:1801–1818
16. Erol M, Keşlioğlu K, Marağh N (2007) Solid-liquid interfacial energy of the solid Mg_2Zn_{11} phase in equilibrium with Zn–Mg eutectic liquid. J Phys Condens Matter 19:176003
17. Liu HY, Jones H (1992) Solidification microstructure selection and characteristics in the zinc-based Zn–Mg system. Acta Metall Mater 40:229–239

Part II
Thermomechanical Processing

Observation of Recrystallization Behavior of Nb-Microalloyed Wide Flange Beams during Hot-Rolling

Bon Seung Koo and Jae Chang Song

Abstract Nb-microalloying has a significant effect on microstructure evolution during hot-rolling. Metallurgical benefits of niobium are associated with the formation of Nb(C, N) in ferrite and austenite, retardation of austenite recrystallization, and further austenite to ferrite phase transformation. Precipitation and microstructure control are therefore important features to achieve better strength and toughness simultaneously. Effects of hot-rolling conditions on strength and toughness are of great interest because the rolling parameters, e.g., temperature, roll force, and cooling rate, contribute to strengthening mechanism of Nb-microalloyed steels. Grain refinement is the most practical way to enhance the niobium-bearing steels by strain-induced precipitation hardening below the non-recrystallization temperature. This rolling research involves residual stress variation associated with parallel-flange section geometry. Therefore, a repetitive hardening–softening mechanism could be an important feature to predict the final mechanical properties. Computational and experimental analysis has been used to determine flow stress according to the temperature change in hot-rolling. The recrystallization behavior is experimentally observed through the multiple rolling.

Keywords Hot-rolling · Niobium · Nitrogen · Wide flange beam · Recrystallization

Introduction

The strength of structural steel is a primary concern to construct high-rise buildings. The number of world high-rise buildings was estimated to 108,924 in 2006, and it represented a nearly 300% increase from 1982 [1]. Steel grades in yield strength

B. S. Koo (✉) · J. C. Song
Hyundai Steel Company, Technical Research Center, 1480 Bukbusaneop-ro, Dangjin, Chungnam 31719, South Korea
e-mail: bsk207@hyundai-steel.com

J. C. Song
e-mail: songjc@hyundai-steel.com

J. Lee et al. (eds.), *Materials Processing Fundamentals 2020*, The Minerals, Metals & Materials Series, https://doi.org/10.1007/978-3-030-36556-1_4

(YS) classes up to 275 MPa prevailed in the 1970s for general construction purposes; higher strength steel materials exceeding YS of 355 MPa have been promoted by the increase in size and height of the building. The steel materials have been replaced slowly, but the desires rapidly spread out to all available higher steel grades. There are always practical challenges to integrating new steel into actual engineering design.

A need for high-strength and high-performance steel is gradually increasing with the latest architectural trends related to high-rise and long-span construction. The field application often requires consistency in quality and good weldability of the supplied steel product. A conventional manufacturing method, i.e., as-rolled condition, therefore has been replaced with a thermo-mechanically controlled process (TMCP) in order to overcome the degradation of weldability [2]. The thermo-mechanically controlled steels have significantly improved strength, toughness, and weldability compared to as-rolled products due to the grain refinement and lower carbon equivalents [3]. In accordance with the construction industry trends, a study has begun to develop a hot-rolled H-shaped beam with a minimum yield strength of 460 MPa. In this study, we described the recrystallization and non-recrystallization behavior of Nb-microalloyed 460 MPa grade structural shapes in relation to the substantial demanding of the higher-grade steels.

Discussion

Materials and Processes

A research subject was a yield strength of 460 MPa hot-rolled H-shaped steel beam. A beam size required for the research was H800 × 300 × 14 × 26 whose web was considered to be wider than regular beams as shown in Fig. 1. The 460 MPa grade beam in Korea is designated to have a minimum yield strength of 460 MPa, a tensile strength in between 570 and 720 MPa, and a maximum yield-to-tensile ratio of 0.85 [4]. Table 1 is the chemical compositions of the 460 MPa grade beam specified in

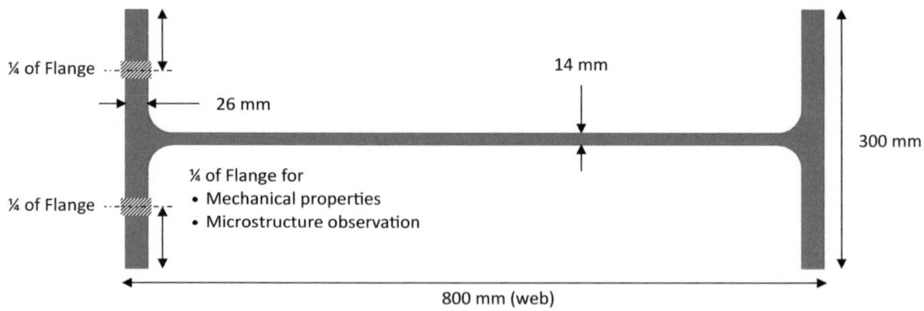

Fig. 1 Dimensions of wide flange beam, H800 × 300 × 14 × 26

Table 1 Chemistry requirements in KS D 3866 standard

KS D 3866	Max. wt%												
	C	Si	Mn	P	S	Cu	Ni	Cr	Mo	V	Nb	Mn/S (min.)	Nb + V	Ceq
460 MPa Gr.	0.2	0.4	1.6	0.035	0.03	0.6	0.45	0.35	0.15	0.11	0.05	20	0.15	0.45

Table 2 Mechanical characteristics in KS D 3866 standard

KS D 3866	Min. (N/mm^2)		Max. (%)		Min. (%)		Min. (J)
	YP	TS	YR		EL		CVN (−5 °C)
	–	–	$t \leq 10$	$t > 10$	$t \leq 40$	$t > 40$	*L*-direction
460 MPa Gr.	460	570–720	87	85	17	20	47

Korean Standard (KS), and Table 2 is the corresponding mechanical properties. The mechanical strength would be enhanced by the adjustment of chemical compositions and optimized rolling conditions.

Niobium was used as a main alloy element in C-Mn-V-0.020Nb-N system. Niobium was assumed to be a key element to satisfy the mechanical testing requirements which required both strengths and yield-to-tensile ratio, as described in KS D 3866 [4]. The metallurgical aspect related to strain-induced precipitation hardening of niobium carbonitrides has been considered enough to achieve the yield and tensile strength [5]. Hot-rolling at a temperature range below recrystallization stop temperature was first tried to strengthen the material. This was the strengthening mechanism of Nb-bearing carbon steel by reducing the thickness of the material and simultaneously diminishing the grain size. Figure 2 is a schematic representation of microstructures which can be formed differently due to the temperature variation in rolling.

Hot-rolling was conducted at Pohang factory of Hyundai Steel. Thermo-mechanically controlled (TMC) process has been operated within a narrow temperature range. Beam blank reheating temperature and start rolling temperature, which corresponded to the initial rolling condition, were handled following by a standard operating procedure. Key control factors of the study were the dissolution temperature of niobium carbide/carbonitride precipitates, the non-recrystallization

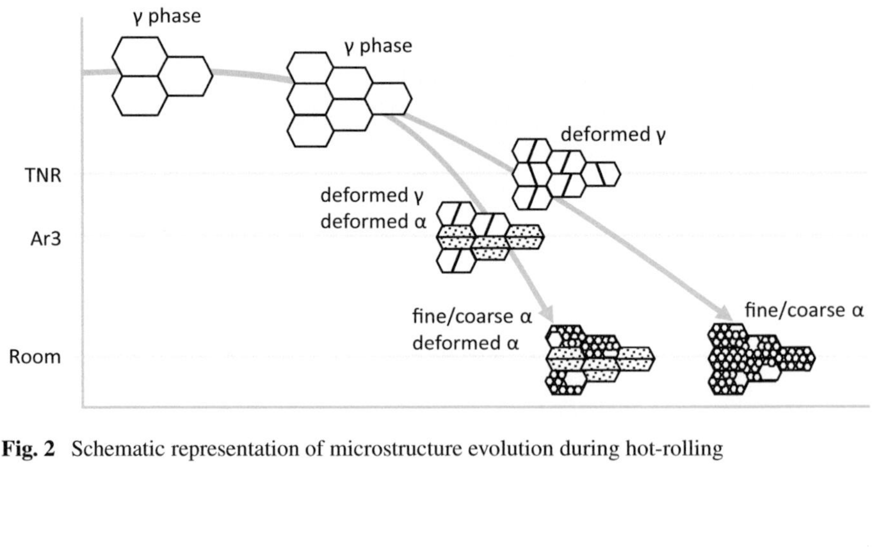

Fig. 2 Schematic representation of microstructure evolution during hot-rolling

Table 3 Controlled rolling conditions for Nb effect

Ctrl.	CASE #1	CASE #2	CASE #3	CASE #4	CASE #5	CASE #6
Recrystallization stop temperature	915 °C					
Finish rolling temperature	760 ± 20 °C			810 ± 20 °C		
Hot-rolling pass	9			9	9	7

temperature (TNR) of the Nb-microalloyed steel, and a finishing rolling temperature. Flow stress analysis was used to measure the degree of dynamic recrystallization during the temperature drop in beam rolling. Table 3 was the 6 different conditions applied to the research. All chemical compositions were the same from CASE #1 to CASE #6 as well as TNR, and the only difference was the finish rolling condition to observe the effect of finishing temperature on the microstructure evolution and material strength. An accumulated reduction ratio below TNR was assumed to be in the range of 30–50%.

Control of nitrogen content was also thought to be another key aspect to enhance mechanical properties in niobium-bearing steels. Nitrogen was proven to an effective strengthening element if it was controlled at an appropriate level [6]. N-binding elements such as Nb, V, Al, and Ti could be added to enhance the performance of steel products through precipitation hardening and grain refining. Additional tests for the nitrogen effect were described in Table 4. All chemical compositions were identically the same as the research cases in Table 3, except nitrogen level and finish rolling temperature. Similar to CASE #4 through #6, the finish rolling temperature for CASE #7 through CASE #10 was 810 ± 20 °C with the 9-pass hot-rolling process.

Mechanical testing was conducted in accordance with KS B 0801 [7]. Basic requirements, e.g., YS, TS, and elongation (EL), were directly measured via tensile testing, and Charpy V-notch impact testing was added to evaluate low temperature toughness. The results were collected over repeated tests for reliability and consistency.

It was important to estimate the influence of niobium addition on the microstructure and mechanical characteristics. The strength increase mechanism by the strain-induced precipitation hardening was supposed to take effect under continuous rolling operations. The hardening procedure below TNR was therefore of great interest in this study. A pre-engineering study should be conducted before the production trial runs to increase the probability of successful implementation.

Table 4 Control of nitrogen content

Ctrl.	CASE #7	CASE #8	CASE #9	CASE #10
Nitrogen level (in ppm)	70	106	135	183

Results

Rolling Reduction Below TNR

We assumed that austenite grain was deformed and recovered repeatedly over TNR, but further recrystallization process would stop occurring below TNR. The temperature of the recrystallization stop was estimated at approximately 915 °C in the given C-Mn-V-Nb-N system, and rolling in non-recrystallization temperature region has caused a substantial increase in strength. The measured temperature profile during hot-rolling was plotted in Fig. 3.

The flow stress profiles for the 6 cases (CASE #1 through CASE #6) are shown in Fig. 4. A roll force applied perpendicular to the flange was obtained reading load cell data. A mean flow stress was calculated back from the roll force using roll bite geometry. Flow stress per each rolling pass was used to analyze the effect of precipitation hardening of Nb(C, N) [8]. There were several intermittent peak values observed in the roll force data; thus, average input signals over stable readings were chosen for the force-to-stress conversion. CASE #1 through CASE #6 showed the variation of the flow stresses, and each graph indicated a meaningful slope change adjacent to the temperature, TNR in the stress vs rolling-pass plots. Thickness reduction occurred above TNR was defined as TYPE I (smooth slope). The strength was slowly increased representing a zigzag pattern due to the repetitive grain recovery after strain-induced grain refinement [9]. A zone, the right-hand side of a “gray” colored bar, was corresponding to the temperature region under TNR. We have thought that Nb(C, N) precipitates resulted in TYPE II (steeper slope) in the non-recrystallization temperature region. The steeper slope was corresponding to an area reduction which

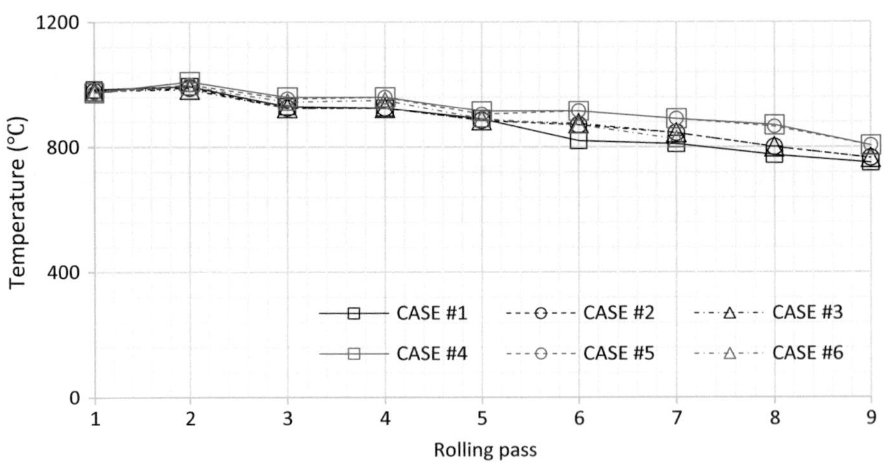

Fig. 3 Temperature profile during hot-rolling (CASE #1 through CASE #6)

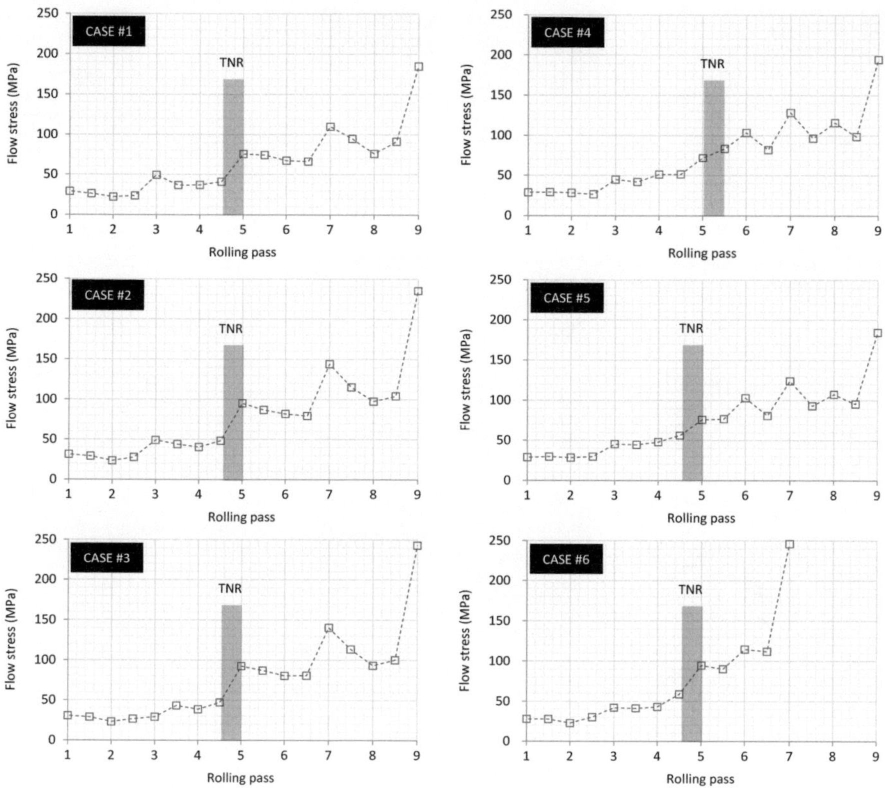

Fig. 4 Flow stress and estimated TNR range during hot-rolling (CASE #1 through CASE #6)

might have occurred with no grain growth. Consequently, the stress was substantially increased due to the microstructure which was continuously refined by strain hardening without recovery. Figure 4 describes that the strain-induced precipitation hardening of Nb(C, N) has come into effect as we expected at the beginning of the research.

The effect of recrystallization/non-recrystallization was also evaluated by visual inspection of the microstructure at a 1/4 point of the flange (as of Fig. 1). The microstructure observation of CASE #1 through CASE #6 was conducted by an optical microscope, and the hot-rolled beam specimens exhibited a ferrite–pearlite microstructure as shown in Fig. 5. Pancake-like flattened ferrite grains existed in CASE #1 through CASE #3, and the finish rolling temperature of 760 ± 20 °C was sufficiently low to include rolling in the ferrite region during the last few steps. The lower finish rolling temperature has therefore initiated to form the microstructure of unrecovered-flattened ferrite in the final stage. In contrast, CASE #4 through CASE #6 resulted in a mixture of equiaxed fine and coarse ferrite grains which had no pancake-like shape. It seemed that hot-rolling with finish rolling temperature of 810 ± 20 °C completed its process in the austenite region, and the deformed austenite

Fig. 5 Microstructure observation after hot-rolling (CASE #1 through CASE #6)

has changed into a fine/coarse equiaxed ferritic microstructure when the temperature dropped below Ar3.

The distinction between two group sets (CASE #1 through CASE #3 and CASE #4 through CASE #6) was the use of different finish rolling temperature and corresponding accumulated rolling reduction ratio below TNR. In order to find a relationship between grain size and accompanying flow stress, the stress profile was plotted as a function of average grain size for six cases as shown in Fig. 6. A correlation coefficient was about 0.71. A data plotting revealed that the stress increases due to the controlled reduction rolling and the final microstructure (grain size and crystallographic texture) had a strong relationship with each other.

Nitrogen Effect

Nitrogen could refine the final grain size and avoid material softening by nitrogen precipitates, i.e., Al-N which prevented grain growth during heat treatment [10]. In high intensity heat welding, the toughness of heat affected zone was enhanced by

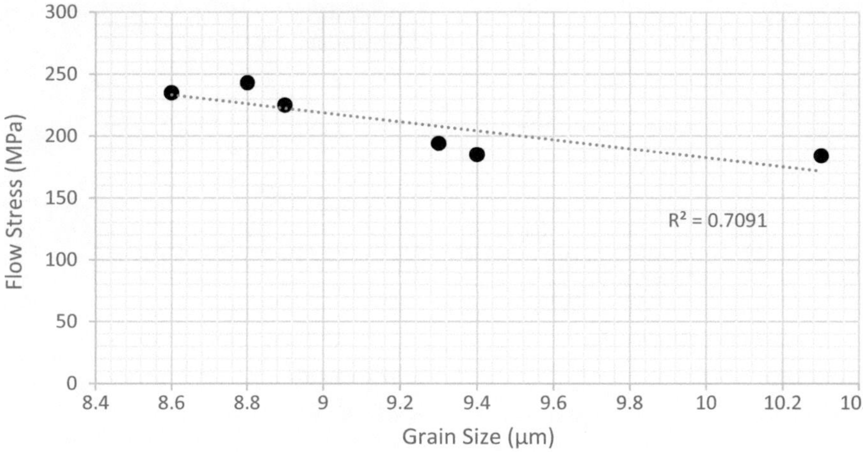

Fig. 6 Grain size variation in flow stress (CASE #1 through CASE #6)

forming Ti-N because grain recovery was a frequent issue decreasing hardness and strength. Change in microstructure and precipitation behaviors due to nitrogen level was considered another key factor influencing material characteristics. Precipitation was associated with the weight percentage of alloying elements; thus, nitrogen level and the resultant grain size were assumed to influence mechanical performance. A 9-pass hot-rolling was conducted at a finish rolling temperature of 810 ± 20 °C, and nitrogen was added in accordance with Table 4. Nitrogen level was increased from 70 to 183 ppm (with an increment of 30 ppm approximately), and microstructure observation through an optical microscope showed that the grain size has been gradually reduced from 15.7 μm down to 10.5 μm as shown in Fig. 7. It seemed that nitrogen precipitates have been acting as ferrite nucleation sites, resulting in the refinement of the microstructure, and the strength enhanced by the increase of the finely dispersed nitrogen precipitates.

The tensile and impact tests of CASE #7 through CASE #10 were also performed to analyze the material characteristic in relation to the nitrogen level. The minimum requirements for TS, YP, and CVN were 460 MPa, 720 MPa, and 47 J at − 5 °C, respectively. Figure 8 describes mechanical testing results taken from a 1/4 point of the flange according to KS specification. A steady increase in tensile and yield strengths was observed while the nitrogen level raised from 70 to 183 ppm. Impact testing result was also proved that nitrogen could be an effective strengthening element up to 183 ppm although there was a relatively slow increase over 135 ppm.

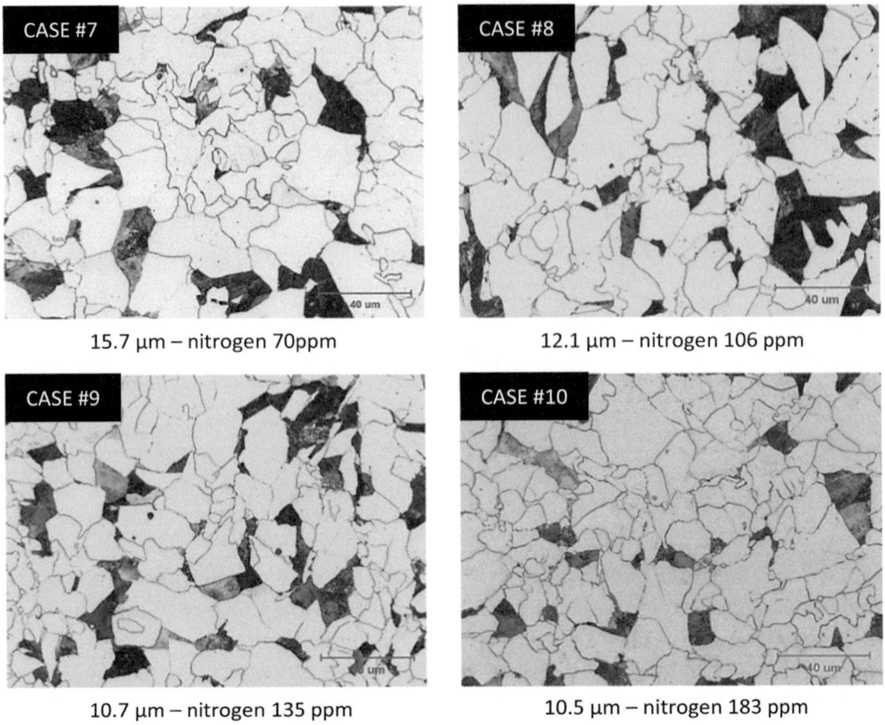

Fig. 7 Grain size variation in nitrogen level (CASE #7 through CASE #10)

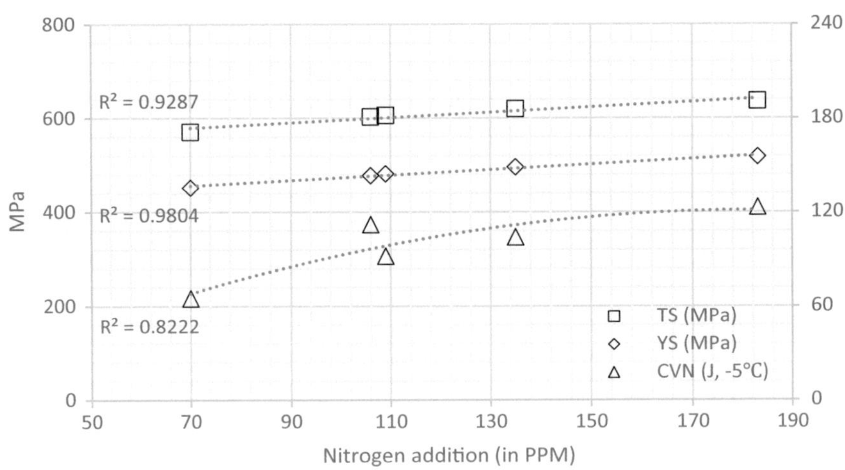

Fig. 8 Change in mechanical properties by nitrogen level

Conclusion

A general principle was to increase strength by precipitation hardening of the niobium element. Rolling conditions have been set up to reveal the characteristics of the alloying element. Mean flow stress analysis per pass confirmed the pattern of increase in strength before and after TNR, and the influence of N precipitates was also examined by tensile testing and microstructure observations. The non-recrystallization behavior during hot-rolling was well-studied in this research, and the strengthening mechanism with alloying elements was investigated as a result. The research results were summarized as below.

(1) Flow stress analysis is an effective way to visualize the dynamic recrystallization behavior of Nb-microalloyed steels during the hot-rolling process.
(2) Pancake-like deformed ferrite can be obtained by rolling in the ferrite region while hot-rolled austenite transforms into equiaxed ferrite due to grain recovery process over Ar3.
(3) With appropriate use, nitrogen can be an effective strengthening element preventing grain growth in austenite.

References

1. Ali MM, Moon KS (2007) Structural developments in tall buildings: current trends and future prospects. Archit Sci Rev 50(3)
2. de Meester B (1997) The weldability of modern structural TMCP steels. ISIJ Int 37(6):537–551
3. Nishioka K, Ichikawa K (2012) Progress in thermomechanical control of steel plates and their commercialization. Sci Technol Adv Mater 13(2)
4. KS D 3866 (2018) Korean Agency for Technology and Standards
5. Li Y, Crowther DN, Mitchell PS, Craven AJ, Baker TN (2004) The effects of vanadium, niobium, titanium and zirconium on the microstructure and mechanical properties of thin slab cast steels 44(6)
6. Bhav Singh B, Siva Kumar K, Madhu V, Arockia Kumar R (2017) Effect of hot rolling on mechanical properties and ballistic performance of high nitrogen steel. Procedia Eng 173:926–933
7. KS B 0801 (2017) Korean Agency for Technology and Standards
8. Dutta B, Sellars CM (1987) Effect of composition and process variables on Nb(C, N) precipitation in niobium microalloyed austenite. Mater Sci Technol 3(197)
9. Silva MBR, Gallego J, Cabrera JM, Balancin O, Jorge AMJ (2015) Interaction between recrystallization and strain-induced precipitation in a high Nb- and N-bearing austenitic stainless steel: influence of the interpass time 637(2015)
10. Razzak MA, Perez M, Sourmail T, Cazottes S, Frotey M (2014) Preventing abnormal grain growth of austenite in low alloy steels. ISIJ Int 54(8):1927–1934

Effects of Heat Treatment Method on Microstructure and Mechanical Properties of Internal Crack Healing in SA 508-3 Steel

Yao Qiu, Ruishan Xin, Jianbin Luo and Qingxian Ma

Abstract The mechanical property is an important index to evaluate the recovery degree of the internal crack zone. This work presents interfacial characteristics and Charpy impact properties of crack healing zones in SA 508-3 steel, healing by hot compression and different healing treatment methods. When the holding time increased, the hardness of the matrix and crack healing zone decreased. The hardness of the matrix was usually higher than the crack healing zone, but the value of differences decreased with the increasing holding time. The percentage recovery of impact property increased when the holding time increased from 0 to 10 h. After holding for more than 10 h, the impact property decreased due to the mixed crystal structure. With the same total holding time, the recovery rate of crack healing slowed down with the increment of heating times. The results show that for specimens with internal cracks, the method of healing at 950 °C and holding for 10 h was enough for crack healing without sacrificing impact property.

Keywords Internal crack healing · Heat treatment · Mechanical properties

Introduction

Methods to achieve internal crack healing in steel have received much attention in recent years, which could offer important economic benefits. Nowadays, the healing methods mainly include heat treatment [1–3], electropulsing method [4, 5], and hot

Y. Qiu · J. Luo · Q. Ma (✉)
Department of Mechanical Engineering, Tsinghua University, Beijing, China
e-mail: maqxdme@mail.tsinghua.edu.cn

Y. Qiu
e-mail: qiuy15@mails.tsinghua.edu.cn

J. Luo
e-mail: luojb@thpx.org

R. Xin
HBIS Group Technology Research Institute, Shijiazhuang, China
e-mail: ruishanxin@163.com

J. Lee et al. (eds.), *Materials Processing Fundamentals 2020*, The Minerals, Metals & Materials Series, https://doi.org/10.1007/978-3-030-36556-1_5

plastic deformation [6–8]. In the actual production of heavy forgings, hot plastic deformation and heat treatment are the most efficient methods. However, the previous studies often focus on the microstructural evolution in the crack healing zone, and quantitative studies have been seldom done on the restoration of mechanical properties.

The evaluation criteria of mechanical properties include hardness evaluation, tensile property evaluation, and impact property evaluation. There have been extensive studies about the tensile properties recovery. For example, Yang et al. [9] studied the tensile properties of twinning-induced plasticity steel healing by the electropulsing method and pointed out that the process of electropulsing treatment could heal the macro-voids completely, which led to the recovery of tensile strength and elongation rate, but the process of annealing did not have the same effect. Zheng et al. [10] conducted tensile tests on the pure Ni sheets with a crack after healing by electropulsing treatment and found out that the tensile strength of the healed sample could be compared to the original materials. The healing efficiency was related to both the fraction of fully healed region and the strength difference in different regions. However, although the effects of healing parameters on the tensile property recovery were demonstrated, little attention has been paid to healing parameters on the impact properties, especially by the methods of hot deformation and heat treatment.

The present paper presents a set of experiments for selecting a proper heat treatment method. On the basis of the different total holding time and different heating times, the microstructure evolution and the impact property recovery were systematically studied. The fracture morphology has also been studied by scanning electron microscope (SEM). Under the analysis of microstructure and mechanical property, a proper heat treatment method was put forward.

Method

The butt-joint method was utilized in this study to preset internal crack in SA 508-3. The chemical compositions of SA 508-3 steel are 0.19 C, 0.22 Si, 1.4 Mn, 0.006 P, 0.006 S, 0.12 Cr, 0.53 Mo, and 0.65 Ni (in wt%) (Fig. 1).

The SA 508-3 steel sample was cut into cylinders with a diameter of 120 mm and a height of 60 mm. The surface of the cylinders was polished to Ra 1.3 μm and cleaned by ethanol before two surfaces were welded along the circumference of the surface. The contact surface was the preset crack zone.

Hot compression test was conducted by an 8MN machine at 950 °C with the compression rate of 20%. Then a series of healing experiments for the specimens after compression were conducted at 950 °C in a batch-type heating furnace for 5 h, 10 h, 15 h, and 20 h, respectively. To study the influence of different heating times, half of the specimens holding for 5 h were cooled to room temperature in the air with the cooling rate of 700 °C/h and then were reheated to 950 °C with the heating rate of 600 °C/h and held for another 5 h, the specimens were named as specimen '5 h

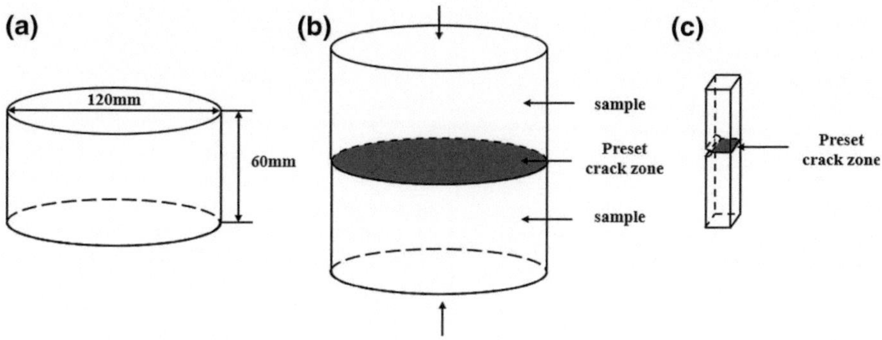

Fig. 1 A schematic illustration of the method to prepare SA 508-3 steel samples with internal pre-cracks **a** the SA508-3 cylinder, **b** butt-joint method, **c** standard Charpy specimens

+ 5 h'. And half of the specimens holding for 10 h were reheated for another 10 h, named as specimen '10 h + 10 h'.

Standard Charpy tests were conducted to study the impact property after internal crack healing by a 300 J pendulum impact testing machine. The Charpy samples, which had the size of 55 mm × 10 mm × 10 mm, were cut along the axial direction. The U notch was set on the middle part of the Charpy samples, and the preset crack zone located at the middle of the u-shaped groove (Fig. 1). To quantitatively evaluate the percentage recovery of impact property, contrastive experiments were performed. The matrix Charpy samples were machined from the SA 508-3 steel cylinders without preset cracks, obtained during the same healing process.

The fractographs were examined using a JSM-7100F scanning electron microscope. The specimens were polished and etched by a picric acid solution to observe the microstructure evolution. The indentation tests were conducted in the crack healing zones and the matrix.

Results and Discussion

Microstructures of Crack Healing Zone After Hot Compression and Post-Heat Treatment

Figure 2a–f shows the microstructures of crack healing zones after holding at 950 °C for 5 h, 10 h, 15 h, 20 h, 5 h + 5 h, 10 h + 10 h, respectively. Obvious remnant crack was observed in Fig. 2a, which indicated that only partially healing occurred when holding for 5 h. When holding for 10 h, it was clear that the nucleation of new grains has not occurred in the crack healing zone, and the absence of new small grains in the crack healing zone implied that the thermal energy was not sufficient to motivate recrystallization on the crack healing zone. However, some grains in the matrix grew

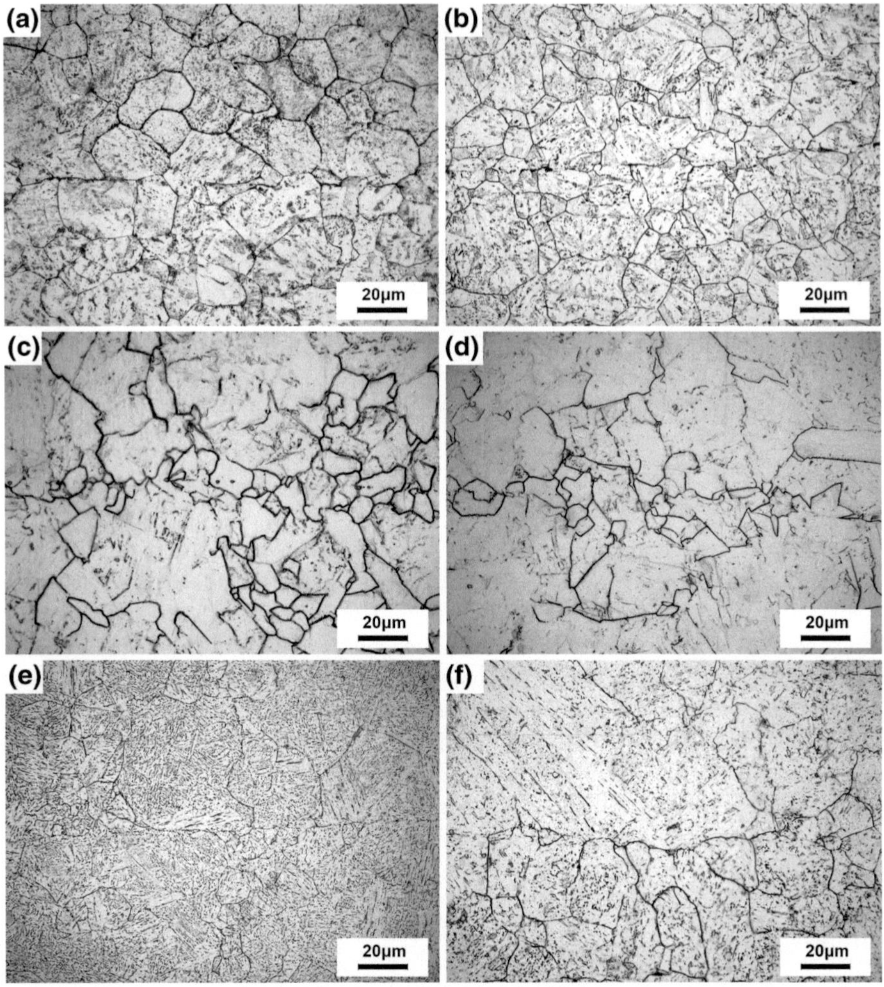

Fig. 2 Microstructure of the crack healing zone after holding at 950 °C for different healing treatment **a** 5 h, **b** 10 h, **c** 15 h, **d** 20 h, **e** 5 h + 5 h, **f** 10 h + 10 h

up and step across the original crack, which segmented the remnant original crack, as shown in Fig. 2b. When the holding time prolonged to 15 and 20 h, small grains were generated along the crack healing zone and replaced the original crack. Meanwhile, it is observed that grain coarsening occurred in the matrix.

When the total holding time was 10 h, the grains in the matrix healing by 5 h + 5 h were bigger than that by 10 h, compared between Fig. 2b, f. In the sample 5 h + 5 h, small grains could be observed in the matrix near the healing zone, and the original crack was still segmented by grownup grains, which shares the same features with the samples 10 h. When the total holding time was 20 h, small grains were hardly observed in sample 10 h + 10 h.

Microhardness of Crack Healing Zone and the Matrix After Hot Compression and Post-Heat Treatment

Indentation hardness tests were carried out in the crack healing zone and the matrix healing by different heat treatments. The hardness measurements were repeated three times, and the average values of hardness were listed in Fig. 3.

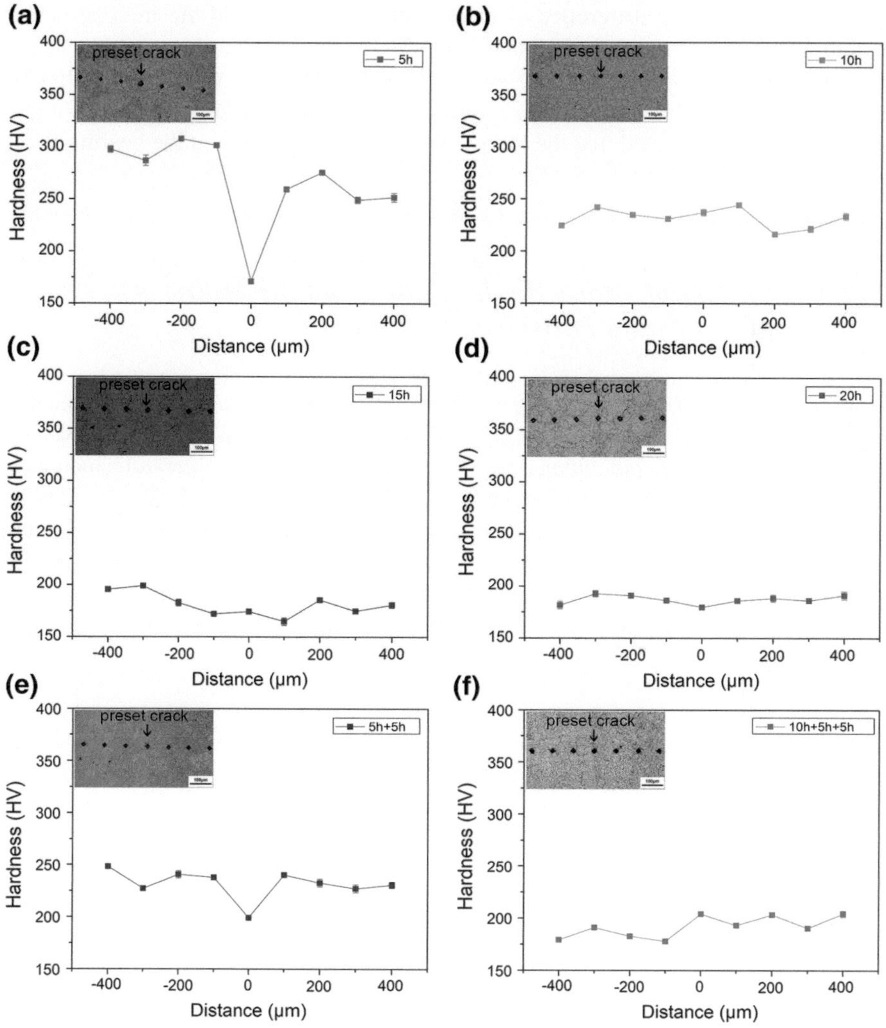

Fig. 3 Distributions of microhardness in the crack healing zone and the matrix after holding at 950 °C for different healing treatment **a** 5 h, **b** 10 h, **c** 15 h, **d** 20 h, **e** 5 h + 5 h, **f** 10 h + 10 h

Figure 3a shows that the matrix has a higher hardness than the crack healing zone when holding time was 5 h. The hardness started to increase sharply within the distance of 100 μm from the interface, which was caused by the incomplete healing of the original crack. When the holding time prolonged, as the recovery degree of internal crack increased, the hardness curve became flattened gradually, and the hardness across the sample became almost the same (Fig. 3b–d). The average hardness in the matrix decreased gradually when the holding time increased.

From the microstructure of the internal crack zone observed in Fig. 2b, e, it could not be figured out the difference of crack recovery degree and the microstructure evolution between the sample 10 h and 5 h + 5 h. Though the hardness was basically the same with the sample holding for 10 h (Fig. 3b), there was still an obvious decrease of the hardness in crack healing zone in the sample holding for 5 h + 5 h (Fig. 3e), which indicated that the recovery degree deceased when the heating times increased.

Impact Property of Crack Healing Zone and the Matrix After Hot Compression and Post-Heat Treatment

Table 1 gives the room-temperature impact properties of crack healing zones at 950 °C for different heat treatment method. It was found that with the increase of total holding time, the percentage recovery increased when temperature increased from 5 to 10 h, and a sharp decline occurred when the holding time increased from 15 to 20 h, which was the result of the grain sizes difference between the matrix and newly formed grains, shown in Fig. 2c and d. With the same total holding time for 10 h, the values of impact absorbed energy in the matrix were basically the same in the samples '10 h' and '5 h + 5 h'. But the sample '10 h' has a higher value of percentage recovery of the impact property than the sample '5 h + 5 h'. For the

Table 1 Impact property of the samples healing at 950 °C by different heat treatment method

Heat treatment method	Total holding time (h)	Impact absorbed energy of the crack healing zone (J)	Error value (J)	Impact absorbed energy of the matrix (J)	Percentage recovery of the impact property (%)
5 h	5	10.11	1.46	47.28	21.4
10 h	10	18.63	3.03	36.11	51.6
15 h	15	12.74	2.08	29.23	43.6
20 h	20	8.02	0.12	21.30	37.7
5 h + 5 h	10	13.42	1.45	36.31	37.0
10 h + 10 h	20	6.89	0.11	21.27	32.3

The table listed the average values of three specimens at each data point samples with the total holding time of 20 h, the impact absorbed energy decreased sharply compared with samples holding for 10 h, but the recovery degrees of impact property were basically the same in the samples '20 h' and '10 h + 10 h', which indicated that when the holding time was long enough, the heat times had little effect on crack healing recovery.

It was observed that the fracture basically occurred along the preset crack. When the holding time was in the range of 5–20 h, the fracture surfaces of the impact samples were shown in Fig. 4. When the holding time was 5 and 10 h, the fracture could be divided into the unhealed region and dimple region (Fig. 4a, b). The existence of the dimple region indicated that effective internal crack healing occurred during the healing treatment. When holding time reached 10 h, the main reason for the percentage recovery enhancement of impact property was correlated with the enlarged dimple region, and the dimples had smaller diameters on the fracture surface in sample '10 h' than sample '5 h'. When holding time was higher than 15 h, the fracture microstructure was characterized by brittle fracture, having a shear character with slip planes (Fig. 4c, d). The internal crack was fully healed, meanwhile, the percentage recovery decreased because of cleavage fracture.

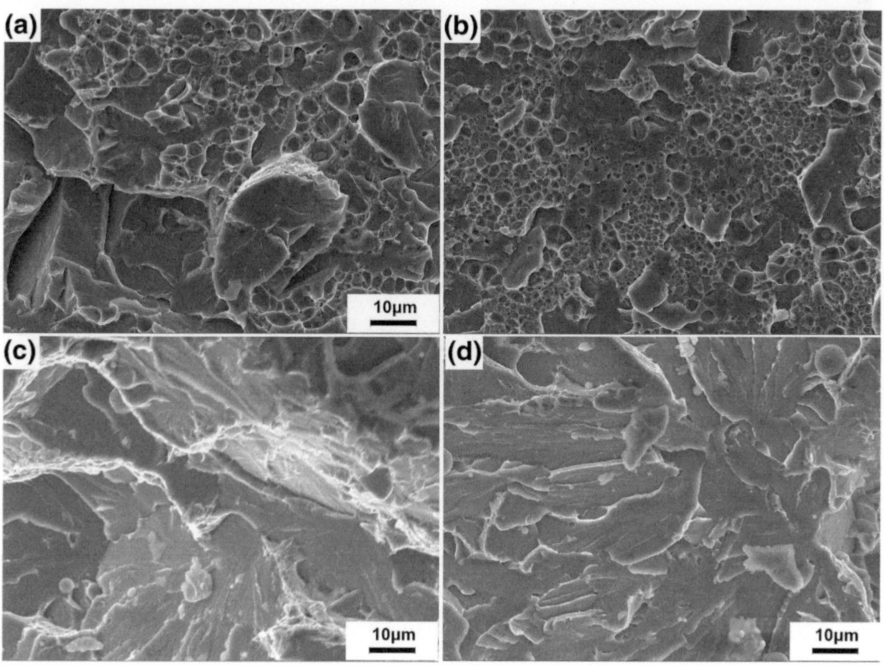

Fig. 4 Microhardness of the crack healing zone and the matrix after holding at 950 °C for different healing treatment **a** 5 h, **b** 10 h, **c** 15 h, **d** 20 h

Summary

1. After hot compression at 950 °C with the compression rate of 20%, the percentage recovery of impact property increased when the holding time increased from 0 to 10 h and decreased when the holding time was more than 10 h.
2. With the same total holding time for 10 h, the increase in heating times slowed down the recovery rate of crack healing. With the same total holding time of 20 h, the impact property decreased sharply and the recovery rate of impact property was basically the same when the heating times increased.
3. For a specimen healing at 950 °C, holding for 10 h and heating only once were enough for crack healing without sacrifice impact property.

Acknowledgements The authors gratefully acknowledge financial support from the National Natural Foundation of China (51775298).

References

1. Zhang HL, Sun J (2014) Diffusive healing of intergranular fatigue microcracks in iron during annealing. Mater Sci Eng, A 382:171–180
2. Wei D, Han J, Jiang ZY, Lu C, Tieu AK (2006) A study on crack healing in 1045 steel. J Mater Process Technol 177:233–237
3. Xin RS, Ma QX, Li WQ (2017) Effect of heat treatment on microstructure and hardness of internal crack healing in a low carbon steel. Key Eng Mater 730:3–7
4. Tang Y, Hosoi A, Morita Y, Ju Y (2013) Restoration of fatigue damage in stainless steel by high-density electric current. Int J Fatigue 56:69–74
5. Hosoi A, Nagahama T, Ju Y (2012) Fatigue crack healing by a controlled high density electric current field. Mater Sci Eng, A 533:38–42
6. Zhao X, Lin X, Chen J, Xue L, Huang W (2009) The effect of hot isostatic pressing on crack healing, microstructure, mechanical properties of Rene88DT superalloy prepared by laser solid forming. Mater Sci Eng, A 504:129–134
7. Yu H, Liu X, Li X, Godbole A (2014) Crack healing in a low-carbon steel under hot plastic deformation. Metall Mater Trans A-Phys Metall Mater Sci 45A:1001–1009
8. Qiu Y, Xin R, Luo JB, Ma QX (2019) Crack healing and mechanical properties recovery in SA 508-3 steel. Materials 12:6
9. Yang CL, Yang HJ, Zhang ZJ (2018) Recovery of tensile properties of twinning-induced plasticity steel via electropulsing induced void healing. Scripta Mater 147:88–92
10. Zheng XG, Shi YN, Lu K (2013) Electro-healing cracks in nickel. Mater Sci Eng, A 561:52–59

Teaching Metal-Forming Processes Using a Laboratory Micro-extrusion Press

Adi Ben-Artzy, Snir Ben-Ze'ev and Nahum Frage

Abstract Metal forming is widely used in the automotive, aviation, and energy industries. As such, teaching metal-forming processes is very important. These days, universities are often not equipped with suitable laboratories where students can perform metal-forming experiments. Instead, the teaching of metal-forming processes is based on simulations and class demonstrations, with no "hands-on" experience provided. An educational extrusion laboratory was established at Ben-Gurion University in Israel. The teaching laboratory contains a micro-hydraulic indirect extrusion press. The students extrude aluminum, as well as magnesium alloys at temperatures up to 650 °C using a variety of dies. The computer-controlled extruder allows online data acquisition of forces, temperature, and movement of the RAM. Students can investigate the effects of the main process parameters on the extrusion forces using different die geometries. Good agreement between measured and calculated extrusion forces was found as well as the influence of extrusion parameters on the microstructure and mechanical properties.

Keywords Metal forming · Teaching · Extrusion · Elavated temperature · Friction · Flow · Theory

Introduction

Extrusion is one of the most common plastic deformation processes. Here, a block of metal (the billet) is forced to flow through a profile-shaped die. This process is widely used in the manufacturing of light alloy products, such as aluminum profiles for doors and window frames, automotive parts, sporting goods, and recreation and outdoor uses. During the extrusion process, a large diameter billet is forced to flow through a die with a small area. In doing so, the billet undergoes severe plastic deformation, which is a driving force for the subsequent recrystallization process. The operation is usually performed at elevated temperatures, such that the flow stress is

A. Ben-Artzy (✉) · S. Ben-Ze'ev · N. Frage
Ben-Gurion University, Be'er Sheva, Israel
e-mail: benartzyadi@gmail.com

J. Lee et al. (eds.), *Materials Processing Fundamentals 2020*, The Minerals, Metals & Materials Series, https://doi.org/10.1007/978-3-030-36556-1_6

low and the strain-hardening coefficient is negligible [1]. Elevated temperature extrusion is usually conducted at the recrystallization temperature (about 0.6 the melting point of the metal comprising the billet). At this temperature, dynamic recovery and recrystallization take place so as to improve the formability of the metal. Elevated temperatures also affect product surface roughness and increase friction between the billet and the container, as well as between the metal and the die [2]. To define extrusion process parameters, one needs to understand their effect on the flow and on the properties of the final product in more detail [3].

Laboratory Work

Experiments described in this paper were conducted by S. Ben Ze'ev, a student in the Department of Materials Engineering at Ben-Gurion University of the Negev, as a part of his graduate thesis. Mr. Ben Ze'ev used equipment located in an extrusion laboratory for the third-year undergraduate students. The role of this laboratory is to enable students to validate theoretical principles and equations taught in class. In the present study, we describe how the students conducted experiments using various parameters to assess the effects of extrusion parameters on extrusion forces and on the microstructure and mechanical properties of extruded parts. The students used different dies (see Fig. 4b) to study the effects of extrusion parameters, such as the extrusion ratio, die angle, and bearing length. Experiments were conducted using aluminum and magnesium alloys.

Analysis of the Extrusion Process

Indirect Extrusion

Indirect extrusion is a form of extrusion in which a hollow stem is used to hold the die (Fig. 1b). The extruded profile exits backward, through the stem. Due to the fact that the billet is static in indirect extrusion, friction between the billet and container

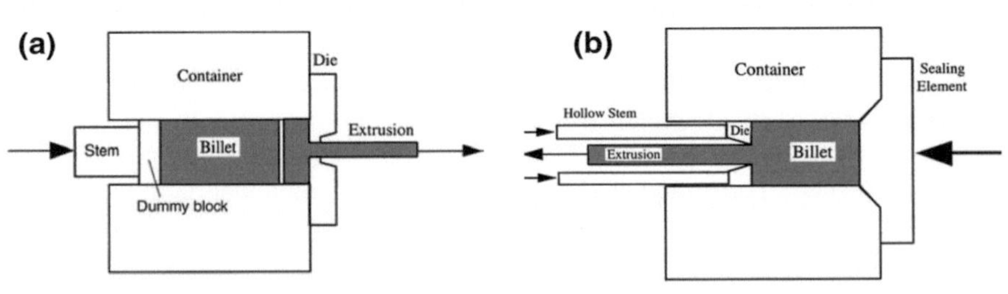

Fig. 1 Direct extrusion (**a**) versus indirect extrusion (**b**) [2]

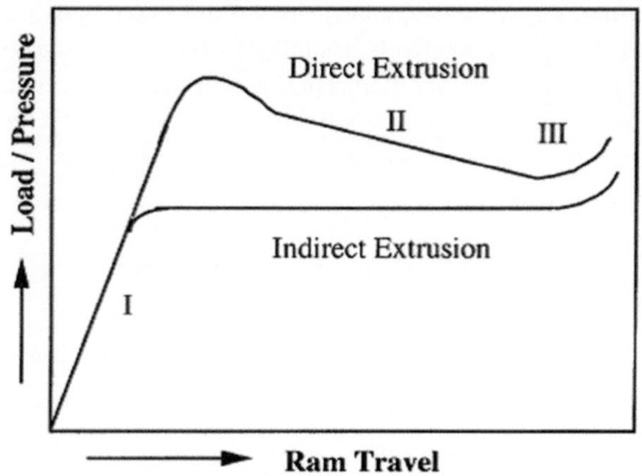

Fig. 2 Processing pressure used in direct and indirect extrusion [2]

is absent. As such, the forces applied are less than those used in direct extrusion [2, 4]. Another advantage of indirect extrusion is that the die is a part of a stem, such that its replacement is very easy and time-saving. Nevertheless, one major disadvantage of the indirect extrusion is that short billets are processed, resulting in the generation of short extruded products.

Differences in the extrusion forces recruited for direct and indirect extrusion are presented in Fig. 2. Region I of the relation portrayed, representing pressure buildup, when the billet is upsetting to an inner diameter of a container, is identical in both processes. Region II represents the extrusion process. In direct extrusion, pressure decreases, whereas in indirect extrusion, pressure remains constant. Finally, Region III represents the end of the extrusion process, when the billet ends and the RAM hits the die [3].

Extrusion Parameters

The extrusion process is characterized by several parameters, with flow stress of the metal (i.e., the resistance of the metal to deformation) being a main parameter that affects the extrusion forces. The flow stress strongly depends on temperature, and therefore, in industry, extrusion is usually performed at elevated temperatures. Friction between the billet, the container, and the die is also determined by temperature. Increasing the temperature reduces extrusion forces while increasing friction, according to empiric Eq. 1 [5]. This equation describes the relation between main extrusion parameters, such as the extrusion ratio ($R = \frac{A_0}{A_f} = \frac{D^2}{d^2}$, where D is the billet diameter and d is the product diameter), the die angle (α), the billet cross-sectional area (Ao), and metal flow stress (σav). The first part of the equation describes the work needed for a change in the cross-sectional area (i.e., the extrusion ratio) of the billet. The second and third parts of the equation define work inside the die and friction forces, respectively. Usually, an extrusion force equation includes an additional component describing the friction within the container. In this work, since extrusion was conducted using an indirect extrusion machine, this component is negligible.

Extrusion force; Flow stress; Extrusion ratio: $R = \frac{D_0^2}{D_f^2}$; Initial billet area; Die angle

$$P = A_0 \bar{\sigma}_{av} \left[\ln(R) + \frac{2}{3}\alpha + \frac{2\ln(R)\mu}{\sin(2\alpha)} \right] \quad (1)$$

Heat Treatment

Before the extrusion process, cast billets must undergo homogenization processing. This is a diffusion-controlled process [6] that has to be carried out at elevated temperature over a long period of time. After the homogenization process, the billets are machined to remove any oxidized surface layer in a procedure called "pealing." After the extrusion process, stress relief and aging treatments are performed. Stress relief is conducted to reduce residual stresses, while aging treatments improve the mechanical properties of the final products. Heat treatment of extruded alloys was done in precise temperature-controlled furnaces. The temperature and duration of such heat treatment were defined by the alloy manufacturer, according to customer requirements [7].

Hardness Tests

The hardness test is a fast, inexpensive, and relatively accurate method for validating the effect of extrusion process on mechanical properties of products. Experimental hardness values can be compared to data reported in the literature to verify the metallurgical condition of an alloy. The correlation between hardness values and tensile strength for various aluminum alloys has been well defined in earlier efforts (Fig. 3 [8]). This empiric graph is sufficient for students to grasp how the process influences the condition of a metal without spending too much time or conducting expensive tensile tests.

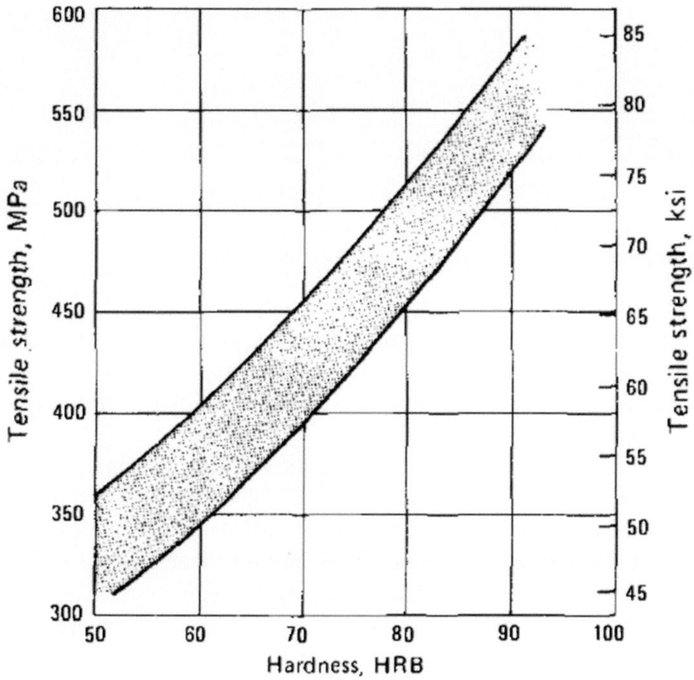

Fig. 3 Correlation between hardness and tensile strength for various aluminum alloys [8]

Experimental Procedures

Experiments were carried out using the micro-hydraulic indirect extrusion machine found in the student laboratory (Fig. 4a). Process parameters, including temperature, RAM movement, and extrusion force, are controlled by a programmable logic controller (PLC). Data from the extrusion process are collected by the computer, allowing the students to subsequently analyze the process.

The temperature of the container is controlled using a thermocouple inserted into the hole in the unit (Fig. 5, part a). The billet (Fig. 5, part b) is heated in a billet

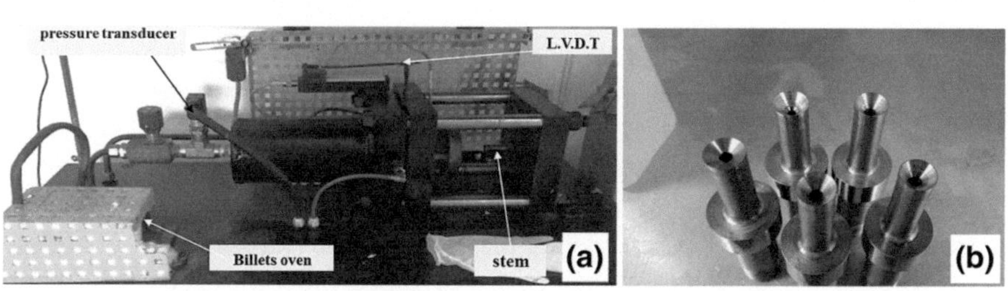

Fig. 4 Hydraulic indirect extrusion micro-press (**a**) and the set of different die stems (**b**)

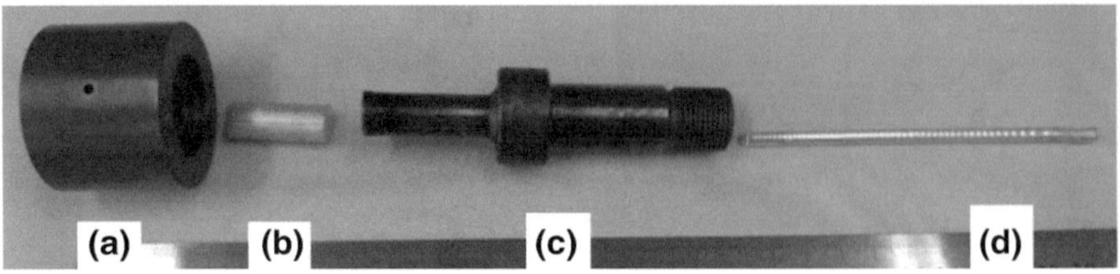

Fig. 5 Container (**a**), Al6063 billet (**b**), stem (**c**) used and the extruded rod (**d**)

oven (Fig. 4a) and inserted into the container. The previously heated RAM (Fig. 5, part c) is then pressed against the billet, and upsetting to the diameter of the hot container, the extrusion process initiates and the metal flows through the stem to yield the extruded rod (Fig. 5, part d).

The extrusion of Al6063 billets was performed at 400 °C using 3, 4.5, and 6 mm flat entry dies. The extrusion ratios for these dies were 20, 10, and 5, respectively, while the inner diameter of the container was 14 mm. Another set of experiments was conducted in the 25–400 °C temperature range to study the effect of temperature. The different extrusions were performed at a constant extrusion ratio of 5 using the 6-mm flat die. The extrusion speed was 13 mm/s. Friction within the die area was reduced using MoS2 lubricant. Extruded samples were annealed at 350 °C for 30 min.

Characterization

Microstructural examination of the extruded samples was conducted by optical microscopy. The extruded samples were examined after grinding and diamond polishing up to 1 μm and etching in a 5% HF solution. Scanning electron microscopy (SEM) and energy-dispersive X-ray spectroscopy (EDS) analyses were undertaken for fractographic studies and the evaluation of cracks. Vickers hardness tests were conducted under a 1000 N load before and after the extrusion process, as well as after annealing heat treatment. Mechanical properties were evaluated at room temperature using tensile tests on the samples according to ASTM with an E8/E8M standard in an Instron 5982 tensile test machine.

Results and Discussion

In this contribution, we report on two test cases, one addressing the influence of the extrusion ratio and the other considering the impact of the extrusion temperature on the extrusion forces, the microstructure, and mechanical properties of extruded parts.

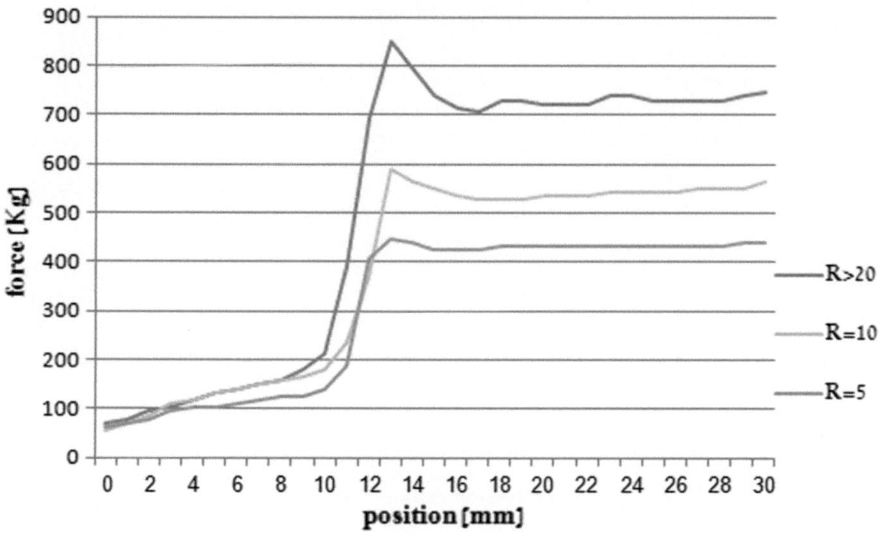

Fig. 6 Extrusion force as a function billet position at various extrusion ratios

The Influence of the Extrusion Ratio on the Extrusion Force

To investigate the effect of the extrusion ratio on extrusion forces, 3, 4.5, and 6 mm flat entry dies were used to achieve extrusion ratios of 20, 10, and 5, respectively. All extrusions were carried out at constant temperature of 400 °C. The influence of the extrusion ratio on the extrusion force can be clearly seen in Fig. 6. The profile of the extrusion curve obtained is a "classic" profile, similar to that generated upon direct extrusion (Fig. 2). The three regions of the extrusion profiles are well defined. The pressure increases can be attributed to the upsetting of the billet within the container during pressure buildup, when the metal is forced to flow through the narrow die opening. The fall in extrusion forces after the peak can be explained by the change from static to dynamic friction coefficiency in the bearing area. The third part of the extrusion graph should be constant. As seen in Fig. 2, such behavior is in contrast to what occurs during direct extrusion, where the force decreases due to a reduction of the friction force, in accordance with the shortness of the billet. The last section of the extrusion graph, when the RAM hits the die, was avoided so as to protect the equipment.

The experimental values of the extrusion forces measured are in good agreement with values calculated using Eq. 1. See Figs. 7 and 11; the 15% difference between the theoretical and experimental results was explained by the slight change in chemical composition of the Al alloy that was used.

The results of a hardness test are presented in Fig. 8. The severe plastic deformation that occurred during the extrusion process caused a rise in residual stress and a remarkable reduction in grain size. Thus, the hardness of the extruded parts is higher

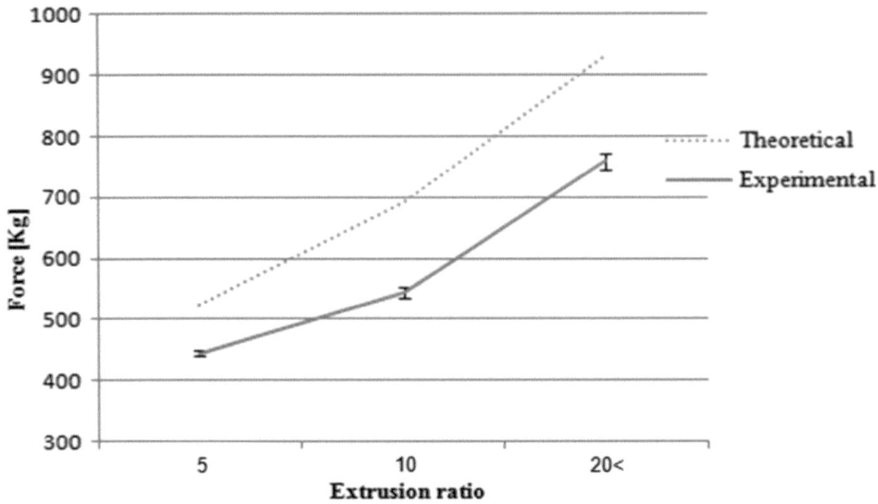

Fig. 7 Comparison of experimental and calculated extrusion force values

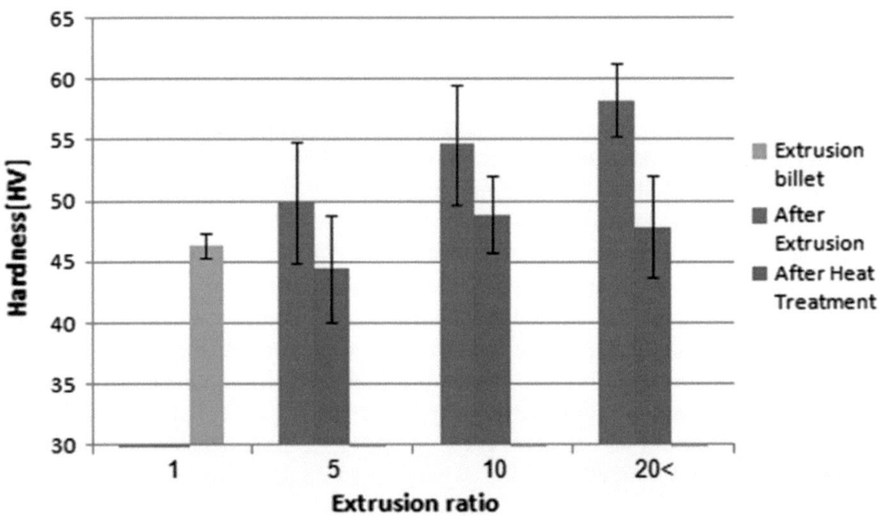

Fig. 8 Effects of the extrusion ratio and heat treatment on hardness

than the hardness of the starting billet. Deformation depends on the extrusion ratio, with hardness increasing as the extrusion ratio grows. The hardness of the extruded billets after annealing decreased yet was still higher than that of the billets before extrusion. This is likely related to the annealing conditions applied (350 °C for 30 min), which differ from the conditions recommended in the literature (450 °C

Fig. 9 Macrostructure of the extruded Al6063 at 400 °C

for 3 h) [8]. The conditions employed were selected because full annealing could "delete" the metallurgical history of the samples.

A longitudinal macro-structural view of the extruded 6063 aluminum alloy is presented in Fig. 9. The microstructure of the sample extruded at 400 °C with extrusion ratio of 20 reflects the dynamic recrystallization of the metal. Nevertheless, due to a fast temperature decrease, only partial recrystallization occurred.

The Effect of Temperature on the Extrusion Process

The flow stress of metals decreases as a function of billet temperature, whereas friction increases. The effect of temperature on the extrusion forces at a constant extrusion ratio equal to 5 is shown in Fig. 10. Maximal extrusion forces were observed at room temperature. The slight increase in force seen in the plateau region can be attributed to strain hardening, while the decrease in extrusion force observed in the plateau region of the 250 °C curve can be associated with added heat generated during extrusion (e.g., deformation can add up to 50 °C). The extrusion forces at 400 °C reflect that the absence of strain hardening at this temperature, as well as of any additional heating as a result of deformation. Calculations of the extrusion force as a function of extrusion temperature (Fig. 11) were made by using the literature values of flow stress for Al6063 at various temperatures [9]. A comparison of the experimental and calculated results demonstrated that a simple set of experiments can serve to clarify the main features of metal deformation.

The effects of extrusion and annealing temperature on hardness are shown in Fig. 12. Students can see that room temperature deformation almost doubled the hardness of the alloy due to a strain-hardening effect. Moreover, the influence of annealing on mechanical properties is clearly demonstrated. One can observe that parts extruded at 400 °C and then annealed attained the hardness of the homogenized

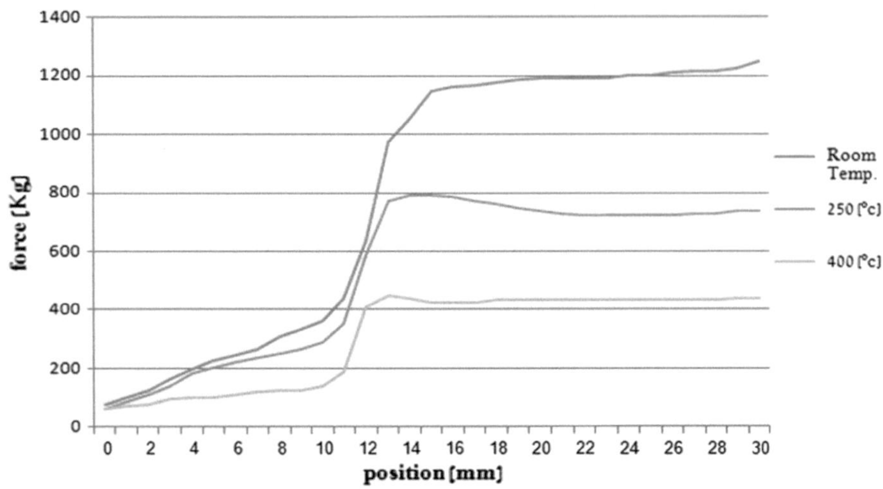

Fig. 10 Influence of temperature on extrusion forces at an extrusion ratio 5 and die angle of $\alpha = 45°$

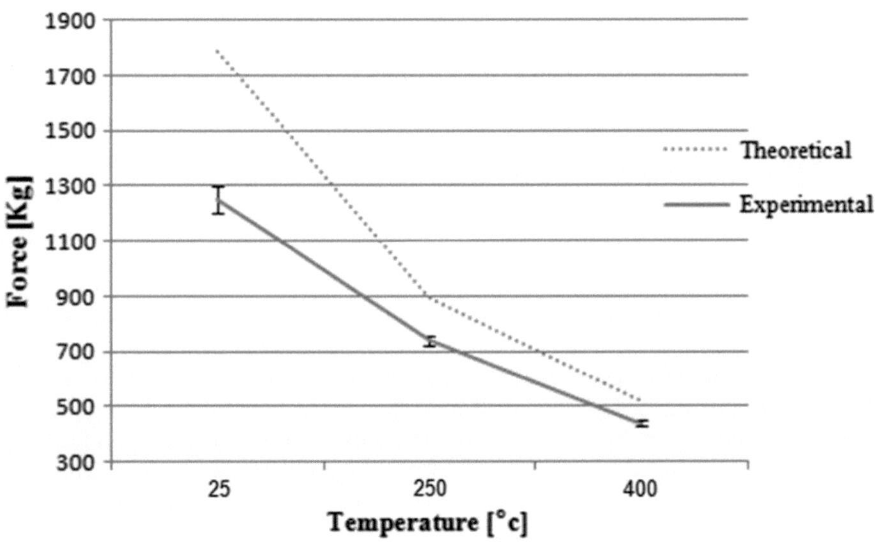

Fig. 11 Comparison of the effect of temperature. Calculated and experimental results are shown

original billet. This can be easily explained by taking into account that extrusion at the recrystallization temperature without strain hardening returns the metal back to the original metallurgical state.

The effect of extrusion temperature on microstructure is demonstrated in Fig. 13. The microstructure of the homogenized original billet displayed coarse grain sizes (about 100–150 μm) due to the high temperature of homogenization (a). The sample

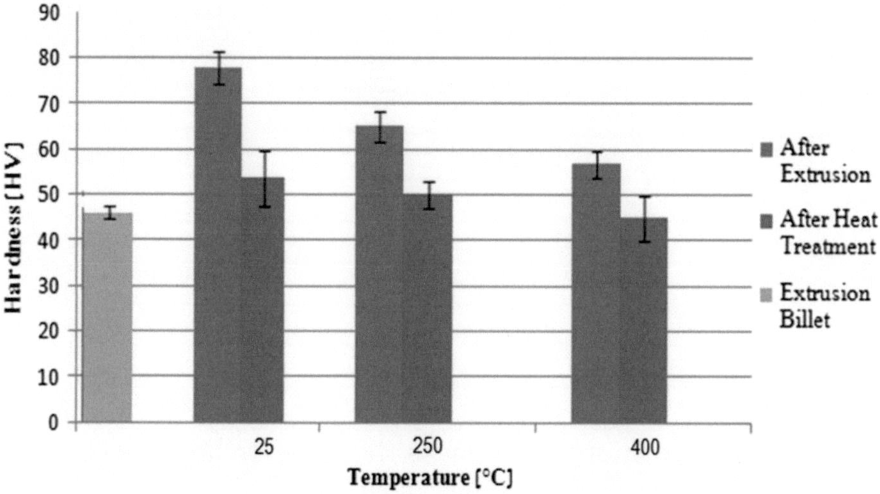

Fig. 12 Effect of extrusion temperature on the hardness of extruded and annealed parts

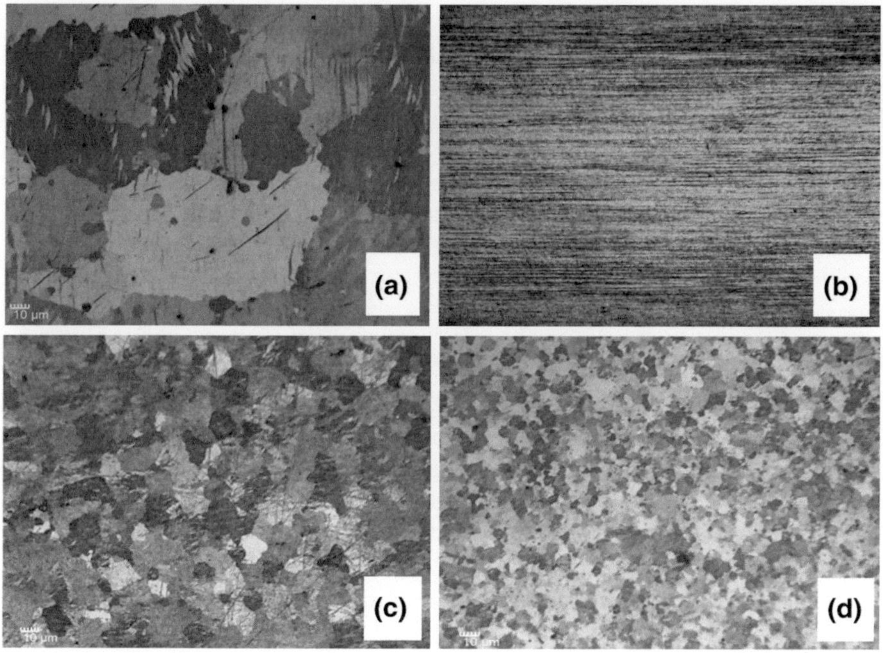

Fig. 13 Microstructure of processed parts with an extrusion ratio of 5. Microstructures of the original billet (**a**), a billet extruded at room temperature (**b**), a billet extruded at 400 °C (**c**), and a billet extruded at 400 °C and fully annealed at 350 °C for 30 min (**d**) are shown

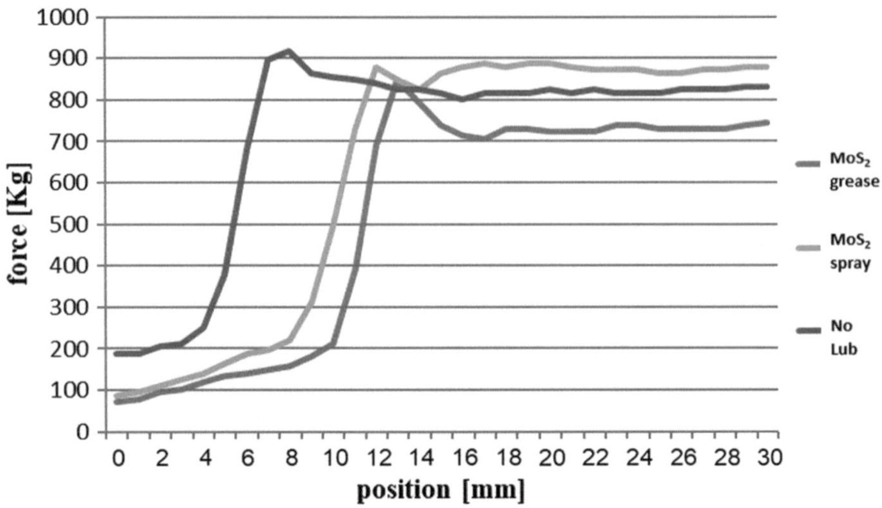

Fig. 14 Influence of lubricants on extrusion forces

extruded at room temperature (b) is highly deformed. A preferred orientation of grains caused by the unidirectional metal flow is detected. The sample extruded at 400 °C (c) displays a partially recrystallized structure with grain sizes of about 20–50 μm. The fully recrystallized microstructure with grain sizes of 5–10 μm is seen with samples extruded at 400 °C and annealed at 350 °C for 30 min (d). This is a very good demonstration of the nature of annealing and recrystallization processes for students.

The Effect of Lubricant on the Extrusion Force

Commonly, friction is the second most important parameter of the indirect extrusion process. The effect of lubricant on the extrusion force is shown in Fig. 14. The experimental results confirm that lubricant affected the extrusion force, especially when the billet was upsetting to the inner diameter of the container.

Conclusions

- An extrusion laboratory was found to be a very important educational tool for teaching metal forming and the effects of heat treatment on the microstructure and mechanical properties of the products.

- The influences of the extrusion ratio and temperature on the indirect extrusion process were investigated by carefully designed extrusion experiments and heat treatments.
- Extrusion at different temperatures clearly demonstrates the nature of the recrystallization process.
- It was established that the extrusion ratio strongly affects the extrusion force and mechanical properties of the final products.
- Good agreement between the extrusion forces measured using different extrusion parameters and the values of forces calculated using an empirical equation was seen.

Acknowledgements The authors thank Alubin-Aluminum Extrusions Ltd. for the extrusion micro-press, Dr. Rinat Katz (Ittah) for significant contributions to this research, Shlomi Levi for metallographic support, and Yoram Rami for technical support.

References

1. Siegert K, Sauer G, Bauser M (2006) Extrusion, 2nd edn. ASM International, Materials Park
2. ASM International (2000) Aluminum Extrusion Technology, 06826G, 2000
3. Thomas MP (2005) Pressure and strain measurement during hot extrusion of Aluminum. Norwegian University of Science and Technology
4. Pradip AS (2000) Aluminum extrusion technology. ASM International, pp 1–29
5. Metals handbook (1998) Forming and forging, vol 14. ASM International, Metals Park
6. Metals handbook (1998) Heat treating, vol 4. ASM International, Metals Park
7. Sourmail T, Opdenacker P, Hopkin G (2000) Annealing twins, Badeshi. University of Cambridge, Cambridge
8. Metals handbook (1991) Heat treating of aluminum alloys, vol 4. ASM International, Metals Park, pp 841–879
9. Metals handbook (1998) Properties and selection: nonferrous alloys, vol 2. ASM International, Metals Park

Investigation and Numerical Modeling of Aluminum Alloys Depending on Different Thermomechanical Processes

B. Güraydin, M. Dinçer, H. Konbul, S. K. İpek, D. Dispinar and A. Karaaslan

Abstract In this study, numerical modelling was applied in order to determine the conductivity values of AA1050, AA3005, AA3104, and AA3105 alloys after different thermomechanical processes. Conductivity difference contour plot graphs of aluminum were performed with different thermomechanical processes. Samples were taken from AA1050, AA3005, AA3104, and AA3105 aluminum alloys and cold rolled with different deformation ratios. Also, samples were annealed with 20 °C intervals for each sample and annealing temperature was applied between 220 and 480 °C. The electrical conductivity change for each annealing process of cold rolled parts was measured. According to the obtained results, sample data were statistically modelled with polynomial regression analysis with Minitab, and with the regression equations electrical resistance and conductivity properties could be predictable with before the thermomechanical process for given aluminum alloys. Experimental results were modelled numerically by regression analysis by using Minitab statistical analysis program.

Keywords Aluminum · Electrical conductivity · Temperature · Twin roll · Numerical modeling · Regression analysis · Rolling · Annealing

B. Güraydin (✉) · M. Dinçer · H. Konbul · A. Karaaslan
Department of Metallurgical and Materials Engineering, Yıldız Technical University, Istanbul, Turkey
e-mail: bdrhngrydn@gmail.com

S. K. İpek
Teknik Alüminyum Sanayi a.Ş, NPD (New Product Development) and Process Development Department, Tekirdag, Turkey

D. Dispinar
Department of Metallurgical and Materials Engineering, Istanbul Technical University, Istanbul, Turkey

J. Lee et al. (eds.), *Materials Processing Fundamentals 2020*, The Minerals, Metals & Materials Series, https://doi.org/10.1007/978-3-030-36556-1_7

Introduction

Aluminum alloys have high resistance, low density, good electrical conductivity, and thermal conductivity as well as corrosion resistance properties [1–3]. Due to these features, it is very common to select these alloys in energy applications where energy transmission is important [1, 4]. Especially, 1xxx series are a special alloy series which are frequently used in electrical transmission cables [5, 6]. Electrical conductivity not only depends on chemical composition but also depends on heat treatment and deformation ratio [7–10]. Plastic deformation causes an increase in dislocation plane and generates residual stress in internal structure of metal [11]. These changes in the internal structure affect the properties of the metal, such as hardness, toughness, and electrical resistivity [12, 13]. In the deformed structure, the average free movement distance of the electrons is reduced. Therefore, resistivity increases whereas electrical conductivity decreases [10]. Recrystallization occurs during all heat treatment processes in cold- or hot-formed aluminum alloys, and grain growth occurs during most of the types of heat treatment processes in multi-crystalline aluminum alloys [14]. During recrystallization process, grain growth varies depending on composition, structure, degree of cold forming, and temperature, and parameters of this process have an important effect on the physical and mechanical properties of aluminum [2, 3, 10, 15]. However, the effect of the alloying elements on electrical conductivity is much greater than that of grain boundary caused by grain growth and the change caused by plastic deformation [1, 7, 9, 17].

The properties of aluminum are also closely related to the purity of the alloy. The most important effect of purity is on electrical and thermal resistance [1, 5, 16]. The alloying elements dissolved in the structure have a negative effect on the electrical conductivity [9]. Because these alloy compositions are located in the lattice as interstitial or substitutional, they form the most effective lattice defects that emit electrons during the conduction of electricity. On the other hand, all these microstructural elements are factors that increase mechanical properties [9, 17]. In this case, the higher-purity aluminum alloys show higher thermal and electrical conductivity [1, 5]. In other words, alloying elements can be regarded reduce the electrical conductivity of aluminum [17].

According to Matthiessen's theory, electrical resistance, and its opposite electrical conductivity depend on many microstructural properties.

$$\rho = \rho 0 + \Delta\rho S + \Delta\rho P + \Delta\rho V + \Delta\rho D + \Delta\rho B \tag{1}$$

These are resistance of pure solvent metal ($\rho 0$), electrical resistance of elements dissolved in solution ($\Delta\rho S$), precipitates ($\Delta\rho P$), voids ($\Delta\rho V$), dislocations ($\Delta\rho D$), and grain boundaries ($\Delta\rho B$) [9].

According to a study by Lipińska et al. [8], the effect of alloying elements in the reduction of electrical conductivity is much higher than grain boundaries and plastic deformation. For instance, Cr, V, Mn, Ti, Mg, Ag, Cu, Zn, Si, Fe, and Ni are effective in reducing the electrical conductivity of aluminium. In particular, Cr, V, and Mn have a significant effect on the electrical conductivity [1, 3, 4, 9]. The number of voids, dislocations, and elements dissolved in the solid solution has a negative effect on the electrical conductivity [16]. Due to the decrease in grain size, the tensile strength of aluminium alloys increases, thus the electrical conductivity [3, 5]. Since the electrical conductivity of aluminium alloy is important for consumers, it is measured after the production process. In this case, it is very important to define the parameters that affect the electrical conductivity [10]. If a regression equation can be defined after heat treatment and cold deformation, electrical conductivity value can be determined before production. The aim of this study is to determine the changes in electricity transmission before the production process of aluminium alloy.

Experimental Studies

In this study, 1xxx and 3xxx series rolled aluminium alloys were produced by twin roll casting method by using the same casting line for all samples. The result of optical emission spectrometer analysis in order to determine the chemical composition of the alloys is shown in Table 1.

The samples with the same width were taken from identical regions. Taguchi test method was used to design this study. One-on-one replacement method is used. The reason for using this method is that it offers the opportunity to examine better the effects of the factors which is causing change on electrical conductivity. Before the experiments were designed, the parameters affecting the electrical conductivity were determined as heat treatment temperature and deformation ratio.

In order to investigate the electrical conductivity of alloys depending on thermo-mechanical processes, different deformation ratios were used varied as 50, 58.30, 66.66, 75, 83.30, 91.60, and 95%. Using different samples at each deformation ratio, heat treatment temperature was increased by 20 °C from 220 to 480 °C and heat-treated at 14 different temperatures for 4 h. Heat treatment was carried out in laboratory-type heat treatment furnace (Nabertherm). The electrical conductivity of the samples was measured by Eddy Current Conductivity Meter (Sigmascope SMP 10). Obtained data were analyzed in Minitab 18.0 program and modelled statistically.

Table 1 Spectral analysis results of AA1050, AA3005, AA3105, and AA3104 alloys

Alloys	Si	Fe	Cu	Mn	Mg	Al
AA1050	0,085	0,249	0,004	0,017	0,004	99,571
AA3005	0,116	0,447	0,073	1,073	0,234	97,924
AA3105	0,120	0,390	0,021	0,377	0,255	98,757
AA3104	0,614	0,868	0,018	1,016	0,938	96,452

Experimental Results

Effect of Heat Treatment Temperature on Electrical Conductivity at Constant Deformation Ratio

The variance of the electrical conductivity with regard to the heat treatment temperature of the aforementioned alloys at the constant deformation ratios ranging from 50 to 95% is shown in Figs. 1, 2, 3, and 4. The data were modeled statistically, and regression graphs were plotted via Minitab 18.0 program. The R-Sq (regression square) and R-Sq (adjective) values shown in the graphs indicate the standard deviation values which confirm the statistical modeling of the data [18, 19].

Discussion

The effect of thermomechanical processes on the electrical conductivity of aluminum alloys is seen obviously. The plastic forming method in the production process increases dislocation, vacancy, grain boundary, and internal stress. After plastic forming, electrical conductivity is reduced due to the decreasing of free movement distance of electrons in the structure [10].

The applied heat treatment processes reduced the internal stresses in the structure of the material and caused the grain growth. Due to the dislocation zones and vacancies in the material are dispersed and the scattering of the electrons is reduced. In this way, the electrical conductivity of the material has increased and the resistivity has decreased [1].

Figure 5, at the deformation ratio of (a) 50%, (b) 66.66%, (c) 83.30%, and (d) 95%, shows the change of the electrical conductivity of different alloys depending on heat treatment temperature. When Fig. 5 is examined, it is clearly seen that the electrical conductivity decreases as the alloying element increases. The electrical conductivity of AA1050 commercial pure aluminum alloy is higher than other aluminum alloys. Figure 5 shows that thermomechanical processes have little effect on the electrical conductivity of commercial pure aluminum. The reason for this was examined in another study [9]. According to this study, in commercial pure aluminum, after the annealing process, a multifaceted restoration mechanism occurs. In commercial pure aluminum, the electrical conductivity changes very little after the deformation because of the multi-directional recovery by heat treatment [16]. Although the electrical conductivity of the 1xxx series is high, it has a low strength due to the fact that it is commercially pure aluminum.

In another study [9], it was found that electrical conductivity was directly proportional to grain size but its effect was less than the effect of alloying elements.

In order to use commercial pure aluminum in electrical conduction, it is necessary to increase the mechanical properties of the alloy. Since alloying reduces electrical conductivity [13], mechanical methods should be applied.

- **AA1050**

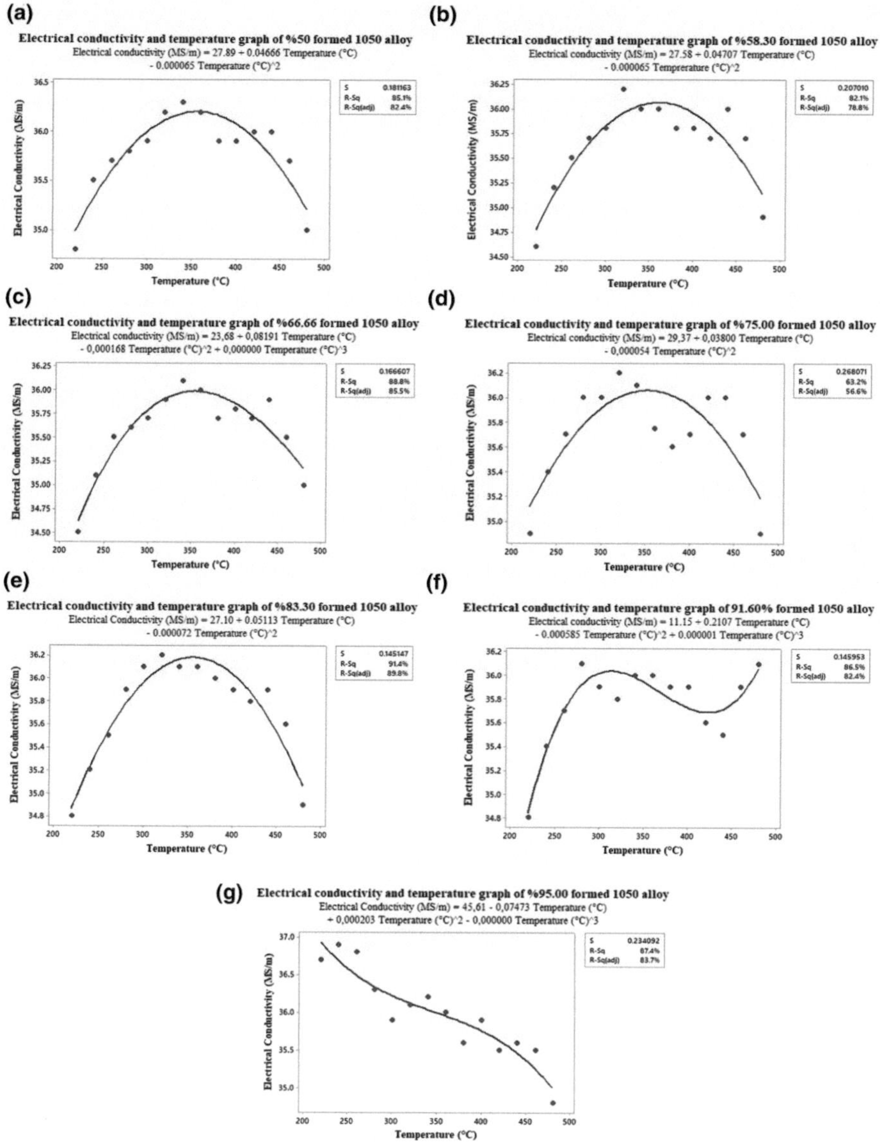

Fig. 1 Regression graphs related to the deformation rate of the varies in electrical conductivity depending on the heat treatment temperature of AA1050 alloy: **a** 50%, **b** 58.30%, **c** 66.66%, **d** 75%, **e** 83.30%, **f** 91.60%, and **g** 95%

- **AA3005**

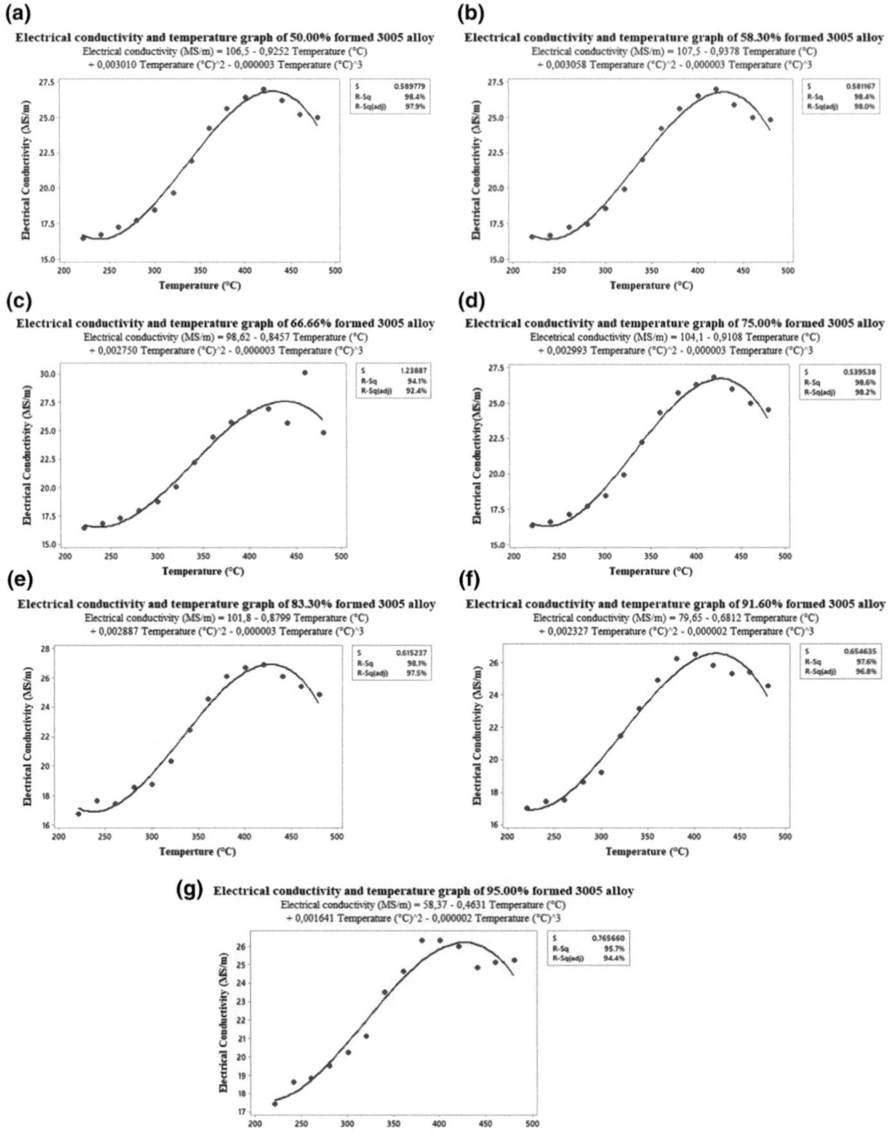

Fig. 2 Regression graphs related to the deformation rate of the varies in electrical conductivity depending on the heat treatment temperature of AA3005 alloy: **a** 50%, **b** 58.30%, **c** 66.66%, **d** 75%, **e** 83.30%, **f** 91.60%, and **g** 95%

- **AA3104**

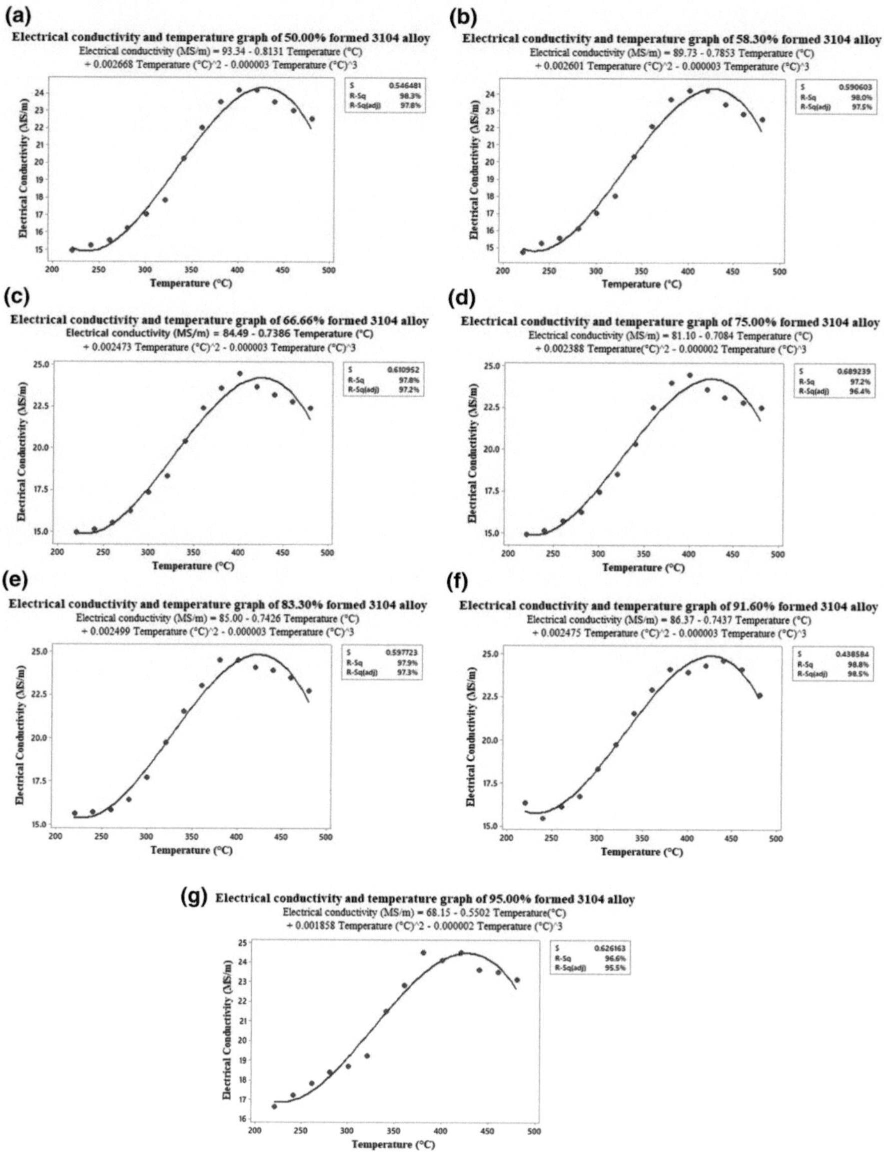

Fig. 3 Regression graphs related to the deformation rate of the variation in electrical conductivity depending on the heat treatment temperature of AA3104 alloy: **a** 50%, **b** 58.30%, **c** 66.66%, **d** 75%, **e** 83.30%, **f** 91.60%, and **g** 95%

- **AA3105**

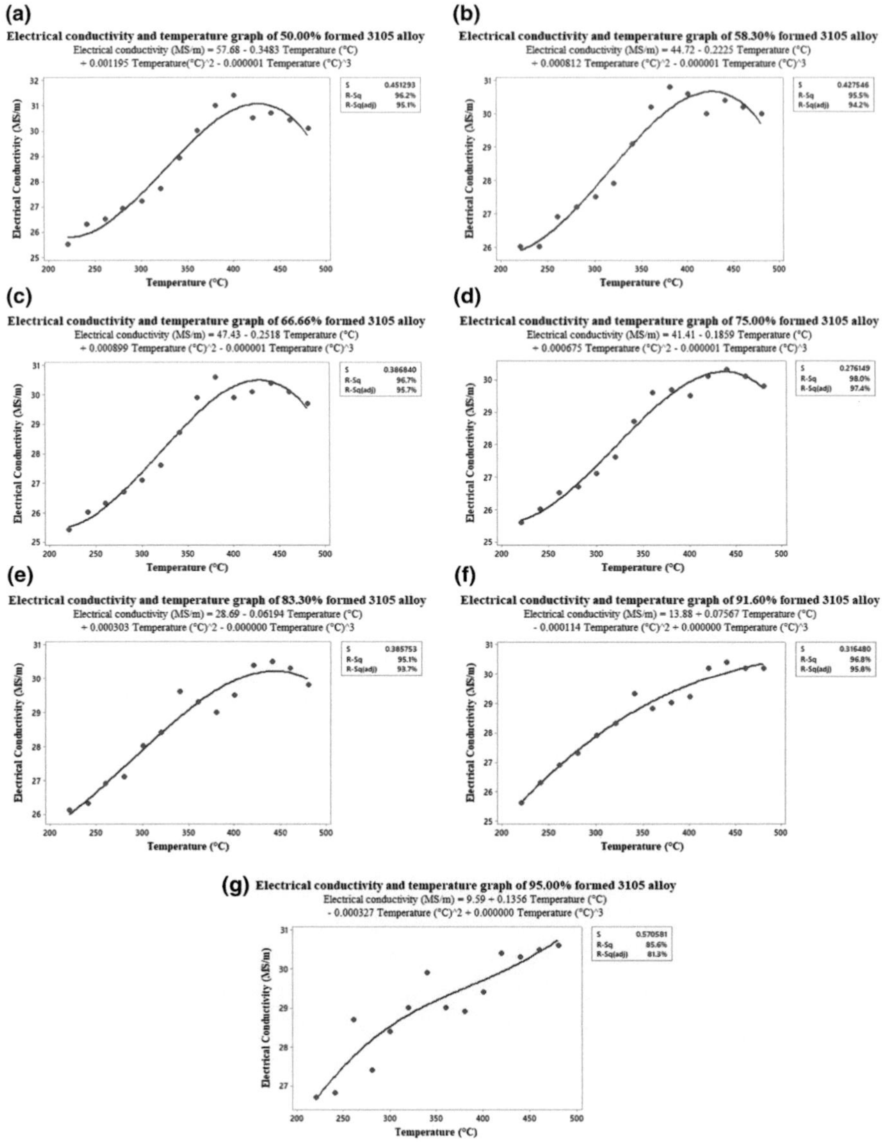

Fig. 4 Regression graphs related to the deformation rate of the variation in electrical conductivity depending on the heat treatment temperature of AA3105 alloy: **a** 50%, **b** 58.30%, **c** 66.66%, **d** 75%, **e** 83.30%, **f** 91.60%, and **g** 95%

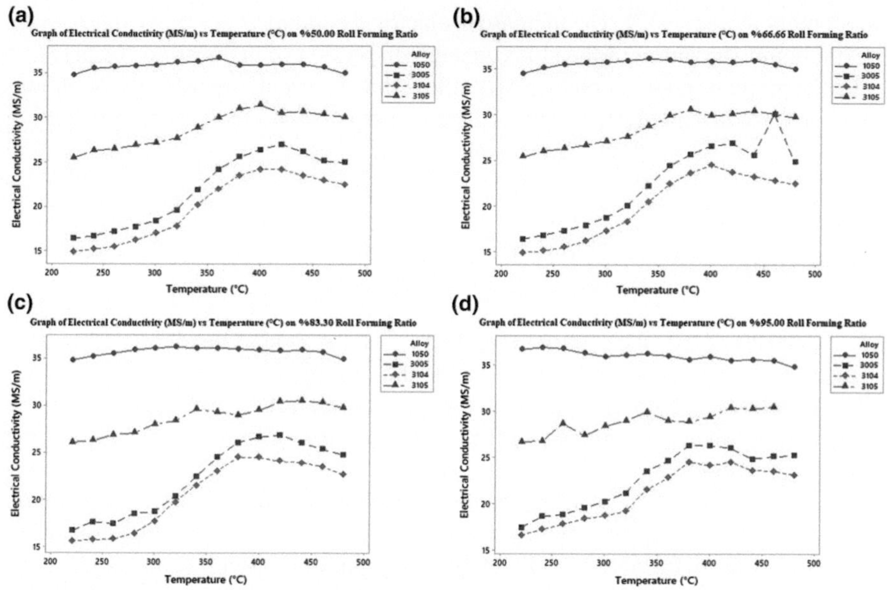

Fig. 5 Variation of electrical conductivity of different alloys depending on heat treatment temperature at same deformation ratios

AA3005, AA3104, AA3105 alloys have lower electrical conductivity because they have lower purity than commercial pure aluminum. It is observed that these alloys have increased electrical conductivity after heat treatment.

Recrystallization occurs during all annealing processes in cold- or hot-formed aluminum alloys. All heat treatment types occur grain growth in polycrystalline aluminum alloys [1]. Because of the recrystallization of these alloys, their electrical conductivity has increased. In a recrystallized structure, defective lattices are recovering [15]. If one sample is subjected to heat treatment with a high plastic forming ratio and another sample subjected to low plastic forming ratio, the sample which has high plastic deformation ratio shows higher electrical conductivity. That is because the recrystallization changes depending on the composition, structure, degree of cold forming, and temperature. Recrystallization temperature decreases as the plastic forming ratio increases [15].

The same alloys with constant deformation ratio—except commercial pure aluminum—have increased electrical conductivity with increase of heat treatment temperature. It is seen that the electrical conductivity of AA3005, AA3104, AA3105 alloys increases in direct proportion with the heat treatment temperature.

Impurities from casting and plastic forming—grain boundary, dislocations, and vacancies—are removed by heat treatment. The resistivity of metals increases as the impurity increases [13]. As a result of the heat treatment, the impurities were removed and an increase in electrical conductivity was observed. Separator grains show recrystallization behavior. As the grain grows due to recrystallization, electrical conductivity increases [7]. The increase in electrical conductivity may be due to grain growth, homogenization of the structure, and dispersion of dislocation zones [7].

The numerical models obtained will allow the electrical conductivity of an aluminum alloy to be calculated based on the processes designed prior to the production process.

Conclusion

- It was observed that the electrical conductivity decreased as the deformation ratio increased and it is directly proportional to the heat treatment temperature.
- The samples with high deformation ratio of the same alloy showed higher electrical conductivity than the samples with low deformation ratio when annealed at the same temperature.
- Samples with high deformation ratios were annealed at the same temperature showing higher electrical conductivity than samples with lower deformation ratios.
- For the same deformation ratio of all alloy samples—without AA1050—the electrical conductivity increased as the applied heat treatment temperature increased.
- As the alloy ratio increases, the electrical conductivity decreases. Therefore, it has been seen that the commercially pure aluminum with the highest electrical conductivity. It has been observed that the electrical conductivity of the commercial pure aluminum has increased slightly with the applied heat treatments.
- Electrical conductivity decreased after 450 °C during heat treatment process. The reason for this decline is regarded as the increased dislocation zones and atomic cavities at high heat treatment temperature.
- It is determined that the numerical model which is formed to determine the electrical conductivity of aluminum alloys is close to the results of the experiment. Thus, it has been observed that the model operates at tight tolerances. The numerical models obtained can be used in the industry.

Acknowledgements The authors are grateful to the TUBITAK (The Scientific and Technological Research Council of Turkey) for the financial support.

References

1. Totten GE, MacKenzie DS (2003) Handbook of aluminum, physical metallurgy and processes, vol. 1
2. Hatch JE (1984) Aluminum: properties and physical metallurgy
3. Lampman AR, Zorc T, Henry S (1990) Properties and selection: nonferrous alloys and special-purpose materials, vol. 2. ASM Handbook
4. Yashpal S, Jawalkar K, Kant CS (2015) A review on use of aluminium alloys in aircraft components. I-Manager's J Mater Sci

5. Wessel JK (2004) The Hand book of advanced material
6. Sivasankaran S (2017) Aluminium alloys: recent trends in processing, characterization, mechanical behavior and applications
7. Huang K, Zhao QL, Li YJ, Marthinsen K (2015) Two-stage annealing of a cold-rolled Al–Mn–Fe–Si alloy with different microchemistry states. J Mater Process Technol 221:87–99
8. Beljajew LAFAI, Rapoport MB (1956) Metallurgie des aluminium
9. Lipińska M, Bazarnik M, Lewandowska P (2016) The influence of severe plastic deformation processes on electrical conductivity of commercially pure aluminium and 5483 aluminium alloy. Archives Civ Mech Eng
10. Vlack LH (1966) Elements of materials science: an introductory text for engineering students
11. Lejeune A, Boudaoud H, Potier-Ferry M, Charpentier I, Zahrouni H (2013) Automatic solver for non-linear partial differential equations with implicit local laws: application to unilateral contact. Int J Numer Methods Eng 94(9):850–867
12. Martins LMC, Padilha JF (2009) Microstructure and texture assessment of Al–Mn–Fe–Si (3003) aluminum alloy produced by continuous and semicontinuous casting processes. J Mater Sci
13. Moldovan F, Popescu P, Miculescu G (2004) Microscopic study regarding the microstructure evolution of the 8006 alloy in the plastic deformation process. J Mater Process Technol
14. Humphreys FJ, Hatherly M (1995) Recrystallization and related annealing phenomena
15. Kassner ME (2008) Fundamentals of creep in metals and alloys, 2nd edn
16. Dhal A, Panigrahi SK, Shunmugam MS (2017) Insight into the microstructural evolution during cryo-severe plastic deformation and post-deformation annealing of aluminum and its alloys. J Alloys Compd 726:1205–1219
17. Qingru Zhao XL, Qian Z, Cui X, Wu Y Optimizing microstructures of dilute Al–Fe–Si alloys designed with enhanced electrical conductivity and tensile strength. J Alloys Compd
18. McNamee R (2005) Regression modelling and other methods to control confounding. Occup Environ Med 62(7):500–506
19. Nemes S, Jonasson JM, Genell A, Steineck G (2009) Bias in odds ratios by logistic regression modelling and sample size. BMC Med Res Methodol 9(1):1–5
20. Rollett A, Humphreys F, Rohrer GS, Hatherly M, Recrystallization and related annealing phenomena, 2nd edn
21. Rometsch PA, Xu Z, Zhong H, Yang H, Ju L, Wu XH (2014) Strength and electrical conductivity relationships in Al–Mg–Si and Al–Sc alloys. Mater Sci Forum 794–796:827–832
22. Rey C, Solas D, Fandeur O (2013) Grain boundaries in cold deformation, in grain boundaries and crystalline plasticity, pp 109–163
23. Nes E (1976) Effect of a fine particle dispersion on heterogeneous recrystallization. Acta Metall 24:391–398
24. Zhong-weiChen LS, Zhao J (2015) Micro textural evolution of different TRC AA8006 alloy sections with homogenization. Int J Miner Metall Mater
25. Mironov SY, Salishchev GA (2001) Effect of grain size and structural homogeneity on the uniformity of deformation of a commercially pure titanium. Fiz Met i Metalloved 92(5):81–88
26. Bay B, Hansen N, Den Riso R (1983) Natl Lab, "influence of small particles and grain boundaries on the deformation structure of aluminium", in deformation of multi-phase and particle containing materials. In: Proceedings of the 4th Riso international symposium on metallurgy and materials science, pp 145–152

Part III
Thermodynamic Modeling

Structure–Thermodynamics Interrelation for the GeO_2 and PdO Containing MgO-Saturated Ferrous Calcium Silicate (FCS) Slag Relevant to E-waste Processing

M. M. Hasan, M. A. Rhamdhani, M. A. H. Shuva and G. A. Brooks

Abstract In black copper smelting, the recovery of the valuable metals from the electronic waste depends on the slag and copper's chemistries. It is therefore important to understand the slag chemistry and its relation to the structure and thermodynamics, and how these are affected by processing conditions. This paper explains the recent work from the authors on the study of the structure of MgO-saturated ferrous calcium silicate (FCS) slags containing Cu_2O, GeO_2, and PdO investigated using Fourier-transform infrared (FTIR) spectroscopy. The effect of slag chemistry on the structure of silicate slag was evaluated. The partition ratio of Ge and Pd in slag and copper was correlated to the structure of the FCS slags. It was found that acidic slags with lower non-bridging oxygen per tetrahedra (NBO/T) are more favourable for Ge-partitioning to copper, and for Pd, the relation is the opposite.

Keywords Basicity · Slag chemistry · Pyrometallurgy · Copper smelting · Slag structure · FTIR spectrometry · Slag viscosity · Thermodynamics · E-waste · Partition ratio

Introduction

Consumers' demand of the electrical and electronic equipment (EEE) is ever-increasing due to the intelligent design and advanced functions of the smart devices [1]. To manufacture the modern EEE, it requires large amount of different valuable metals. Underground ores are the primary source of those valuable metals, which are limited in higher grade. Processing plants will require higher capacity and capability to process lower grade ores which are often complex and expensive to process [2]. Mining ores are also associated with environmental and ecological disturbances in the mine area [3]. On the other hand, waste EEE are highly rich with different valuable elements, sometimes higher in concentration from their respective ores. For example, gold content in waste mobile phones is seven times higher than from its

M. M. Hasan · M. A. Rhamdhani (✉) · M. A. H. Shuva · G. A. Brooks
Swinburne University of Technology, Melbourne, Australia
e-mail: arhamdhani@swin.edu.au

J. Lee et al. (eds.), *Materials Processing Fundamentals 2020*, The Minerals, Metals & Materials Series, https://doi.org/10.1007/978-3-030-36556-1_8

ore [4]. About 44.7 million metric tonnes of E-waste were generated in 2016, valuing USD 64.7 billion, only 20% of which were recycled, and the rest are discarded mostly in landfills [5].

Previous research showed that the recovery of the valuable materials from the copper melt depends on the thermodynamics of the elements at the smelting condition [6, 7]. It is evident that the thermophysical and thermodynamic properties such as viscosity, surface tension, specific volume, sulphide capacity, and mixing free energy of oxides are dependent on the structure of the silicate slags [8–12]. These properties determine the success of the metal production process. Therefore, it is important to find the thermodynamics–structure–property relations to ensure the optimization of the process.

Here, structure of silicate means the relative abundance of different silicate units, *i.e.* the fully polymerized units ($Si_2O_4^0$: 3D network) and other lesser polymerized units ($Si_2O_7{}^{6-}$: sheets, $Si_2O_6{}^{4-}$: chains, $Si_2O_7{}^{6-}$: dimers and $Si_2O_8{}^{8-}$: monomers) [13, 14]. These units are differentiated by the number of non-bridging oxygen present per silicate tetrahedral unit (NBO/T) and referred to as Q^n (n is the number of bridging oxygen) [15]. The higher the number of bridging oxygen, the larger the silicate unit. Slag structure can be measured by different spectroscopic techniques, such as Raman spectroscopy, Fourier-transform infrared (FTIR) spectroscopy, nuclear magnetic resonance (NMR) spectroscopy, X-ray photoluminescence spectroscopy (XPS), and others [15–23]. Park introduced the term, degree of polymerization. This represents the fraction of highly polymerized structural unit, Q^4, in the silicates and can also be measured as Q^3/Q^2 [8].

Several attempts were made to correlate slag structure with the thermodynamic properties of the system. Park related sulphide capacity with the fraction of free oxygen in CaO–SiO_2–MO (M=Mn and Mg) melts at 1500–1650 °C [8]. The entropy of mixing (ΔS_{mix}) of Li_2O–SiO_2 and Na_2O–SiO_2 binary systems was calculated by Yano et al. [24]. Park [8] also calculated the molar excess free energy of the components in CaO–SiO_2–MO (M=Mn and Mg) melts at 1873 K using FactSage™ 6.3 and correlated the data with the structure of the slags, presented as Q^3/Q^2. The authors described the relation of slag structure with thermodynamics of Ge at copper smelting conditions in a recent article [25]. The distribution ratio of the elements in copper and slag was related to slag structure and processing conditions through semi-empirical equations.

Although there have been several studies that try to relate the structure to thermophysical property (especially viscosity), there is still lack of information on the structure–thermodynamic property at conditions relevant to black copper smelting. This paper explains the recent studies by the authors on the structure of Cu–Ge/Pd equilibrated MgO-saturated FCS slags using FTIR spectroscopy.

Experimental Conditions

Slags were prepared from silicon (IV) oxide (99.5% purity, Alfa Aesar), calcium carbonate (99.5% purity, Alfa Aesar), iron (>99.8% purity, Alfa Aesar), iron (III) oxide (99.99% purity, Alfa Aesar) powders, germanium oxide (99.99% purity, Sigma-Aldrich), and palladium oxide (99.99% purity, Sigma-Aldrich). $CaCO_3$ powder was dried at 500 °C for 5 h to remove moisture and then calcined at 1100 °C for 4 h in a muffle furnace to produce CaO. Powders of required ratio were mixed in a ball mill for 36 h and then melted at 1300 °C at partial pressure of oxygen of 10^{-8} atm (obtained by controlling CO/CO_2 flowrate) in a vertical tube furnace (Nabertherm, Germany). High purity magnesia crucibles (97.4% MgO, Ozark Technical Ceramics, USA) were used for melting the samples.

The details of the experimental conditions with schematic of the furnace setup are shown in previous articles [26–28]. For equilibration of slag–copper reaction, sufficient melting time was provided. In this research, Fe/SiO_2 ratio varied from 0.81 to 1.16 and basicity of the slag $B = (\%CaO + \%MgO)/\%SiO_2$ varied from 0.49 to 0.89. The experimental conditions used for melting are listed in Table 1. It is worth to mention here that the samples were cooled by purging of argon (grade 5, 99.995% purity) and followed by lowering to the cooler zone of the furnace. It has been suggested in the literature that there is no significant difference between the spectra of quenched sample and in situ at high temperature [29, 30].

For analysis, the obtained slag samples were broken to pieces and then ground to powders. The bulk composition of the samples was determined using an inductively coupled plasma atomic emission spectrometer (ICP-AES). Fourier-transform infrared (FTIR) spectrometer equipped with a Nicolet iD5 attenuated total reflection unit (Thermo Fisher Scientific, USA) was used to collect the vibrational spectra. FTIR spectra acquisition was carried out at room temperature in the wavenumber range from 200 to 1600 cm^{-1}. For each sample, 100 scans were taken at a resolution of 4.0 cm^{-1}. The obtained spectra were deconvolved using PeakFit v4.12 software using Gaussian functions. The percentage abundance of different silicate structural

Table 1 Experimental conditions and slag compositions used in the present study

Temperature (°C)	p_{O_2} (atm)	Slag Composition (mass pct)					Fe_T/SiO_2	Basicity (B)
		Fe_T	SiO_2	CaO	MgO	Cu_2O		
1300	10^{-8}	26.97	33.37	9.57	6.67	1.57	0.81	0.49
		32.82	33.14	9.91	6.49	1.61	0.99	0.49
		37.30	32.28	9.65	6.41	1.58	1.16	0.49
		37.29	35.32	5.58	6.27	1.61	1.06	0.34
		32.55	32.81	11.53	6.46	1.63	0.99	0.55
		32.88	33.08	14.27	7.40	1.56	0.99	0.65
		30.23	30.82	19.10	8.64	1.54	0.98	0.89

units (Q^n) was calculated in the same way reported in other references [25]. Position of the silicate unit peaks is iterated with the initial input of 860 cm^{-1} for (Q^0), 910 cm^{-1} for (Q^1), 970 cm^{-1} for (Q^2), and 1100 cm^{-1} for Q^3 bands shift [14].

Results and Discussion

Structure Analysis of the Equilibrated FeO_x–CaO–SiO_2–MgO–Cu_2O–GeO_2/PdO Slag

The FTIR spectra of equilibrated FeO_x–CaO–SiO_2–MgO–Cu_2O–GeO_2/PdO slags were analysed to evaluate the effect of compositional variation on the silicate structure of slags. Selected results from oxygen potential and temperature study have already been published in previous article [25]. In this article, the function of Fe/SiO_2 ratio and basicity on the structure of the slags were studied.

Effect of Fe/SiO_2

In Fig. 1a, b respectively, the composition of silicate structure from Ge study and Pd study at varied Fe/SiO_2 ratio is shown. It is observed from Fig. 1a that the relative abundance of Q^2 and Q^3 units decreases and Q^0 unit increases with increasing Fe/SiO_2 ratio from 0.81 to 1.16 for GeO_2 containing slags. It can be deduced that the highly polymerized Q^3 (Si_2O_5-sheet) and Q^2 (SiO_3-chain) units depolymerize to Q^0 (SiO_4-monomer) and Q^1 (Si_2O_7-dimer) unit as Fe/SiO_2 ratio increases in the GeO_2 containing slag. For the PdO-containing slags, similar depolymerization action is observed for varying Fe/SiO_2 ratio in the equilibrated slag from 0.8 to 0.99. Fe/SiO_2 ratio beyond that value resulted in inverse action with more polymerized units in the

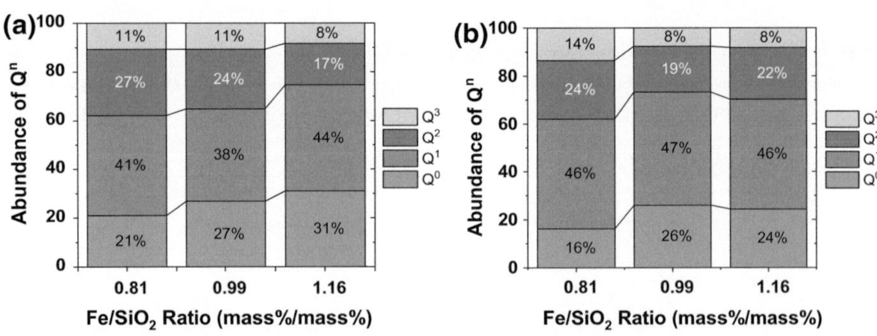

Fig. 1 Abundance of structural units in FeO_x–CaO–SiO_2–MgO–Cu_2O–(GeO_2/PdO) equilibrated slag as a function of Fe/SiO_2 ratio from **a** Ge study and **b** Pd study ($p_{o_2} = 10^{-8}$ atm and $T =$ 1300 °C)

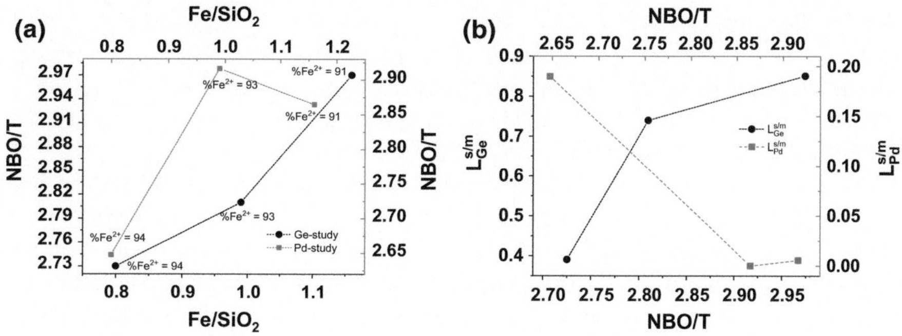

Fig. 2 Relationship of NBO/T with **a** Fe/SiO_2 ratio and **b** distribution ratio of Ge and Pd in slag and metal for different Fe/SiO_2 ratio at $p_{O_2} = 10^{-8}$ atm and $T = 1300$ °C

structure. Therefore, it can be summarized that the added Fe-species acts as network breaker in the tested systems except when added beyond the Fe/SiO_2 ratio of 0.99 in the PdO-containing slag.

The relation of NBO/T with Fe/SiO_2 ratio in FeO_x–CaO–SiO_2–MgO–Cu_2O–GeO_2/PdO slags at temperature of 1300°C and p_{O_2} of 10^{-8} atm is shown in Fig. 2a. Increasing Fe/SiO_2 ratio from 0.81 to 1.16 resulted in increased NBO/T of the GeO_2-containing slag system. For PdO-containing slags, NBO/T increased with increasing Fe/SiO_2 ratio from 0.81 to 0.99 and decreased when Fe/SiO_2 ratio of the slag increased to 1.16. Therefore, in summary, iron oxide behaves as network modifier in GeO_2-containing slag and as amphoteric oxide in PdO-containing slag system.

The distribution ratio of the elements is plotted against NBO/T of the slag in Fig. 2b. Increase in NBO/T of the slag system resulted in increased partitioning of Ge in the slag phase and vice versa for Pd. This suggests that iron addition depolymerizes the silicate structures in the GeO_2 containing FCS slag, where GeO_2 acted as network former capturing the released free oxygen by iron oxide. For PdO study, the scenario is exactly opposite, where PdO acted as network modifier. Ge^{4+} has higher ion-oxygen attraction ($z/r^2 = 26$) than Si^{4+} ($z/r^2 = 25$) and therefore acts as network former, while Pd^{2+} with lower ion-oxygen potential ($z/r^2 = 5.4$) acts as network modifier in the tested silicate systems.

Fe can be present in the slag as Fe^{2+} or Fe^{3+}. Fe^{2+} is a well-known network breaker in silicate slags. Fe^{3+} on the other hand acts as network former when the $Fe^{3+}/(Fe^{2+} + Fe^{3+})$ ratio is greater than 0.5 and as network breaker when the ratio is less than 0.3 [9, 31]. The concentration of Fe^{3+} in the iron-containing slag is related to p_{O_2} of the system. Generally, with increasing p_{O_2}, the concentration of the Fe^{3+} species increases. In this research, the $Fe^{3+}/(Fe^{2+} + Fe^{3+})$ ratio was calculated using FactSage 7.2 and found to have a value of less than 0.1. Approximately 90% of the Fe is in the form of Fe^{2+}; therefore, the role of the Fe in the slag can be taken as network breaker.

Effect of Basicity

The abundance of different silicate structural units with varying basicity of the FeO_x–CaO–SiO_2–MgO–Cu_2O–GeO_2/PdO slags is shown in Fig. 3a, b, respectively. The effect of basicity solely in this case is not that clear.

However, increasing basicity of the slags resulted in increased NBO/T in both slag systems (Fig. 4a). This is expected, as increased basicity means there is more source of free oxygen in the system that can break some of the bridging bonds in silicate network. In Fig. 4b, the distribution ratio of Ge and Pd is plotted as a function of NBO/T at a fixed temperature and p_{o_2}. It is observed that partitioning of Ge in slag phase increases with increasing NBO/T of the slag, which was expected for an acidic element. In case of Pd-containing system, the element partitioning increases to the slag phase with increasing NBO/T to a certain limit and then decreases. The higher partitioning of Pd at higher NBO/T can be explained by the relatively basic character of Pd^{2+}, owing to its lower ion-oxygen potential ($z/r^2 = 5.4$) than Si^{4+} ($z/r^2 = 25$).

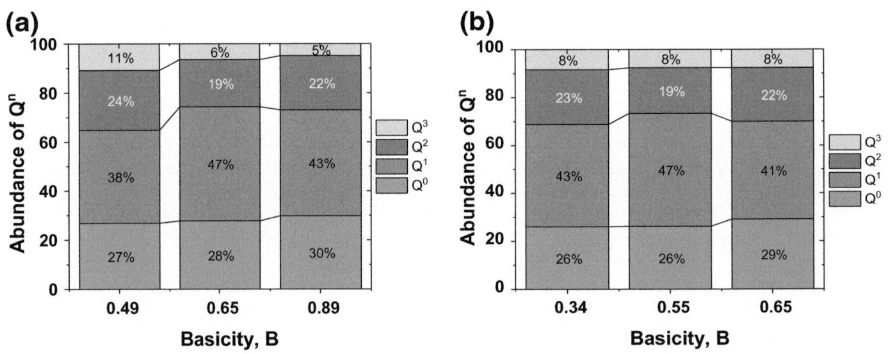

Fig. 3 Abundance of silicate structural units in FeO_x–CaO–SiO_2–MgO–Cu_2O–(GeO_2/PdO) slags with different basicity; **a** Ge study and **b** Pd study ($p_{o_2} = 10^{-8}$ atm and $T = 1300$ °C)

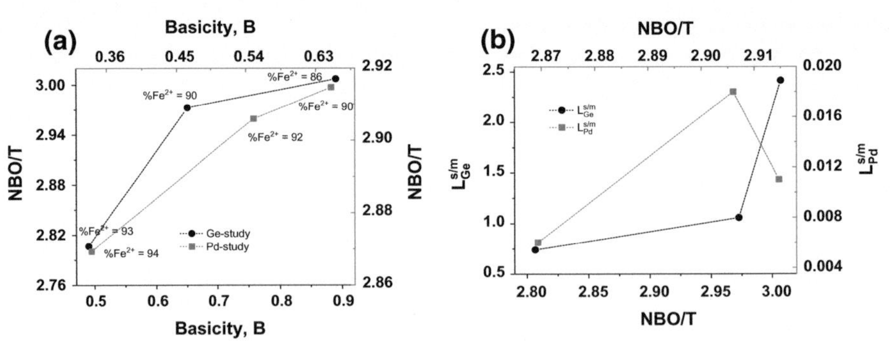

Fig. 4 Relationship of NBO/T with **a** basicity and **b** distribution ratio of Ge and Pd in slag and metal at $p_{o_2} = 10^{-8}$ atm and $T = 1300$ °C

Oxygen Speciation from FTIR Spectra

Park et al. described about the estimation of NBO/Si values in the slag using a mass balance of Si and O atom [8, 17, 32, 33]. In this study, similar technique to calculate NBO/Si using the following Eqs. (1) and (6) was adopted.

$$\frac{\text{NBO}}{\text{Si}}(\text{measured}) = \sum_{n=1}^{4} n \cdot f_{Q^{4-n}} \tag{1}$$

where $f_{Q^{4-n}}$ represents the area fraction of silicate structure units.

The theoretical NBO/Si of silicates can be obtained from Eq. (2).

$$\frac{\text{NBO}}{\text{Si}} = \frac{\sum 2X_{\text{MO}}}{X_{\text{SiO}_2}} \tag{2}$$

Here, mole fraction of oxide is shown as X_{MO} and M=Ca, Mg, and Fe. The value of NBO/Si of FeO_x–CaO–SiO_2–MgO–Cu_2O–(GeO_2/PdO) slags was calculated according to Eq. (2).

The calculated and measured value of NBO/Si is shown in Fig. 5. The graph is divided into two regions: first is the 'Acidic Region' where the measured value of NBO/Si is higher than the calculated value of NBO/Si and the opposite for the other region known as 'Basic Region' [8, 17, 33]. It can be seen from Fig. 5 that the calculated value of NBO/Si is significantly greater than measured NBO/Si. The overestimated value of NBO/Si (calc.) suggests that silicon network is not perfectly depolymerized. Therefore, free oxygen (O^{2-}) interacts with M^{2+} ions and forms a low-strength ionic bond instead of reaching equilibrium state according to Eq. (3).

$$O^0 + O^{2-} = O^- \tag{3}$$

where O^- is non-bridging and O^0 is bridging oxygen, and K is the equilibrium constant of Eq. (3) and given as Eq. (4).

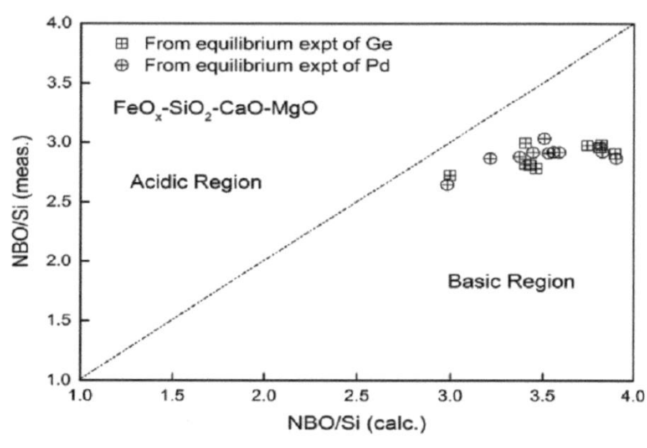

Fig. 5 Relationship between the measured and calculated value of NBO/Si

$$k = \frac{[O^-]^2}{[O^0] \cdot [O^{2-}]} \tag{4}$$

The measured value of the NBO/Si is lower than the calculated value of the NBO/Si, which suggests that there is free network modifier from the depolymerization reaction, which is called 'excess base'. The mole fraction of oxygen species in the slag can be estimated from simple mass balance. The fraction of three types of oxygen speciation in the slag can be calculated from Eqs. (5) to (8) [8, 17, 32, 33].

$$X_{O^{2-}} = \frac{\text{excess base}}{2X_{SiO_2} + \sum X_{MO}} \tag{5}$$

$$X_{O^-} = (1 - X_{O^{2-}}) \cdot \frac{(\text{NBO/T})_{\text{meas.}}}{(\text{NBO/T})_{\text{meas.}} + \sum_{n=1}^{4} \frac{4-n}{2}\left(f_{Q^{4-n}}\right)} \tag{6}$$

$$X_{O^0} = (1 - X_{O^{2-}}) \cdot \frac{\sum_{n=1}^{4} \frac{4-n}{2}\left(f_{Q^{4-n}}\right)}{(\text{NBO}/T)_{\text{meas.}} + \sum_{n=1}^{4} \frac{4-n}{2}\left(f_{Q^{4-n}}\right)} \tag{7}$$

Here, M indicates the metals (Fe and Mg for this case) and excess base is calculated using Eq. (8).

$$\text{excess base} = \sum X_{MO} - X_{SiO_2} \cdot \sum_{n=1}^{4} \frac{n}{2}\left(f_{Q^{4-n}}\right) \tag{8}$$

The calculated oxygen fraction of free and bridging oxygen in the FeO_x–CaO–SiO_2–MgO–Cu_2O–(GeO_2/PdO) slags as a function of mole fraction SiO_2 and CaO at fixed $p_{O_2} = 10^{-8}$ atm and 1300 °C is shown in Fig. 6a, b, respectively. In Fig. 6a, the relative concentration of free oxygen decreases, while bridging oxygen increases with increasing silica contents and it will disappear when silica contents exceed a certain composition. It supports the formation of large silicate anionic network

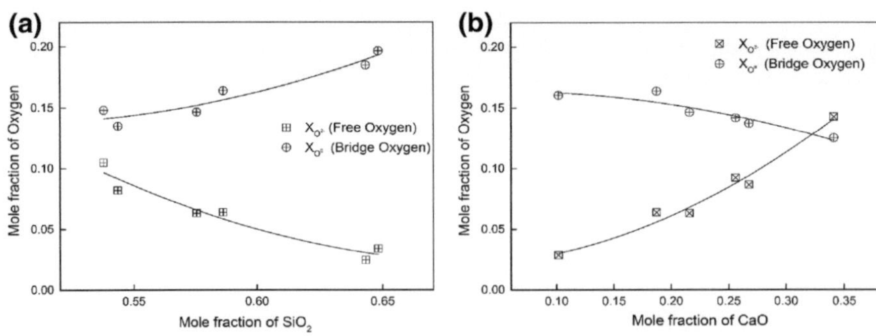

Fig. 6 Mole fraction of oxygen as a function of mole fraction of **a** SiO_2 and **b** CaO in the FeO_x–CaO–SiO_2–MgO–Cu_2O–(GeO_2/PdO) slag ($p_{O_2} = 10^{-8}$ atm and $T = 1300$ °C)

structure by linking $[SiO_4]^{4-}$ tetrahedra units together when it reaches that critical composition [34]. On the other hand, it is seen from Fig. 6b that the fraction of free oxygen increases, while bridging oxygen decreases with increasing CaO contents. This indicates the breaking of large silicate anionic network structure by introducing more Ca^{2+} cations. Thus, the analysis of the structure of silicate melts is important to investigate the structure-dependent property changes and for deducing the scientific theories for those changes related to the concentration of different oxygen species in the slag.

Summary

The FTIR spectra of the FeO_x–CaO–SiO_2–MgO–Cu_2O–(GeO_2/PdO) slags were measured to investigate the effect of slag composition on the silicate structure. The structure of slags was interpreted based on the abundance of different silicon tetrahedral units and the NBO/T calculated from that. The summary of the study is explained below:

(1) Ge^{4+} with higher ion-oxygen ratio than Si^{4+} behaves as acidic species while Pd^{2+} with lower ion-oxygen attraction value acts as basic species in the tested FeO_x–CaO–SiO_2–MgO–Cu_2O–GeO_2/PdO slag system.
(2) Acidic slags with lower non-bridging oxygen per tetrahedra (NBO/T) are more favourable for Ge-partitioning to copper, and for Pd, the relation is opposite.
(3) The speciation of oxygen was calculated in the FeO_x–CaO–SiO_2–MgO–Cu_2O–GeO_2/PdO slag system. It is shown that slag structure analysis can be used to calculate the concentration of different oxygen species in the slag.

Acknowledgements This work was part of M. A. H. Shuva Ph.D. study supported by the Swinburne University Postgraduate Research Award (SUPRA) and the Wealth from Waste Research Cluster, a collaborative program between the Australian Commonwealth Scientific and Industrial Research Organisation (CSIRO), Swinburne University of Technology, University of Technology Sydney, Monash University, University of Queensland, and Yale University.

References

1. Lifespan of consumer electronics is getting shorter, study finds, https://www.theguardian.com/environment/2015/mar/03/lifespan-of-consumer-electronics-is-getting-shorter-study-finds. The Gurdian, 3 March, 2015, accessed on 10 August, 2018
2. Peckham V (2018) These 10 mines will set the copper price for the next decade. MINING.com, http://www.mining.com/these-10-mines-will-set-the-copper-price-for-the-next-decade/, 3 Nov 2015. Accessed on 10 Aug 2018
3. Keeling A, Sandlos J (2009) Environmental justice goes underground? Historical notes from Canada's northern mining frontier. Environ Just 2(3):117–125
4. Rhamdhani MA et al (2015) More from less, generating wealth from lower grade and urban metal/ore sources. In: Advanced Materials Research. Trans Tech Publ

5. University UN (2018) World e-waste rises 8% by weight in 2 years as incomes rise, prices fall: UN-backed report. https://www.sciencedaily.com/releases/2017/12/171213143714.htm?utm_medium=cpc&utm_campaign=ScienceDaily_TMD_1&utm_source=TMD. 13 December, 2017. Accessed on 10 Aug 2018
6. Nakajima K et al (2011) Thermodynamic analysis for the controllability of elements in the recycling process of metals. Environ Sci Technol 45(11):4929–4936
7. Shuva M et al (2016) Thermodynamics data of valuable elements relevant to e-waste processing through primary and secondary copper production: a review. J Clean Prod 131:795–809
8. Park JH (2013) Effect of silicate structure on thermodynamic properties of calcium silicate melts: quantitative analysis of Raman spectra. Met Mater Int 19(3):577–584
9. Mills KC (1993) The influence of structure on the physico-chemical properties of slags. ISIJ Int 33(1):148–155
10. Maroufi S et al (2016) Diffusion coefficients and structural parameters of molten slags. In Advances in molten slags, fluxes, and salts: proceedings of the 10th international conference on molten slags, fluxes and salts 2016. Springer
11. Min DJ, Tsukihashi F (2017) Recent advances in understanding physical properties of metallurgical slags. Met Mater Int 23(1):1–19
12. Park Y, Min DJ (2017) A structural study on the foaming behavior of CaO–SiO_2–MO (MO=MgO, FeO, or Al_2O_3) ternary slag system. Metall Mater Trans B 48(6):3038–3046
13. Brooks GA, Hasan MM, Rhamdhani MA (2019) Slag basicity: What does it mean? In: Jiang T et al (eds) 10th International symposium on high-temperature metallurgical processing. The minerals, metals and materials series. Springer, Cham, pp 297–308
14. Virgo D, Mysen B, Kushiro I (1980) Anionic constitution of 1-atmosphere silicate melts: implications for the structure of igneous melts. Science 208(4450):1371–1373
15. Mysen BO, Virgo D, Kushiro I (1981) The structural role of aluminum in silicate melts: a Raman spectroscopic study at 1 atmosphere. Am Miner 66(7–8):678–701
16. Li J, Shu Q, Chou K (2014) Structural study of glassy CaO–SiO_2–CaF_2–TiO_2 slags by Raman spectroscopy and MAS-NMR. ISIJ Int 54(4):721–727
17. Park JH (2012) Composition–structure–property relationships of CaO–MO–SiO_2 (M=Mg^{2+}, Mn^{2+}) systems derived from micro-Raman spectroscopy. J Non-Cryst Solids 358(23):3096–3102
18. Sun Y et al (2015) FTIR, Raman and NMR investigation of CaO–SiO_2–P_2O_5 and CaO–SiO_2–TiO_2–P_2O_5 glasses. J Non-Cryst Solids 420:26–33
19. Wang L et al (2016) Raman structure investigations of CaO–MgO–Al_2O_3–SiO_2–CrO_x and its correlation with sulfide capacity. Metall Mater Trans B 47(1):10–15
20. Sohn I et al (2012) Influence of TiO_2 on the viscous behavior of calcium silicate melts containing 17 mass% Al_2O_3 and 10 mass% MgO. ISIJ Int 52(1):158–160
21. Farges F, Brown GE (1997) Coordination chemistry of titanium (IV) in silicate glasses and melts: IV. XANES studies of synthetic and natural volcanic glasses and tektites at ambient temperature and pressure. Geochimica et Cosmochimica Acta 61(9):1863–1870
22. Farges F, Brown GE, Rehr JJ (1996) Coordination chemistry of Ti (IV) in silicate glasses and melts: I. XAFS study of titanium coordination in oxide model compounds. Geochimica et Cosmochimica Acta 60(16):3023–3038
23. Chaskar V, Richards G, McCammon C (1993) A mössbauer study of the behavior of iron cations in iron oxide-containing melts at 1400 °C. Metall Trans B 24(1):101–111
24. Yano T, Shibata S, Maehara T (2006) Structural equilibria in silicate glass melts investigated by Raman spectroscopy. J Am Ceram Soc 89(1):89–95
25. Shuva M et al (2018) Structural analysis of germanium (ge)-containing ferrous calcium silicate magnesia slag for applications of black copper smelting. In: TMS annual meeting and exhibition. Springer
26. Shuva M et al (2016) Thermodynamics behavior of germanium during equilibrium reactions between FeO_x–CaO–SiO_2–MgO slag and molten copper. Metall Mater Trans B 47:2889–2903
27. Shuva M (2017) Analysis of thermodynamics behaviour of valuable elements and slag structure during e-waste processing through copper smelting. Ph.D. thesis, Swinburne University of Technology

28. Shuva M et al (2017) Thermodynamics of palladium (Pd) and tantalum (Ta) relevant to secondary copper smelting. Metall Mater Trans B 48(1):317–327
29. Mysen BO, Virgo D (1980) Solubility mechanisms of carbon dioxide in silicate melts; a Raman spectroscopic study. Am Miner 65(9–10):885–899
30. Mysen BO, Richet P (2018) Silicate glasses and melts. Elsevier
31. Mysen BO (1990) Relationships between silicate melt structure and petrologic processes. Earth-Sci Rev 27(4):281–365
32. Park JH, Min DJ, Song HS (2002) FTIR spectroscopic study on structure of CaO–SiO_2 and CaO–SiO_2–CaF_2 Slags. ISIJ Int 42(4):344–351
33. Park JH (2013) Structure–property relationship of CaO–MgO–SiO_2 slag: quantitative analysis of Raman spectra. Metall Mater Trans B 44(4):938–947
34. Zhang L, Jahanshahi S (2013) Review and modeling of viscosity of silicate melts: part I. Viscosity of binary and ternary silicates containing CaO, MgO, and MnO. Metall Mater Trans B 29(1):177–186

A Model for the Interaction of Fe with MgO–14.5 wt% C Refractory Under Flash Ironmaking Conditions

Rahul Sarkar and Hong Yong Sohn

Abstract In this work, the interaction of Fe with MgO–14.5 wt% C refractory under the conditions relevant to the novel flash ironmaking technology (FIT), which has been developed at the University of Utah, has been studied. Oxidation of carbon occurred and the formation of magnesiowustite ($Mg_xFe_{1-x}O$) solid solution took place as a result of the interaction between Fe and MgO (in the presence of O_2). A kinetic model for the growth of magnesiowustite was developed based on the counterdiffusion of Fe^{2+} and Mg^{2+} cations, and experiments were conducted with Fe powders and MgO–14.5 wt% C refractory in the temperature range 1200–1400 °C under flash ironmaking atmospheres. Analyses of samples using SEM-EDX and EPMA confirmed the formation of magnesiowustite and using the experimentally determined composition profiles the values of interdiffusion coefficient, averaged over the composition range ($\bar{D}_{Fe-Mg}$), were calculated. The activation energy for the solid-state diffusion was calculated to be 377 kJ/mol.

Keywords Flash ironmaking · Magnesia-carbon refractory · Solid-state diffusion · Magnesiowustite

Introduction

A novel flash ironmaking technology (FIT) has been developed at the University of Utah based on the direct reduction of iron ore concentrate with a reductant gas such as hydrogen, natural gas, or coal gas in a flash furnace. This technology aims at overcoming the shortcomings of the conventional blast furnace (BF) technology. First of all, the FIT bypasses the problematic pelletization/sintering and cokemaking steps required in the BF technology since iron ore concentrates in the form of powders can be directly used in this process. Also, since a gaseous reducing agent is used instead of solid carbon, there is no need for the cokemaking step. These factors greatly reduce the energy consumption and CO_2 emissions, and the FIT can operate with 30% less energy input and 39–51% less CO_2 emissions as compared to the BF technology

R. Sarkar (✉) · H. Y. Sohn
Department of Metallurgical Engineering, University of Utah, Salt Lake City, Utah 84112, USA
e-mail: rahulsarkarmet.89@gmail.com

J. Lee et al. (eds.), *Materials Processing Fundamentals 2020*, The Minerals, Metals & Materials Series, https://doi.org/10.1007/978-3-030-36556-1_9

[1, 2]. Moreover, since this process can be operated over a wide temperature range, it is possible to produce iron either in solid or liquid [3–5].

In the development of FIT and its proposed scale-up, the choice and design of refractories are expected to play a pivotal role. Since the refractories are meant to sustain high temperatures, they are often quite expensive and failure in the refractories results in the loss of production and equipment time. Sometimes, the type of refractories used affects the quality of the manufactured product. Economic factors greatly impact these issues, and the refractory most suitable for the FIT might not necessarily be the one that lasts the longest but rather the one which gives the best balance between the initial cost of installation and its performance [6]. For identifying a suitable refractory for the FIT, understanding the interactions between iron, iron oxides, and selected candidate refractories under flash ironmaking conditions is necessary. This article presents a kinetic model for the interaction of iron (Fe) powder with MgO–14.5 wt% C refractory under flash ironmaking conditions.

Theory

Thermodynamics

Carbon Oxidation

Under flash ironmaking conditions, oxidation of carbon takes place from the MgO–C refractory as a result of *direct burnout of carbon* by O_2 (from oxidizing gases such as CO_2 and H_2O) as per the reaction

$$C(s) + \frac{1}{2}O_2(g) = CO(g) \tag{1}$$

For Reaction (1), the standard Gibbs free energy (ΔG_1^0) and equilibrium constant (K_1) are given by [7]

$$\Delta G_1^0 = -114400 - 85.77T \left(\frac{\mathrm{J}}{\mathrm{mol}}\right) \tag{2}$$

$$K_1 = \frac{p_{CO}}{a_{C(s)} \cdot p_{O_2}^{1/2}} \tag{3}$$

Taking the activity of solid carbon as unity, the oxidation of carbon by Reaction (1) can take place only if the oxygen partial pressure (p_{O_2}) is higher than that the corresponding equilibrium value given by Eq. (3). For p_{CO} values typically relevant to flash ironmaking, the p_{O_2} values are much higher than the equilibrium values given by Eq. (3) at flash ironmaking temperatures. Therefore, *direct burnout of carbon* under flash ironmaking conditions per Reaction (1) takes place. Other probable mechanisms of carbon oxidation from the MgO–C refractory are available in the literature [7, 8] but in this case the *direct burnout of carbon* is the most probable mechanism, as described in detail elsewhere [9].

Interaction of Fe and MgO in the Presence of O_2 (From Oxidizing Gases CO_2 and H_2O)

The system that needs to be considered for studying the interaction between Fe and MgO–C refractory in the presence of O_2 (from oxidizing gases such as CO_2 and H_2O) is the Fe–MgO–O system because carbon from the magnesia-carbon refractory oxidizes. Of the various iron oxides, the most important in the present case is FeO because Fe^{2+} is the primary diffusing species for iron [10, 11]. Thus, the phases formed in this system as a result of interaction under FIT conditions were determined from the FeO–MgO phase diagram. To account for the effect of gas atmospheres, the pseudo-binary FeO–MgO phase diagrams for p_{O_2} values relevant to FIT were considered. In this system, the only solid solution phase that forms is magnesiowustite ($Mg_xFe_{1-x}O$). Thus, under FIT conditions, the only phase that is expected to form as a result of interactions between iron and MgO from the magnesia-carbon refractory is the magnesiowustite phase.

Kinetic Modeling

In this section, a kinetics model using solid-state diffusion is developed for the growth of magnesiowustite formed as a result of interaction between Fe and MgO in the presence of O_2.

Model Assumptions

The following assumptions are first made:

i. The reactions occur isothermally.
ii. Thermodynamic equilibrium is maintained at the iron (Fe)-magnesiowustite phase boundary.
iii. The magnesiowustite formed is completely compact and does not contain any pores.
iv. The molar volume of magnesiowustite (V_{MW}) is assumed to be a linear function of its composition.

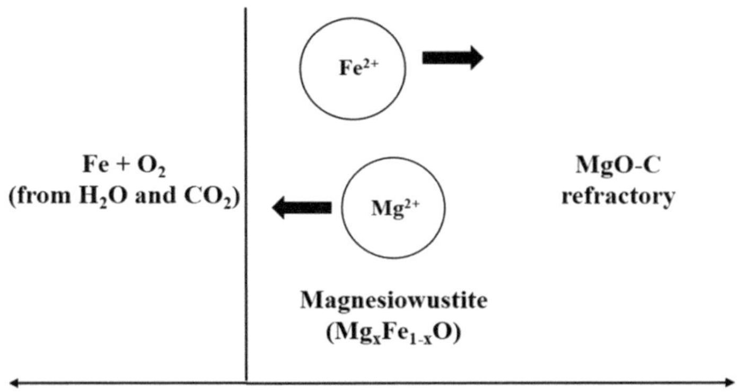

Fig. 1 Schematic representation of the interaction mechanism between Fe and MgO–C refractory under flash ironmaking conditions

Interaction Mechanism

A number of possible mechanisms for the growth of magnesiowustite are possible but the solid-state diffusion in the FeO–MgO system can best be described by the counter diffusion of Fe^{2+} and Mg^{2+} cations through a relatively rigid oxygen ion lattice of the rock salt structure of magnesiowustite [12–15]. Thus, in this work, the counterdiffusion of Fe^{2+} and Mg^{2+} cations through magnesiowustite is chosen as the dominant interaction mechanism. A schematic representation of this mechanism is shown in Fig. 1.

Model Development

The FeO–MgO system is known to follow the regular solution model [16]. Therefore, in this case, the volume change associated with the mixing process of FeO and MgO is zero. This postulation, along with the assumption of linear dependence of molar volume of magnesiowustite (V_{MW}) on its composition, gives

$$V_{MW} = x \cdot V_{MgO} + (1 - x) \cdot V_{FeO} \tag{4}$$

where V_{MgO} and V_{FeO} are the molar volumes of MgO and FeO, respectively.

Moreover, the difference between V_{MgO} and V_{FeO} is less than 10% at the experimental temperatures [17]. Hence, V_{MW} is expected to be a weak function of the magnesiowustite composition, and in this model, V_{MW} is assumed to be a constant over the entire range of magnesiowustite composition.

The quantitative treatment to this problem is done by performing a shell mass balance of species i (where i can be Fe^{2+} or Mg^{2+}) over a control volume of unit cross-sectional area and width Δz in the magnesiowustite layer. In the limit $\Delta z \to 0$, this gives

$$-\frac{\partial N_i}{\partial z} = \frac{\partial C_i}{\partial t} \tag{5}$$

The relationship between N_i and diffusive molar flux (J_i) is given by:

$$N_i = J_i + X_i \cdot N_T = J_i + X_i \cdot \left(N_{\mathrm{Fe}} + N_{\mathrm{Mg}}\right) \tag{6}$$

where X_i is the mole fraction of cation i with respect to total cations in the magnesiowustite, and N_T is the total molar flux expressed as

$$N_T = N_{\mathrm{Fe}} + N_{\mathrm{Mg}} \tag{7}$$

J_i is given by Fick's first law for binary diffusion as

$$J_i = -\bar{D}_{\mathrm{Fe-Mg}} \cdot \frac{\partial C_i}{\partial z} \tag{8}$$

In Eq. (8), $\bar{D}_{\mathrm{Fe-Mg}}$ is the interdiffusion coefficient in the Fe^{2+} – Mg^{2+} binary system. As reported by Liermann and Ganguly [18] in the case of Fe^{2+} and Mg^{2+} counterdiffusion, the dependence of $\bar{D}_{\mathrm{Fe-Mg}}$ on composition is weak. In this work, $\bar{D}_{\mathrm{Fe-Mg}}$ represents the average value of the interdiffusion coefficient over the entire magnesiowustite composition range.

At $z = 0$, there is no motion of Mg^{2+} cation due to the negligibly small solubility of Mg^{2+} ions in pure Fe. Therefore,

$$N_{\mathrm{Mg}}\big|_{z=0} = 0 \tag{9}$$

Applying Eq. (6) for Fe^{2+} cations at $z = 0$ and using Eqs. (8) and (9), we get:

$$N_{\mathrm{Fe}}\big|_{z=0} = -\bar{D}_{\mathrm{Fe-Mg}} \cdot \left.\frac{\partial C_{\mathrm{Fe}}}{\partial z}\right|_{z=0} + X_0 \cdot N_{\mathrm{Fe}}\big|_{z=0} \tag{10}$$

where X_0 is the mole fraction of Fe^{2+} cation with respect to total cations at $z = 0$.

Rearranging Eq. (10) and using Eq. (9), we get

$$\left(N_{\mathrm{Fe}} + N_{\mathrm{Mg}}\right)\big|_{z=0} = \frac{-\bar{D}_{\mathrm{Fe-Mg}}}{1 - X_0} \cdot \left.\frac{\partial C_{\mathrm{Fe}}}{\partial z}\right|_{z=0} \tag{11}$$

Now using Eq. (5), we get

$$-\frac{\partial N_T}{\partial z} = -\frac{\partial\left(N_{\mathrm{Fe}} + N_{\mathrm{Mg}}\right)}{\partial z} = \frac{\partial\left(C_{\mathrm{Fe}} + C_{\mathrm{Mg}}\right)}{\partial t} = \frac{\partial C}{\partial t} = 0 \tag{12}$$

since C, the total concentration, expressed as

$$C = C_{\mathrm{Fe}} + C_{\mathrm{Mg}} = 1/V_{\mathrm{MW}} \tag{13}$$

is not a function of time. It follows, therefore, that the total flux, $N_T = N_{\mathrm{Fe}} + N_{\mathrm{Mg}}$ is not a function of z and is a function of t alone. Thus for all z, Eq. (11) must hold true. In other words, for all z

$$N_T = N_{\mathrm{Fe}} + N_{\mathrm{Mg}} = \frac{-\bar{D}_{\mathrm{Fe-Mg}}}{1 - X_0} \cdot \left.\frac{\partial C_{\mathrm{Fe}}}{\partial z}\right|_{z=0} \tag{14}$$

Substituting Eqs. (14) and (8) in Eq. (6) (for $i = \mathrm{Fe}^{2+}$), we get

$$N_{\mathrm{Fe}} = -\bar{D}_{\mathrm{Fe-Mg}} \cdot \frac{\partial C_{\mathrm{Fe}}}{\partial z} - X_{\mathrm{Fe}} \cdot \frac{\bar{D}_{\mathrm{Fe-Mg}}}{1 - X_0} \cdot \left.\frac{\partial C_{\mathrm{Fe}}}{\partial z}\right|_{z=0} \tag{15}$$

Differentiating Eq. (15) with respect to z and using Eq. (5), we get

$$\frac{\partial N_{\mathrm{Fe}}}{\partial z} = -\frac{\partial}{\partial z}\left(\bar{D}_{\mathrm{Fe-Mg}} \cdot \frac{\partial C_{\mathrm{Fe}}}{\partial z} + X_{\mathrm{Fe}} \cdot \frac{\bar{D}_{\mathrm{Fe-Mg}}}{1 - X_0} \cdot \left.\frac{\partial C_{\mathrm{Fe}}}{\partial z}\right|_{z=0}\right) = -\frac{\partial C_{\mathrm{Fe}}}{\partial t} \tag{16}$$

Substituting,

$$C_{\mathrm{Fe}} = \frac{X_{\mathrm{Fe}}}{V_{\mathrm{MW}}} \tag{17}$$

and rearranging, we get

$$\begin{aligned} \frac{\partial X_{\mathrm{Fe}}}{\partial t} &= \frac{\partial}{\partial z}\left(\bar{D}_{\mathrm{Fe-Mg}} \frac{\partial X_{\mathrm{Fe}}}{\partial z} + X_{\mathrm{Fe}} \cdot \frac{\bar{D}_{\mathrm{Fe-Mg}}}{1 - X_0} \cdot \left.\frac{\partial X_{\mathrm{Fe}}}{\partial z}\right|_{z=0}\right) \\ &= \bar{D}_{\mathrm{Fe-Mg}} \frac{\partial^2 X_{\mathrm{Fe}}}{\partial z^2} + \frac{\bar{D}_{\mathrm{Fe-Mg}}}{1 - X_0} \cdot \left.\frac{\partial X_{\mathrm{Fe}}}{\partial z}\right|_{z=0} \cdot \frac{\partial X_{\mathrm{Fe}}}{\partial z} \end{aligned} \tag{18}$$

Choosing the origin (i.e., $z = 0$) at the point of contact between Fe and MgO, the boundary conditions for the concentration profiles are given by

$$\text{at } z = 0 \quad \left.\begin{aligned} X_{\mathrm{Fe}} &= X_0 \\ X_{\mathrm{Mg}} &= 1 - X_0 \end{aligned} \quad \text{for all } t > 0\right\} \tag{19}$$

$$\text{as } z \to \infty \quad \left.\begin{aligned} X_{\mathrm{Fe}} &= 0 \\ X_{\mathrm{Mg}} &= 1.0 \end{aligned} \quad \text{for all } t > 0\right\} \tag{20}$$

Introducing, dimensionless variables:

$$\left.\begin{array}{c}\eta = \frac{z}{2\sqrt{\bar{D}_{\text{Fe-Mg}}t}} \\ \varphi = \frac{X_{\text{Fe}}}{X_0}\end{array}\right\} \tag{21}$$

After substituting η and φ, Eq. (18) becomes

$$\varphi'' + 2(\eta - \lambda)\varphi' = 0 \tag{22}$$

where

$$\lambda = -\frac{1}{2} \cdot \frac{X_0}{1 - X_0} \cdot \varphi'\big|_{z=0} \tag{23}$$

The boundary conditions in terms of dimensionless variables now are:

$$\begin{array}{c}\text{at } z = 0 \ \varphi = 1 \\ z \to \infty \ \varphi = 0\end{array} \tag{24}$$

Using these boundary conditions, Eq. (22) can be integrated to obtain the concentration profile in terms of dimensionless variables as

$$\varphi = \frac{1 - \text{erf}(\eta - \lambda)}{1 + \text{erf}\lambda} \tag{25}$$

The relation between X_0 and λ is obtained as

$$X_0 = \frac{1}{1 + (\sqrt{\pi}(1 + \text{erf}\,\lambda) \cdot \lambda \cdot \exp \lambda^2)^{-1}} \tag{26}$$

From Eq. (25), we get:

$$\text{erfc}(\eta - \lambda) = (1 + \text{erf}\,\lambda) \cdot \varphi \tag{27}$$

Back-substituting the dimensional variables X_{Fe}, z and t and rearranging

$$\frac{z}{2\sqrt{\bar{D}_{\text{Fe-Mg}}\,t}} = \text{erfc}^{-1}\left[(1 + \text{erf}\,\lambda).\frac{X_{\text{Fe}}}{X_0}\right] + \lambda = \theta \tag{28}$$

Using the experimental conditions, X_0 is calculated as follows:

We assume that FeO in $Mg_xFe_{1-x}O$ is in equilibrium with Fe and O_2 at $z = 0$ as per the reaction

$$\text{Fe(s)} + \frac{1}{2}\text{O}_2\text{(g)} = \text{FeO}_{\text{in MW}} \tag{29}$$

$$K_2 = \frac{a_{FeO}|_{z=0}}{p_{O_2}^{1/2}} \tag{30}$$

where $a_{FeO}|_{z=0}$ is the activity of FeO in magnesiowustite at $z = 0$ and K_2 is the equilibrium constant for Reaction (29). Using Eq. (30), $a_{FeO}|_{z=0}$ can be calculated from a knowledge of K_2 from Barin [19] and p_{O_2} (set by experimental conditions). Once $a_{FeO}|_{z=0}$ is known, X_0 can be calculated as

$$X_0 = \frac{a_{FeO}|_{z=0}}{\gamma_{FeO}} \tag{31}$$

where γ_{FeO} is the activity coefficient of FeO in magnesiowustite, calculated using FactSage Version 7.2 [20]. Once X_0 is known, λ can be calculated using Eq. (26). Thereafter, if the composition profile for any time t is known, $\bar{D}_{Fe-Mg}$ can be calculated using Eq. (28) from the plot of function θ versus z.

Experimental Work

Excess amounts of Fe powder were applied on both sides of an MgO–C refractory sample placed inside an alumina crucible. The alumina crucible containing the Fe powder and MgO–C was then placed on a carved out fire brick and the entire assembly was inserted into the hot zone of a horizontal tubular furnace. After a specific interaction time as required for the particular experiment, the assembly was quickly removed from the hot zone of the furnace, and the sample along with the crucible was quenched in water. Experiments were conducted at three different temperatures, viz. 1200, 1300, and 1400 °C, and at each temperature, experiments were carried out for at least three different interaction times. The gas atmospheres were so maintained that metallic Fe was stable at the experimental temperatures. A gas composition of $X_{H_2} = 0.50$, $X_{CO} = 0.25$ and $X_{CO_2} = 0.25$ satisfied this condition. Hence, this gas composition was used as the initial gas composition for calculating the equilibrium gas compositions using HSC version 5.1 [21] and the equilibrium gas compositions were then used for determining the flow rates of H_2, H_2O, CO, and CO_2. Figure 2 shows the p_{O_2} values for Fe–MgO–C experiments.

Results and Discussion

Interaction Mechanism

As discussed in Sect. 2.1.1, the carbon in the MgO–C refractory gets oxidized and it is only the MgO in the MgO–C refractory that interacts with Fe. As a result of

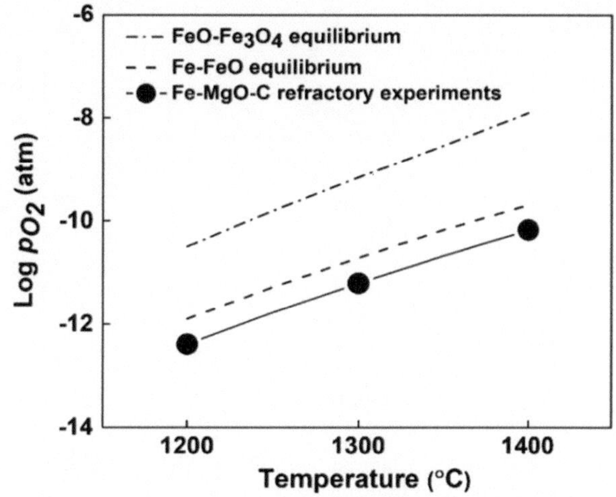

Fig. 2 p_{O_2} versus temperature plot for Fe–MgO–C refractory experiments compared with the corresponding values for Fe–FeO and FeO–Fe_3O_4 equilibria

interaction between Fe, MgO and O_2 (from oxidizing gases viz. H_2O and CO_2), the formation of magnesiowustite solid solution ($Mg_xFe_{1-x}O$) takes place at all experimental temperatures. Figure 3a shows the SEM micrograph of reacted MgO–C cross section after interaction with iron for 16 h at $T = 1300$ °C and $p_{O_2} = 6.28 \times 10^{-12}$ atm. (henceforth referred to as Fe–MgO–C-1300-1). As evident from Fig. 3a, three separate regions can be distinguished from the SEM micrograph shown in Fig. 3a. These were the iron (Fe) on the left, the MgO–C refractory on the right and the magnesiowustite ($Mg_xFe_{1-x}O$) in between the two. The composition of magnesiowustite

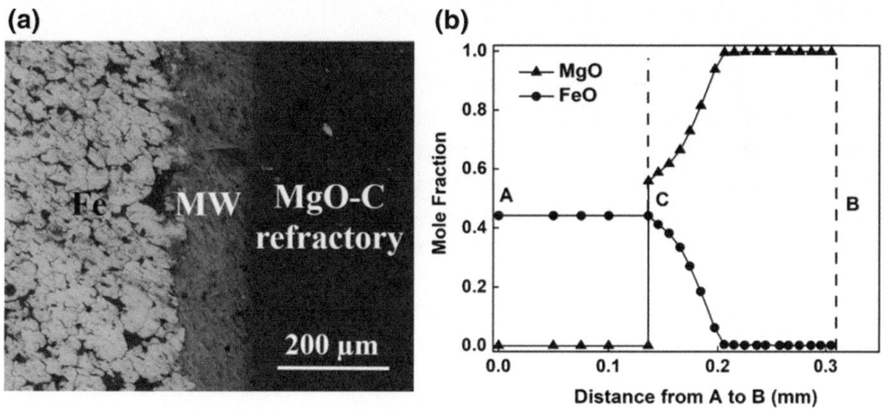

Fig. 3 **a** SEM micrograph of reacted MgO–C refractory cross section after interaction with iron for 16 h at $T = 1300$ °C and $p_{O_2} = 6.28 \times 10^{-12}$ atm. [MW: Magnesiowustite ($Mg_xFe_{1-x}O$)]. **b** Concentration profiles for FeO and MgO across the reacted MgO–C refractory cross section for the same sample (point A is in pure Fe region, point B is deep inside MgO–C refractory, and point C is Fe–MW phase boundary)

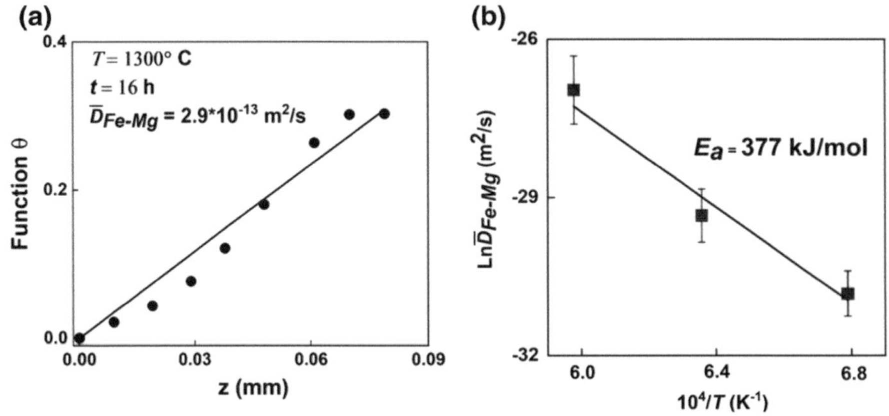

Fig. 4 **a** θ versus z plot for sample Fe–MgO–C-1300-1. **b** Arrhenius plot for interdiffusion coefficient $\left(\bar{D}_{\text{Fe-Mg}}\right)$ in magnesiowustite

($Mg_xFe_{1-x}O$) varied across the reacted refractory cross section and accurate determination of the magnesiowustite composition as a function of distance was done using EPMA. Figure 3b shows the composition profiles obtained from an EPMA line scan for the reacted MgO–C cross section for the same sample. The scan is performed from point A which is in the pure Fe region to a point B which is deep inside the MgO–C refractory.

Determination of Interdiffusion Coefficient ($\bar{D}_{Fe-Mg}$)

Once the composition profiles are obtained, the interdiffusion coefficient ($\bar{D}_{\text{Fe-Mg}}$) can be obtained using the kinetics model already described in Sect. 2.2.3. Using the experimental conditions, X_0 is first calculated and then λ is calculated using Eq. (26). Then the function θ is plotted as a function of z, and $\bar{D}_{\text{Fe-Mg}}$ is calculated from the slope of the curve. Figure 4a shows the plot of θ versus z for the same sample. In this case, $\bar{D}_{\text{Fe-Mg}}$ is calculated as 2.9×10^{-13} m^2/s. At each temperature, the $\bar{D}_{\text{Fe-Mg}}$ values for different interaction times were averaged and plotted in Fig. 4b as function of inverse of temperature. The activation energy (E_a) for $\bar{D}_{\text{Fe-Mg}}$ was calculated as 377 kJ/mol.

Conclusions

The following are the conclusions from this work:

1. It has been shown through thermodynamic calculations that oxidization of carbon from MgO–C refractory occurs, and the formation of magnesiowustite

($Mg_xFe_{1-x}O$) takes place as a result of Fe–MgO–O_2 interaction under flash ironmaking conditions.
2. A kinetics model for the growth of magnesiowustite ($Mg_xFe_{1-x}O$) has been developed by taking into account the counter diffusion of Fe^{2+} and Mg^{2+} ions through the magnesiowustite.
3. Using the kinetics model and the compositional profiles obtained from EPMA line scans, the values of interdiffusion coefficient ($\bar{D}_{Fe-Mg}$) were determined at 1200, 1300, and 1400 °C. The activation energy (E_a) for $\bar{D}_{Fe-Mg}$ was 377 kJ/mol.

References

1. Pinegar HK, Moats MS, Sohn HY (2012) Flow sheet development, process simulation and economic feasibility analysis for novel suspension ironmaking technology based on natural gas: part 1—flow sheet and simulation for ironmaking with reformerless natural gas. Ironmak Steelmak 39(6):398–408
2. Choi ME, Sohn HY (2010) Development of green suspension ironmaking technology based on hydrogen reduction of iron oxide concentrate: rate measurements. Ironmak Steelmak 37(2):81–88
3. Wang H, Sohn HY (2007) Hydrogen reduction kinetics of magnetite concentrate particles relevant to a novel flash ironmaking process. Metall Mater Trans B 2013 44(1):133–145
4. Sohn HY (2007) Suspension ironmaking technology with greatly reduced energy requirement and CO_2 emissions. Steel Times Int 31(a):68–72
5. Mohassab-Ahmed MY, Sohn HY (2013) Application of spectroscopic analysis techniques to the determination of slag structures and properties: effect of water vapor on slag chemistry relevant to a novel flash ironmaking technology. JOM 65(11):1559–1565
6. Hubble DH (1998) Steel plant refractories. In: Fruehan RJ (ed) The making, shaping and treating of steel 1998. The AISI Steel Foundation; AIST, Pittsburgh, pp 159–226
7. Smets S, Parada S, Weytjens J, Heylen G, Jones PT, Guo M et al (2003) Behaviour of magnesia-carbon refractories in vacuum-oxygen decarburisation ladle linings. Ironmak Steelmak 30(4):293–300
8. Pickering GD, Batchelor JD (1972) Carbon–MgO reactions in BOF refractories. Bull Am Ceram. Soc. 50:611–614
9. Sarkar R (2019) Interactions of iron, wustite and slags with selected refractory materials under flash ironmaking conditions. PhD [dissertation]. University of Utah, Salt Lake City, Utah, USA
10. Sarkar R, Sohn HY (2019) Interaction of iron with alumina refractory under flash ironmaking conditions. Metall Mater Trans B 50(4):2063–2076
11. Sarkar R, Sohn HY (2019) Interaction of ferrous oxide with alumina refractory under flash ironmaking conditions. Ceram Int 45(12):15417–15428
12. Rigby EB, Cutler IB (1965) Interdiffusion studies of the system Fe_xO–MgO. J Am Ceram Soc 48(2):95–99
13. Carter RE (1961) Mechanism of solid-state reaction between magnessium oxide and aluminum oxide and between magnessium oxide and ferric oxide. J Am Ceram Soc 44(3):116–120
14. Sata N, Goto KS (2006) Oxygen pressure dependence of cation inter diffusivity in magnesiowustite solid solutions. J Am Ceram Soc 65(3):158–162
15. Bygden J, DebRoy T, Seetharaman S (1994) Dissolution of MgO in stagnant CaO–FeO–SiO_2 slags. Ironmak Steelmak 21:318–323
16. Wu P, Eriksson G, Pelton AD, Blander M (1993) Prediction of the thermodynamic properties and phase diagrams of silicate systems-Evaluation of the FeO–MgO–SiO_2 systems. ISIJ Int. 33(1):26–35

17. Robie RA, Bethke PB (1962) Molar volumes and densities of minerals. Unites States Department of the Interior Geological Survey, pp 4–21
18. Liermann HP, Ganguly J (2002) Diffusion kinetics of Fe^{2+} and Mg in aluminous spinel-experimental determination and applications. Geochim Cosmochim Acta 66(16):2903–2913
19. Barin I (1993) Thermodynamical data of pure substances, vol I. VCH Publishers, New York, NY
20. Bale CW, Bélisle E, Chartrand P, Decterov SA, Eriksson G, Gheribi AE et al (2016) FactSage thermochemical software and databases 2010–2016. Calphad (Sept) 54: 35–53. www.factsage.com
21. Outokumpu Research Oy, Pori, Finland, A. Roine, HSC Chemistry, Version 5.2. www.hsc-chemistry.net

Process of Thermal Decomposition of Lithium Carbonate

Lei Shi, Tao Qu, Dachun Liu, Yong Deng, Bin Yang and Yongnian Dai

Abstract In recent years, the methods of lithium preparation by metallothermic reduction of its oxide in negative pressure have been developed. Since Li_2CO_3 was considered as an important raw material for the preparation of Li_2O, it is important to clarify the decomposition and melting mechanisms of Li_2CO_3. The behaviors of decomposition of lithium carbonate under argon, carbon dioxide, and negative pressure were studied by thermogravimetric behavior. Results showed that the decomposition of Li_2CO_3 can be divided into two steps and the mass loss under different atmospheric conditions is different. The first decomposition temperature of carbonate was 1000 K in argon gas. Decomposition of lithium carbonate was a complex process, including the multiple reactions such as melting of lithium carbonate, dissolution of Li_2O and CO_2 in Li_2CO_3, and adsorption of CO_2 in Li_2O. The first step of the decomposition is reduced, and the second step of the decomposition is increased in carbon dioxide atmosphere or negative pressure, compared with the condition of the argon gas.

Keywords Lithium · Lithium oxide · Lithium carbonate · Thermal decomposition · TG-DSC

Introduction

Lithium and its compounds have been widely applied in various branches of industry due to its excellent performance, in medicine, glass, refrigeration, ceramics, atomic energy, metallurgical industry, military industry, etc. The development of effective technologies or the improvement of existing technologies for obtaining metal lithium

---

L. Shi · T. Qu (✉) · D. Liu · Y. Deng · B. Yang · Y. Dai
National Engineering Laboratory of Vacuum Metallurgy, Kunming University of Science and Technology, Kunming 650093, People's Republic of China
e-mail: qutao_82@126.com

Key Laboratory of Vacuum Metallurgy for Non-Ferrous Metal of Yunnan Province, Kunming 650093, People's Republic of China

State Key Laboratory of Complex Nonferrous Metal Resources Clean Utilization in Yunnan Province, Kunming 650093, People's Republic of China

J. Lee et al. (eds.), *Materials Processing Fundamentals 2020*, The Minerals, Metals & Materials Series, https://doi.org/10.1007/978-3-030-36556-1_10

is a topical problem now. At present, the electrolytic process and thermal reduction process are the main methods of industrial production of lithium in the world. The metallothermic method is more environmentally safe and less waste-producing compared with the electrolytic method.

In recent years, the methods of obtaining lithium by metallothermic reduction of its oxide in negative pressure have been developed. Yang bin et al. [1] extracted metallic lithium from Li_2CO_3 by negative pressure silicothermic reduction. Fan et al. [2] reported that lithium was obtained by negative pressure iron thermal reduction. The methods of obtaining lithium by the aluminothermic reduction from its aluminate in negative pressure have been studied at the Moscow Institute of Steel and Alloys [3, 4]. In addition, the reductants involve carbon, calcium carbide, aluminum–silicon alloy and silicon–iron alloy [5, 6], etc.

In general, Li_2O is obtained from the form of Li_2CO_3, which is subsequently decomposed during the melting process [7, 8]. The main metallothermic reduction processes according to the following reaction are:

$$Li_2CO_3(l) = Li_2O(s) + CO_2(g) \tag{1}$$

$$yLi_2O(s) + xM(s) = 2yLi(s) + M_xO_y(s) \tag{2}$$

The two processes are combined in one apparatus, namely the decomposition of lithium carbonate and the reduction of lithium oxide by reductant. Therefore, it is important to clarify the decomposition and melting mechanisms of Li_2CO_3. However, the behavior of lithium carbonate during heating is a subject of controversial reports such as crystallographic transition, melting point, and the decomposition temperature of Li_2CO_3. According to Pasierb's [9, 10] description, it was reported that the melting points of Li_2CO_3 were 998, 973, 985, 996 K [11, 12], and 1001 K. Bazhenov [3] had reported that Li_2CO_3 was melted at 1004 K. KIM [10] had reported that Li_2CO_3 was melted at 993 K. On the other hand, the decomposition temperature of Li_2CO_3 was reported as 1173 K [6], 1001 K [9], and 1043 K [13]. The diffusion of gas and its escape from the reacting mass are prevented by the fusion of lithium carbonate powder on reaching the melting temperature because the decomposition of lithium carbonate is accompanied by the escape of CO_2 from the reactants [14, 15]. In addition, the melting temperature of lithium oxide produced in the course of reaction is as high as about 1840 K, and Li_2O is a solid phase at the decomposition temperature of Li_2CO_3. This also impedes the escape of CO_2 [16].

The aim of the present work was to investigate dissociation of lithium carbonate in argon gas, carbon dioxide, and negative pressure. The behaviors of Li_2CO_3 at different temperatures were studied, respectively, by simultaneous thermal analysis.

Experimental

Analytical grade of Li_2CO_3 was used as the raw material in experiments. The simultaneous thermal analysis was NETZSCH STA449F3. The analyzer enabled simultaneous analysis of the TG and DSC measurements with a detection accuracy of 1 μg. Experiments on the decomposition of pure Li_2CO_3 were conducted in a different atmosphere, such as a flow of argon gas, a flow of CO_2 gas, or negative pressure atmosphere. The gaseous flow rate was maintained at 50 mL/min. The sample mass was in the range of 5–10 mg. The pure Li_2CO_3 was contained in a zirconia crucible. The samples were heated to the desired temperature at a heating rate of 2, 5, 10, and 20 K min^{-1}.

Results and Discussion

The Dissociation of Lithium Carbonate in Argon Gas

The results of the thermal decomposition of pure Li_2CO_3 in argon gas with different heating rates are shown in Figs. 1, 2, 3, 4, and 5. If Li_2CO_3 decomposes according to the reaction given in Eq. (1), the theoretical mass loss would be 59.5% (residual mass: 40.5%). It can be seen from figures that the decomposition of pure Li_2CO_3 can be divided into two steps. According to the TG/DTG curve, the two steps of decomposition were not continuous and the mass loss at each step is also different at different heating rates. The mass loss of first step was about 30%, and the mass

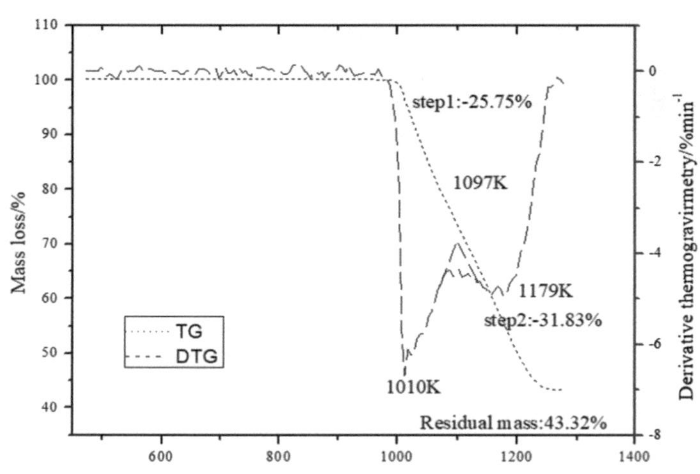

Fig. 1 TG results for lithium carbonate in argon gas/20 K min^{-1}

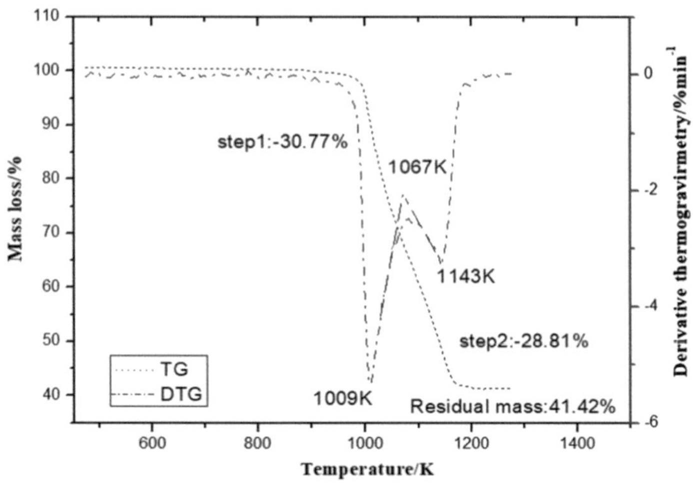

Fig. 2 TG results for lithium carbonate in argon gas/10 K min^{-1}

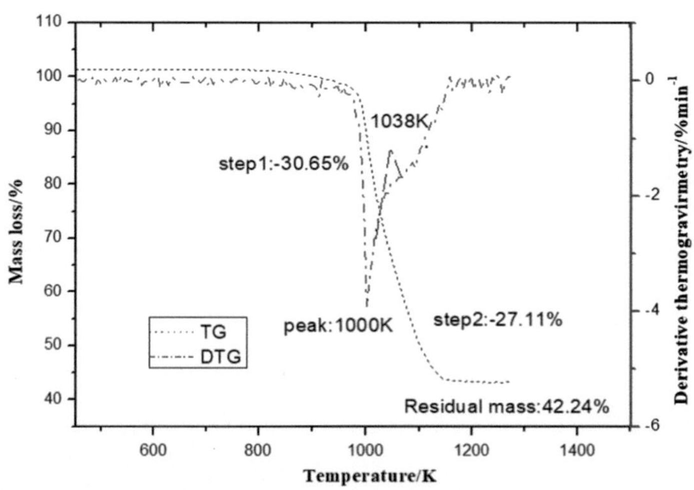

Fig. 3 TG results for lithium carbonate in argon gas/5 K min^{-1}

loss of second step increased with the increase in heating rate. It is also seen that the temperature at which lithium carbonate decomposes completely increased with the increase in heating rate, and the total mass loss approximated continuity 59.5%. Z Cancarevic and his peers have studied the relevant equilibria $M_2O + M_2CO_3 = M_4CO_4$ (M alkali metal) at various thermodynamic conditions using computational techniques [17, 18]. Recent studies showed that during exposure to air, molten Li_2CO_3 decomposes into Li_2O and CO_2 [19–21]. In our study, we consider that the decomposition of carbonate is divided into two responses as follows:

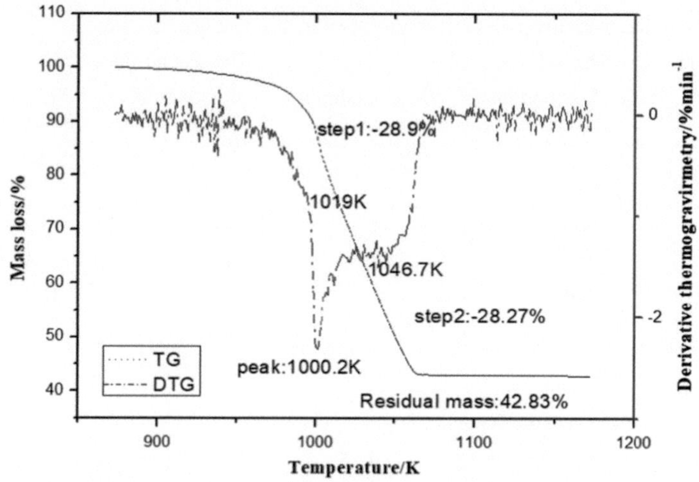

Fig. 4 TG results for lithium carbonate in argon gas/2 K min^{-1}

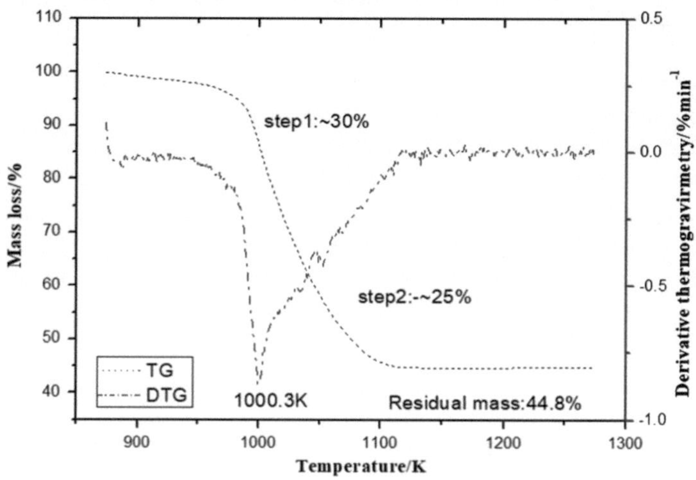

Fig. 5 TG results for lithium carbonate in argon gas/1 K min^{-1}

$$2Li_2CO_3(l) = Li_4CO_4(l) + CO_2(g) \tag{3}$$

$$Li_4CO_4(l) = 2Li_2O + CO_2(g) \tag{4}$$

The theoretical mass loss of two steps would be 30 pct. This value is in line with our assumption. Table 1 shows the maximum mass loss rates of decomposition at different heating rates. It can be seen from the table that the loss of first decomposition

Table 1 Maximum mass loss rate of decomposition in different heating rates

Heating rate ($K\ min^{-1}$)	The first stage		The second stage	
	Maximum mass loss temperature (K)	Maximum mass loss rate (% min^{-1})	Maximum mass loss temperature (K)	Maximum mass loss rate (% min^{-1})
20	1010	−6.92	1179	−4.99
10	1009	−5.88	1143	−3.28
5	1000	−3.93	1092	−1.43
2	1000.2	−2.44	1046.7	−1.34
1	1000.3	−0.88	1054.7	−0.39

Table 2 Onset and the peak of TG and DSC

Heating rate ($K\ min^{-1}$)	20	10	5	2	1
TG onset (K)	1004.1	997.6	991	992.5	989.3
DSC onset (K)	997.9	991.1	990	989.1	984.6
DSC peak (K)	1010.6	1002.2	997.6	98.4	997.2

can achieve the equilibrium state as the decrease of the heating rates. We consider that the first decomposition temperature of carbonate was 1000 K.

Table 2 shows the onset and the peak of TG and DSC. It can be seen that TG extrapolation starting point (the temperature of mass loss start) is smaller than DSC extrapolation starting point(the temperature of start endothermic). It illustrates that the melting temperature of lithium carbonate is inferior to the decomposition temperature of lithium carbonate. The reaction is shown in Eq. (1). The first step of decomposition is associated with the melting process. As the heating rate increases, the temperature of lithium carbonate melting is increased (984–997.9 K).

The Dissociation of Lithium Carbonate in Carbon Dioxide

It is known that the process of destruction of lithium carbonate was restricted by the saturation pressure of carbon dioxide. Figure 6 illustrates the thermogravimetric behavior of Li_2CO_3 under an atmosphere of CO_2. And the lithium carbonate decomposition has been diminished. The curve shows a first mass loss between 990 and 1041 K and a second mass loss between 1119 and 1265 K. The maximum temperature and mass loss of the first step are 1010.8 K and 4.03%. The maximum temperature and mass loss of the second step are 1187.5 K and 53.44%. The first step of mass loss decreased significantly, and the main dissociation occurred in the second step. There are obvious boundaries between two steps, decomposition from 1041 to 1119 K. And there is almost no mass loss in the middle (lithium carbonate without decomposition). Under experimental conditions of temperature, the quality of the residue

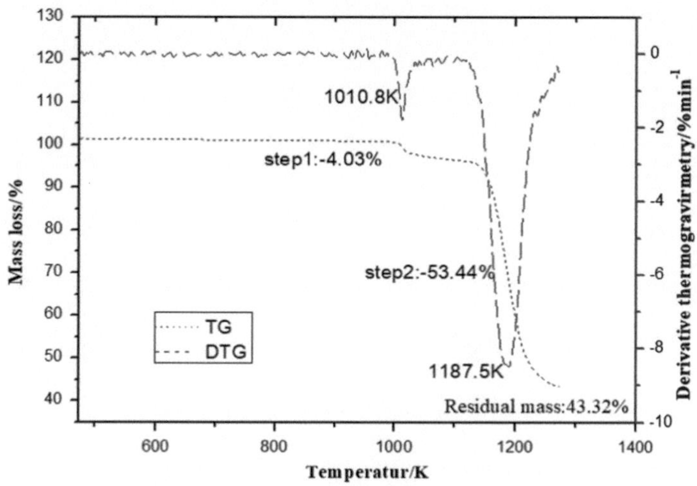

Fig. 6 TG results for lithium carbonate in carbon dioxide/10 K min^{-1}

is 43.32% (far greater than 40.5%), shows that lithium carbonate decomposition is not completely. The decomposition of lithium carbonate in a carbon dioxide atmosphere is affected by the pressure of carbon dioxide, which hinders lithium carbonate decomposition. While Li_2CO_3 is known to release CO_2 efficiently in the dissociation reaction, a question remains as Li_2O can absorb CO_2 easily in the dissociation reaction under moderate temperature and/or pressure conditions [22, 23]. At the beginning of the decomposition, lithium carbonate was becoming more stable in the presence of Li_2O and CO_2. When the temperature rises to 1117 K, lithium carbonate could be further decomposed. We can also observe the discontinuous phenomenon in "chemical sorption of carbon dioxide (CO_2) on lithium oxide (Li_2O)" [24]. This is why all the masses are different, although they are all related to the absorption of carbon dioxide.

The Dissociation of Lithium Carbonate in Negative Pressure

It is known that the diffusion limitation of the process may be eliminated under conditions of low pressure by permanently evacuating carbon dioxide being formed. The pressure was maintained at 20 kPa. Under negative pressure condition, the reaction temperature decreases obviously as it is shown in Fig. 7. The curve shows a first mass loss between 940 and 1039 K and a second mass loss between 1039 and 1140 K. The boundaries of the decomposition of two steps were more apparent in comparison with the argon gas and more continuous in comparison with the carbon dioxide gas. Then, it can be assumed that the outer layer of the particle will decompose in the first place [25]. Lithium oxide that forms here must be in the liquid, rather than solid,

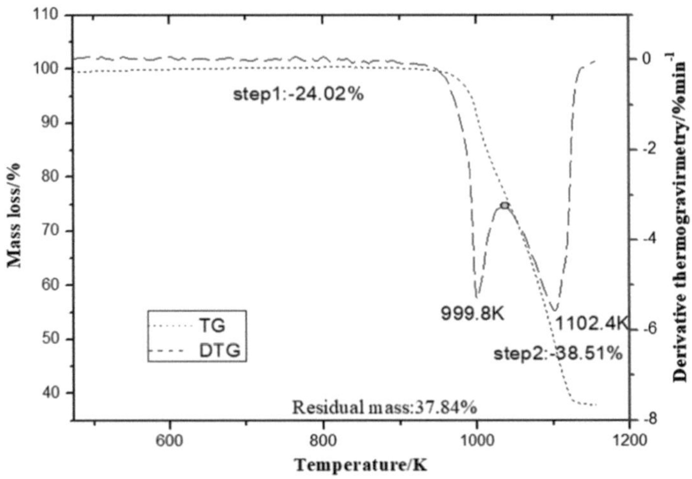

Fig. 7 TG results for lithium carbonate in negative pressure/10 K min^{-1}

which should facilitate the diffusion of carbon dioxide that forms during the reaction in negative pressure.

Table 3 shows the maximum mass loss rate of decomposition in a different atmosphere in 10 K min^{-1}. From the table, we can see that the first step of the decomposition is reduced and the second step of the decomposition is increased in carbon dioxide atmosphere or negative pressure, compared with the condition of the argon gas.

The temperature of lithium carbonate decomposition was reduced in negative pressure; on the contrary, the decomposition temperature of the lithium carbonate was increased in carbon dioxide atmosphere, even reached up to 1187.5 K.

Table 3 Maximum mass loss rate of decomposition in a different atmosphere/10 K min^{-1}

Atmosphere	The first stage			The second stage		
	Maximum mass loss temperature (K)	Maximum mass loss rate (% min^{-1})	Mass loss (%)	Maximum mass loss temperature (K)	Maximum mass loss rate (% min^{-1})	Mass loss (%)
Argon gas	1005.8	−5.88	30.77	1147.2	−3.28	28.81
Carbon dioxide	1010.8	−1.70	4.03	1187.5	−8.47	53.44
Negative pressure	999.8	−5.30	24.02	1102.4	−5.58	38.51

Conclusions

The decomposition of pure Li_2CO_3 can be divided into two steps, and the losing mass of each step is not the same in different atmosphere conditions. The first decomposition temperature of carbonate was 1000 K in argon gas. Lithium carbonate decomposition is a complex process, including the multiple reactions such as lithium carbonate melting, dissolution of Li_2O and CO_2 in Li_2CO, adsorption of CO_2 in Li_2O. The first part of the decomposition can be reduced, and the second part of the decomposition can be increased in carbon dioxide atmosphere or negative pressure, compared with the condition of the argon gas.

Acknowledgements The present project was financially supported by the National Natural Science Foundation of China Project (Grant No. 51604133) and the Academician Free Exploration Fund of Yunnan Province, China (Grant No. 2018HA006).

References

1. Yang B (1999) Study on extracting metallic lithium form Li_2CO_3 by vacuum metallurgy. Yunnan Science and Technology Press, Kunming
2. Fan FX (2012) Research of vacuum thermal reduction preparation of metallic lithium iron. Master Thesis of Kunming University of Science and Technology, Kunming
3. Bazhenov AA, Miklushevskii VV, Vatulin II, Kropacheva EN, Bidylo AP (2010) Study of the process of dissociation of lithium carbonate in the presence of aluminum powder. Russian J Non-Ferrous Metals 51(1):44–48
4. Kulifeev VK, Vatulin II, Tarasov VP, Miklushevskii VV (2004) Technology of producing lithium metal by aluminothermic reduction of lithium aluminates. Russian J Non-Ferrous Metals 45(11):6–14
5. Dai YN, Yang B (2008) Vacuum metallurgy of non-ferrous metal materials. Metallurgical Industry Press, Beijing
6. Di YZ, Feng NX, Dong WW, Peng JP, Wang YW (2009) Study on thermal decomposition of Li_2CO_3 in production of lithium by vacuum thermal reduction process. Nonferrous Metals (Extractive Metallurgy) 6
7. Olivares RI, Chen C, Wright S (2012) The thermal stability of molten lithium–sodium–potassium carbonate and the influence of additives on the melting point. J Sol Energy Eng 134(4):041002
8. El-Shobaky GA, Ibrahim AA (1987) Solid-solid interactions between ferric oxide and lithium carbonate and the thermal stability of the lithium ferrites produced. Thermochem Acta 118:151–158
9. Pasierb P, Gajerski R, Rokita M, Rekas M (2001) Studies on the binary system Li_2CO_3–$BaCO_3$. Physica B 304(1–4):463–476
10. Pasierb P, Gajerski R, Komornicki S, Rękas M (2001) Structural properties and thermal behavior of Li_2CO_3–$BaCO_3$ system by DTA, TG and XRD measurements. J Therm Anal Calorim 65(2):457–466
11. Ahamad L, Rakshit SK, Parida SC, Naik YP, Rao GR, Kulkarni SG, Gadkari SC (2013) Solid-state synthesis and heat capacity measurements of ceramic compounds $LiAlSiO_4$, $LiAlSi_2O_6$, $LiAlSi_3O_8$, and $LiAlSi_4O_{10}$. J Therm Anal Calorim 112(1):17–23
12. Licht S (2012) Stabilization of STEP electrolyses in lithium-free molten carbonates. arXiv preprint arXiv:1209.3512

13. Kim JW, Lee HG (2001) Thermal and carbothermic decomposition of Na_2CO_3 and Li_2CO_3. Metall Mater Trans B 32(1):17–24
14. Berbenni V, Milanese C, Bruni G, Girella A, Marini A (2013) Synthesis of Li_2SnO_3 by solid state reaction and characterization by TG/DSC, XRPD, and MTDSC. J Therm Anal Calorim 113(2):763–767
15. Surzhikov AP, Pritulov AM, Lysenko EN, Sokolovskii AN, Vlasov VA, Vasendina EA (2011) Dependence of lithium–zinc ferrospinel phase composition on the duration of synthesis in an accelerated electron beam. J Therm Anal Calorim 110(2):733–738
16. Timoshevskii AN, Ktalkherman MG, Emel'kin VA, Pozdnyakov BA, Zamyatin AP (2008) High-temperature decomposition of lithium carbonate at atmospheric pressure. High Temp 46(3):414–421
17. Čančarevič Ž, Schön JC, Jansen M (2006) Alkali metal carbonates at high pressure. Zeitschrift für anorganische und allgemeine Chemie 632(8–9):1437–1448
18. Cancarevic Z, Schön JC, Jansen M (2006) Existence of alkali-metal orthocarbonates at high pressure. Z Anorg Allg Chem 632(12–13):2084
19. Kaplan V, Wachtel E, Lubomirsky I (2011) Conditions of stability for ($Li_2CO_3 + Li_2O$) melts in air. J Chem Thermodyn 43(11):1623–1627
20. Kaplan V, Wachtel E, Lubomirsky I (2014) CO_2 to CO electrochemical conversion in molten Li_2CO_3 is stable with respect to sulfur contamination. J Electrochem Soc 161(1):F54–F57
21. Kaplan V, Wachtel E, Lubomirsky I (2012) Titanium carbide coating of titanium by cathodic deposition from a carbonate melt. J Electrochem Soc 159(11):E159–E161
22. Duan Y, Sorescu DC (2009) Density functional theory studies of the structural, electronic, and phonon properties of Li_2O and Li_2CO_3: application to CO_2 capture reaction. Phys Rev B 79(1):014301
23. Kaplan V, Wachtel E, Gartsman K, Feldman Y, Lubomirsky I (2010) Conversion of CO_2 to CO by electrolysis of molten lithium carbonate. J Electrochem Soc 157(4):B552–B556
24. Mosqueda HA, Vazquez C, Bosch P, Pfeiffer H (2006) Chemical sorption of carbon dioxide (CO_2) on lithium oxide (Li_2O). Chem Mater 18(9):2307–2310
25. Ktalkherman MG, Emelkin VA, Pozdnyakov BA (2009) Production of lithium oxide by decomposition lithium carbonate in the flow of a heat carrier. Theor Found Chem Eng 43(1):88–93

The Chemical Stability of MoS_2 in Chloride Eutectic Molten Salt

Cheng Lv, Jianxun Song, Yusi Che, Yongchun Shu and Jilin He

Abstract Molten salt electrolysis is used for the preparation of metal with high purity in a one-step procedure. When dealing with metal sulfides, a modified FFC (replacing oxides with sulfides) process shall prevent the emission of sulfur oxides if a protective atmosphere is provided. In this paper, the chemical stability of MoS_2 in chloride eutectic molten salt system (NaCl–KCl, $CaCl_2$–KCl, CsCl–KCl, and LiCl–KCl) was studied to explore the reaction mechanism of MoS_2 in molten salt. Moreover, the influence of adding K_2S was also investigated. MoS_2 tablets were immersed into the molten salt for 8 h and the SEM graph of the samples was observed focusing on the boundary between the tablet and solid salt. MoS_2 powder was also used to investigate existence of the intermediate product by XRD through the same process. The solubility of MoS_2 in each molten salt system was also determined.

Keywords Chemical corrosion · Molten salt · Molybdenum sulfide · Electrolysis

Introduction

Refractory metals are a group of metals that possess a high melting point, which accounts for their great performance under high-temperature conditions. Molybdenum is one of the metals that hold lots of potential. Considering its pyrometallurgy, a fact is Mo exists in the nature in form of molybdenite (MoS_2), which leads the traditional Mo metallurgy process containing both calcination and H_2 or carbothermic reduction [1]. Not only a long procedure is required, the emission of SO_2 is inevitable for the process, which accounts for severe pollution to the environment. Thus, a clean and short process for molybdenum metallurgy is needed. With the help of molten salt, we were able to get a wide electrochemical window, which allows the direct electrolysis of some rare metals (FFC process) [2]. According to the previous studies, Mo can be extracted directly from MoS_2 through electrolysis method in molten salt [3].

C. Lv · J. Song (✉) · Y. Che · Y. Shu · J. He
Henan Province Industrial Technology Research Institute of Resources and Materials, Zhengzhou University, Zhengzhou 450001, China
e-mail: jianxun.song@zzu.edu.cn

J. Lee et al. (eds.), *Materials Processing Fundamentals 2020*, The Minerals, Metals & Materials Series, https://doi.org/10.1007/978-3-030-36556-1_11

Among various molten salt systems available, chloride molten salts are always favorable, for the low cost in practical productivity. However, even with all the pros that molten salt system shall have, cons are that different molten salt systems hold a variety of ions, which may lead the electrochemical process to be variable [4]. Thus, it is necessary to investigate both chemical stability and physical stability of MoS_2 in various molten salts. One thing to be noticed is that the corrosion happens between molten salt and MoS_2 will change the reduction pathway, which directly results from the ions' interaction and the formation of complexes [5]. Different molten salt systems may lead to different reduction steps [6], where the polarization force of cations varies from salts to salts. There are two main impacts that a polarized system has on the electrolysis process: firstly, the interaction between ionized ions from the salt and ions produced during the electro-reduction process; secondly, the formation of complexes, which may stabilize the ions with intermediate valence. Another basis for selecting electrolyte is the diffusion rate and solubility, which allows the electrolysis process to be carried out efficiently with stable performance. Reports came that NaCl–KCl eutectic molten salt system is more suitable for sulfide electrolysis, for a high mass transportation coefficient of S^{2-} in such system, resulting in a high efficiency of current [7]. Moreover, the form of S^{2-} existing in the system may to some degree effect the pathway behind the electrolysis process of MoS_2, as the CaO had been proved to play an important role in the $CaTiO_3$-forming procedure [8]. Thus, K_2S was added to molten salts to investigate the influence of S^{2-}on the formation of the intermediate product.

With the purpose of exploring the influence of composition of the molten salt on the stability of the MoS_2, systems including NaCl–KCl, LiCl–KCl, $CaCl_2$–KCl, and CsCl–KCl were taken into consideration in this work. The corrosion without electrochemical effect was studied by observing the interaction surface directly, where a sintered pellet was directly immersed into the molten salt system at the working temperature, and was kept for long enough to give time for corrosion to reveal. Scanning electron microscope (SEM) was used to observe the interface between the tablet and molten salt, and the composition form was analyzed by X-ray diffraction (XRD).

Experimental

Preparation of MoS_2 Pellet

MoS_2 powder with the purity of 99.95% is used in this study. The powder was firstly die-pressed into tablets by using a mould with an inner diameter of 12.5 mm and then, in order to gain enough mechanism strength, was sintered at 930 °C with continuous Ar flow for 8 h. Pressure of tablet making was employed as a main factor, so tablets under 10 MPa, 20 MPa, 30 MPa, 40 MPa, and 50 MPa were made, respectively. Along with the MoS_2 corrosion experiment in various molten salts, tablets with different pressure were evaluated by checking the performance of the tablets made under different conditions.

Table 1 Salt composition, polarization powers, and working temperature

Salt	Mole ratio	Polarization power	Temperature (°C)
LiCl–KCl	59:41	1.24	400
$CaCl_2$–KCl	1:3	0.90	650
NaCl–KCl	1:1	0.75	750
CsCl–KCl	5:3	0.42	630

Molten Salt Preparation

In this study, NaCl–KCl, LiCl–KCl, $CaCl_2$–KCl, and CsCl–KCl were employed for investigating the composition influence of the salt, which were chosen for their different ion radius, resulting in different polarization power. The polarization power calculated for each system is listed in Table 1. It is known that the chemical reactions happen with the breakage of the old chemical bond, and the formation of new ones, and that is when polarization power takes charge. Depending on the different polarization powers in each system, different existential forms of elements are expected, resulting in various ways of electrochemical reactions.

When it came to eutectic salt preparation, steps were taken as follows. Firstly, various salts were weighed according to the above ratios. Then, experiments were carried out in a vertical electrothermal furnace, where Ar flow was used to provide a dehydrated atmosphere for a better preparation of eutectic salt with high purity. With aims of removing the moisture and obtaining homogeneous eutectic salt, molten salts were preprocessed under the temperature shown in Table 1 for 6 h in crucibles covered with lids.

Corrosion Experiment

There are mainly two parts in this work: (1) checking stability of MoS_2 in molten salt and (2) finding the final form of the MoS_2 to reveal if any chemical reaction happens throughout the molten salt corrosion. Thus, after the preparation of the binary eutectic salt, the corrosion testing was carried out. To be used as the electrode, MoS_2 tablets are required to possess enough mechanical strength to maintain a compact form throughout the electrolysis; otherwise, the electrochemical process will not be able to proceed continuously. However, it is suggested that the dissolution of reactant in the electrolyte shall assure the reaction to perform steadily. In order to figure out the appropriate condition, several aspects of stability were considered, including both physical and chemical. At the same time, the solubility of MoS_2 in several different eutectic molten salts was estimated.

To check the stability of MoS_2 tablets, a sintered MoS_2 tablet was placed above eutectic salt at room temperature. The eutectic salt was obtained through processes mentioned above. A pellet was placed above the cooled-to-solid salt and was expected to be immersed into the salt after the salt was melted. With the application of Ar, a protective atmosphere was employed, preventing MoS_2 from oxidation during the corrosion procedure. After cooling the molten salt together with the tablet, the sample was embedded in resin for further observation of the boundary between MoS_2 and salt. By polishing the sample, the interface of the supposed corrosion area was revealed and was observed by SEM (FEI QUANTA 200). Due to the strong tendency of water absorption, samples must be preserved carefully in vacuum desiccator, away from any form of moisture. In order to judge the physical stability of the MoS_2 tablet, the assistant of SEM is required. If no obvious corrosion was found, we may draw the conclusion that the effect of physical dissolution is little.

When it comes to chemical stability, MoS_2 powders were applied in each molten salt system. Theoretically, thinner particles may be able to provide more superficial area than a compact form, which leads to a higher reaction chance and a more notable result. In this study, MoS_2 powders were scattered over the solid salt, prepared by the pre-melt process, and it was expected to be mixed up when the salt reached the setting temperature. It was kept for 8 h and then cooled to room temperature. Samples were washed by deionized water, soluble salt fully dissolved into the water, leaving insoluble powders in the beaker. To separate them, a filter was used. Chloride salts were easy to dissolve in the water, leaving the product of MoS_2 powder alone undissolved. A filter was then used to separate the sample of the experiment, and after drying up in an oven, XRD (PANalytical Empyrean) tests were taken respectively.

Solubility Experiment

The solubility of MoS_2 in each eutectic salt system was measured in this work. A sintered MoS_2 tablet was immersed into the molten salt after the salt was melted, and was kept for a certain time picked from the experimental goal, which is 3 h, 6 h 9 h, 12 h, and 15 h, respectively. Through the whole procedure, tablet's mass loss was measured by comparing the weight before and after the immersing process. At the same time, considering the possibility of error coming from the volatilization of MoS_2 throughout the dissolution process, an undipped MoS_2 tablet, used as a reference, was also set into the oven. Due to the pellets' absorption of salt, the MoS_2 tablet shall gain weight right after the immersing process, and a washing process using deionized water together with ultrasonic washer was employed. The application of ultrasonic vibration was employed for 5 min each time to fully wash out the soluble impurities, mainly salts, from the tablet. After the salt is dissolved into the solution, deionized water was used to replace the solution of salt, leaving the insoluble MoS_2 to be dried and weighed. Through the procedure above, a dry MoS_2 tablet after dissolution in salt is obtained. Because of the application of ultrasonic washer, the tablet may suffer from powdering to some degree, but since the whole procedure was operated in a beaker, the powders together with the remaining tablet were all collected and weighed. However, the replacement of solution was not 100%, in afraid of losing MoS_2 powder for the powder is scattered when pouring away the solution of salt. In a way like this, after replacing deionized water and ultrasonic washing for three times, little salt remaining is still foreseeable.

Results and Discussion

Pellet Preparations

Once immersed in molten salt, the particles of MoS_2 powder, particularly a fine one, will experience two forces, buoyancy and interfacial tension, that tend to prevent the small particles from contacting each other and hinder the further reduction process. Used as the electrode, a dense phase is needed, so the first step is to press the MoS_2 powders into tablets and then obtain a relatively compact form after sintering. To study the impact of pressure on tablets making, MoS_2 tablets with different pressure were made to find a proper condition for the following experiments. MoS_2 tablets under 10 MPa, 20 MPa, 30 MPa, 40 MPa, and 50 MPa were made respectively as shown in Fig. 1. Because of the stickiness of MoS_2 powder, the presence of uneven surface is common in tablets made from MoS_2. Layered structure of MoS_2 may also account for the exfoliation. However, the sintering procedure played an important part in the pellet making, for the raw tablets turn into powders easily.

The result above shown in Fig. 1 reveals that: Firstly, under the condition of 10 MPa, the tablet broke into pieces after the sintering, which reveals the lack of physical strength. Meanwhile, with the pressure over 20 MPa, MoS_2 pellets shall survive the sintering process with a contact form. Secondly, the shrink of the MoS_2 tablets is not obvious, and samples made over 40 MPa possess a denser form and better mechanical properties. Finally, in order to attach to the electrode, a hole is needed to fix the tablet, so drilling tests were carried out on the pellets respectively, and the result shows that pressure over 40 MPa did not crack after drilling.

According to the results above, 40 MPa was chosen for MoS_2 tablets making in the following experiments.

Fig. 1 MoS_2 tablets made under, **a, f** 10 MPa, **b, g** 20 MPa, **c, h** 30 MPa, **d, i** 40 MPa, **e, j** 50 MPa before and after sintering

SEM Graph of MoS_2 Tablets and Solid Salt

The boundary between MoS_2 and eutectic salt was captured by SEM. Samples should be kept very carefully because of the salt's tendency of water absorption. Here, SEM graphs mainly focused on the edge of the MoS_2 tablets, which carry the information about the corrosion condition.

For the NaCl–KCl system, experiments both with and without K_2S (1.5 wt%) addition were carried out, and in order to get images with high quality and contrast, backscattered electron (BSE) was used, which is shown in Fig. 2. As shown in Fig. 2, the boundary between the tablet and solid eutectic salt is notable and no corrosion is observed. With the presence of K_2S, the edge is still clear, which means the influence of K_2S addition is not notable. The crack shown in the picture is mainly due to the water absorption of salt, and in Fig. 2a, an obvious absorption of salt is shown in MoS_2 tablet, leaving a long crack at the polished surface. Results shown in Fig. 2d–f are similar, which allows us to conclude that the addition of K_2S does not improve the corrosion process in NaCl–KCl eutectic salt.

Figure 3 shows the boundary between the MoS_2 and the CsCl–KCl eutectic salt. From the images in Fig. 3, it can be found that the edge of the tablet is clear under SEM, and the holes appear in the image result from the water absorption of solid salt after the sample was prepared. However, with the addition of K_2S, the edge became blurry; since this might result from the polishing process, EDS analysis was employed. By means of EDS, a distinguishable line was found and the boundary of MoS_2 tablet was revealed. The sample without the addition of K_2S was also analyzed

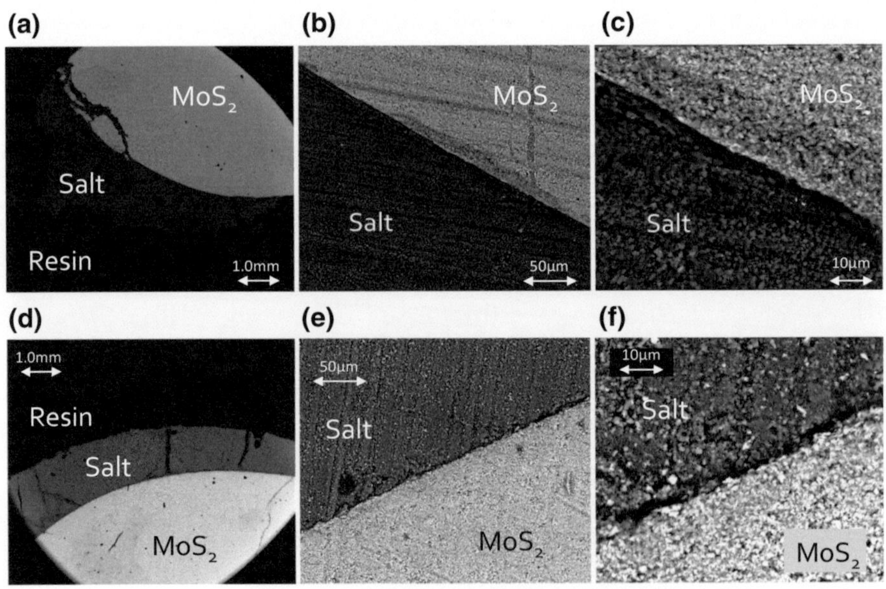

Fig. 2 SEM images of MoS_2 tablet, **a–c** in NaCl–KCl and **d–f** in NaCl–KCl–K_2S

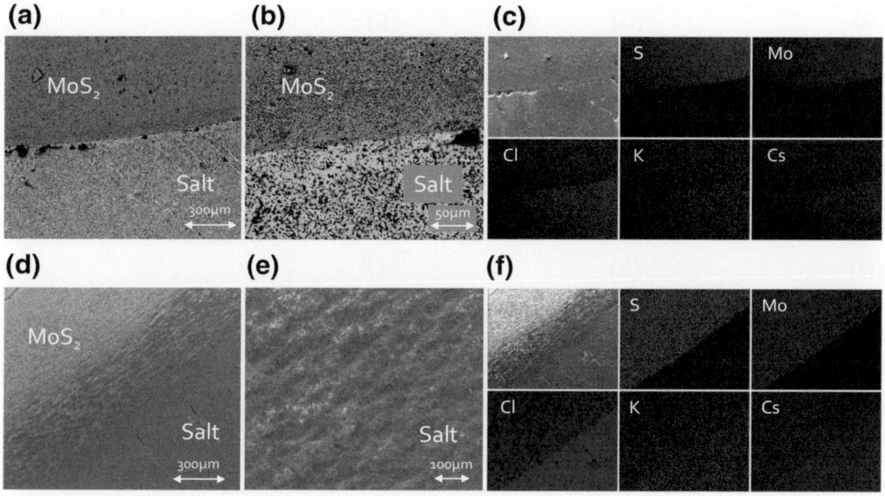

Fig. 3 SEM and EDS images of MoS_2 tablet in solid, **a–c** CsCl–KCl and **d–f** CsCl–KCl–K_2S

by SEM and EDS, showing a similar result of the former one, no intermediate product or transition area appeared. Here, we can draw the conclusion that the MoS_2 tablet is also physically stable in CsCl–KCl and CsCl–KCl–K_2S system, though there might be salt absorption, but the pill is intact after the experiment.

In Fig. 4, MoS_2 tablet in LiCl–KCl and LiCl–KCl–K_2S eutectic salt system also shows a distinguishable boundary, while some salt may be carried onto the surface of the tablet due to the polishing process or the salt absorption of the tablet. Though the

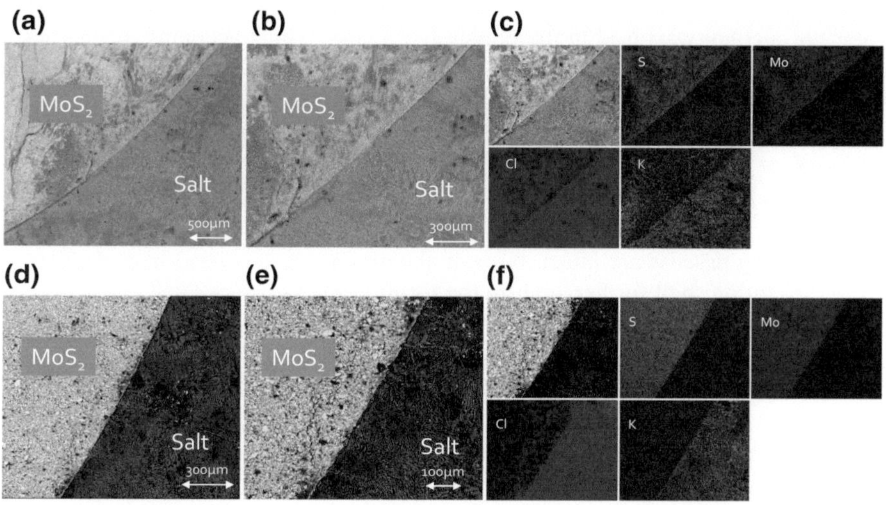

Fig. 4 SEM and EDS images of MoS_2 tablet in solid, **a–c** LiCl–KCl and **d–f** LiCl–KCl–K_2S

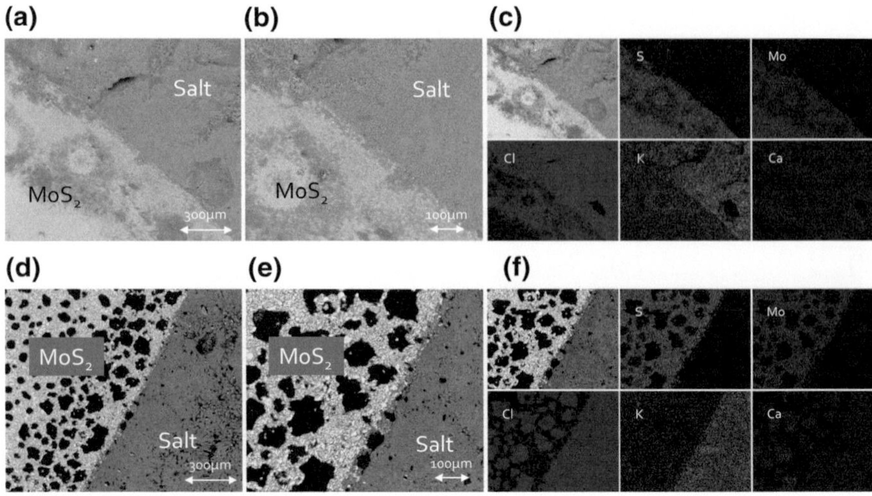

Fig. 5 SEM and EDS images of MoS_2 tablet in solid, **a–c** $CaCl_2$–KCl and **d–f** $CaCl_2$–KCl–K_2S

characterization of element Li is not available for EDS, due to its light atom mass, other elements shall help draw a line between MoS_2 and the salt. The result shows that LiCl–KCl eutectic salt has great tendency of water absorption, resulting in a height difference between MoS_2 tablet and solid salt, which were at the same altitude after polishing process. Since the tablet stayed intact after the immersing process and a distinguishable line was found, LiCl–KCl and LiCl–KCl–K_2S are both suitable for further electrolysis tests.

MoS_2 tablet was also studied in $CaCl_2$–KCl and $CaCl_2$–KCl–K_2S system as the result is shown in Fig. 5. Because of the strong power of water absorption, neither water nor alcohol was used in the polishing process, which accounts for the black spots in Fig. 5. From the result of EDS analysis, $CaCl_2$ was thought to be responsible for black dots. However, with the tendency of water absorption, the edge of the tablet in Fig. 5 is not very clear. Further XRD test was carried out to determine whether an intermediate component was formed or not.

Each batch of experiment contains both with and without K_2S addition. With the help of EDS, a clear boundary in each system is shown. All the tablets collected after the immersing process are intact, but salt absorption can be found in each system. X-ray diffraction shall further analyze the form of product.

XRD Pattern of the Product

In order to check the formation of samples after corrosion experiment, MoS_2 powders were used to examine its existing form in each molten salt. Here, in this section, the XRD patterns of MoS_2 product after mixing up and reacting with different eutectic molten salt systems were presented.

The experiment was firstly carried out in NaCl–KCl eutectic salt with different K_2S addition. XRD patterns were obtained as shown in Fig. 6. In spite of the concentration of K_2S, peaks of MoS_2 were found in every pattern, and any intermediate product was not detected. It means that MoS_2 is stable in molten NaCl–KCl. Also, the addition of K_2S did not have any effect on the dissolution of MoS_2 by forming a different product. Similar results were obtained in LiCl–KCl and $CaCl_2$–KCl molten salt systems, where molybdenite was considered the main form of product as shown in Fig. 7.

While dealing with CsCl–KCl system, a different phenomenon was found: The solution of the salt with the addition of K_2S is orange, while the solution of CsCl–KCl stayed clear. In order to check the component that accounts for the different color, XRD analysis was carried out. According to the XRD patterns shown in Fig. 8, the main form of the product is still MoS_2, which means that MoS_2 did not get involved with any chemical reaction under such condition. When attempting to identify the phenomenon of the color changes, rotatory evaporator is employed to prepare samples for XRD test. As shown in Fig. 8c, the main solutes of the orange-colored dissolution are still KCl and CsCl, which should not be responsible for the change of color, and further studies shall be carried out.

From the results obtained, no evidence for the existence of intermediate component was found in the salt systems chosen, which allows concluding that MoS_2 gets involved with no chemical reaction under our experimental conditions in every eutectic salt with or without the addition of K_2S, and no chemical reaction between MoS_2 and chloride salts happened.

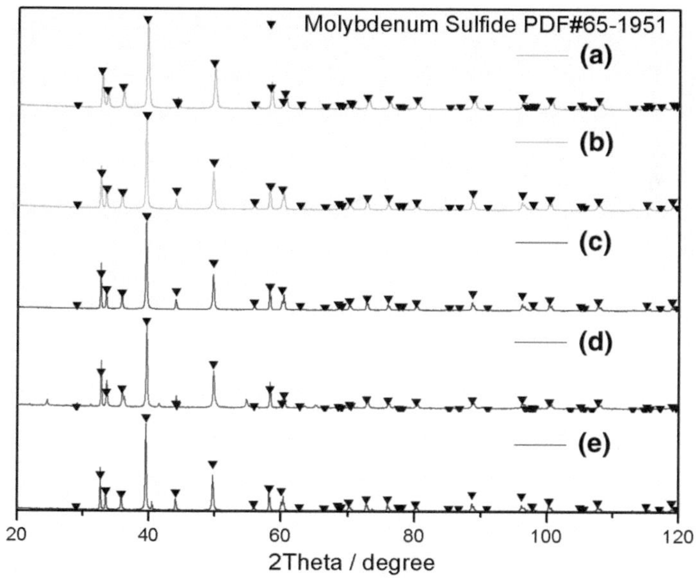

Fig. 6 XRD patterns of powders after immersing in NaCl–KCl molten salt with **a** 0 wt%, **b** 0.5 wt%, **c** 1.0 wt%, **d** 1.5 wt%, **e** 3.5 wt% K_2S addition

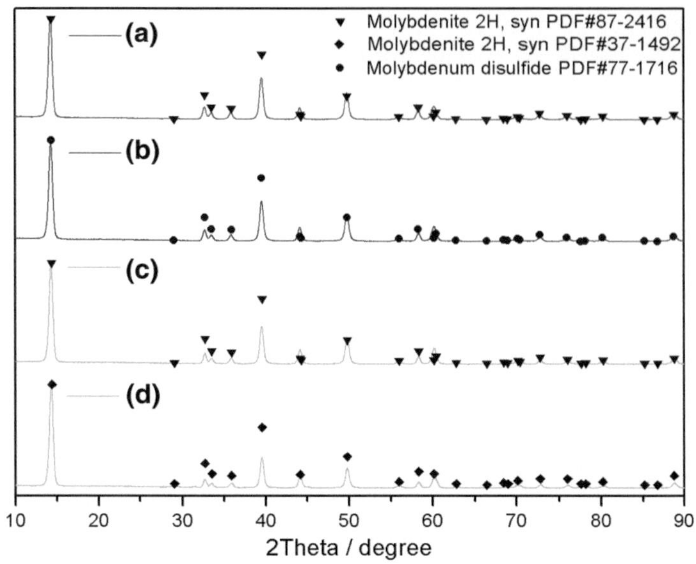

Fig. 7 XRD patterns of powders after immersing in **a** LiCl–KCl, **b** LiCl–KCl–K_2S, **c** $CaCl_2$–KCl, **d** $CaCl_2$–KCl–K_2S

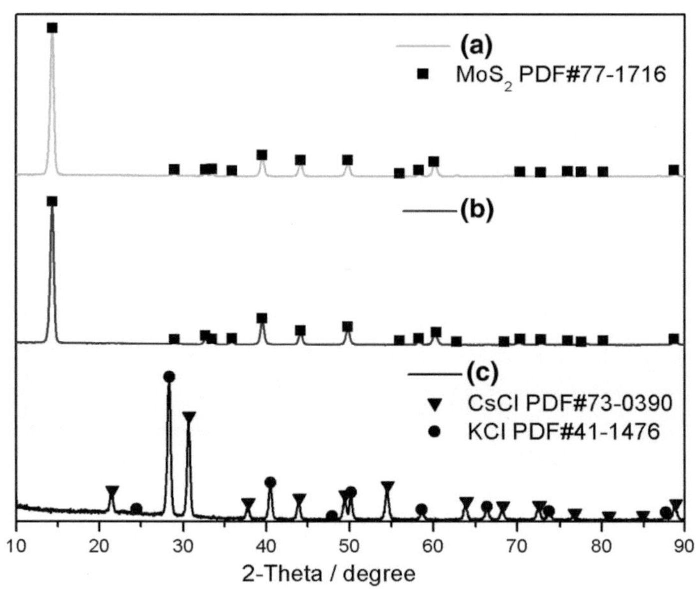

Fig. 8 Powders after immersing in **a** CsCl–KCl molten salt, **b** powders after immersing in CsCl–KCl–K_2S, **c** sample of the orange-colored solution obtained by rotatory evaporating

Solubility of MoS_2

In this section, the solubility of MoS_2 in each chosen eutectic salts was measured for the purpose of studying the status of MoS_2, as an electrode, under the operating temperature. The solubility of MoS_2 was measured by the mass loss while immersed in the salt. Throughout the experiment, MoS_2 tablets in different molten salts all presented a different degree of salt absorption, so a washing process was employed, preventing error. However, since the washing process was crude, salt elimination may not proceed perfectly. The solubility of MoS_2 in chosen molten salt systems is shown in Fig. 9.

Through the sintering process, the die-pressed MoS_2 tablets suffer from the sublimation, resulting in nearly 6% mass loss. However, in further corrosion experiments, no more mass loss was observed in the following experiments where a sintered pellet was introduced as a reference to check whether sublimation still had impact on the dissolving process. From Fig. 9, the solubility of MoS_2 in both LiCl–KCl and $CaCl_2$–KCl is little, due to the calculation method, and errors are introduced due to the remaining salt or the accuracy of the electro scale. It is revealed that MoS_2 is more likely to dissolve in CsCl–KCl eutectic salt, and with the dissolving time of 6 h, the solubility of MoS_2 in the CsCl–KCl eutectic salt shall come to a saturation at approximately 4.1 wt%. When it comes to NaCl–KCl, the saturation solubility is around 2.1 wt%, reached in time of 12 h.

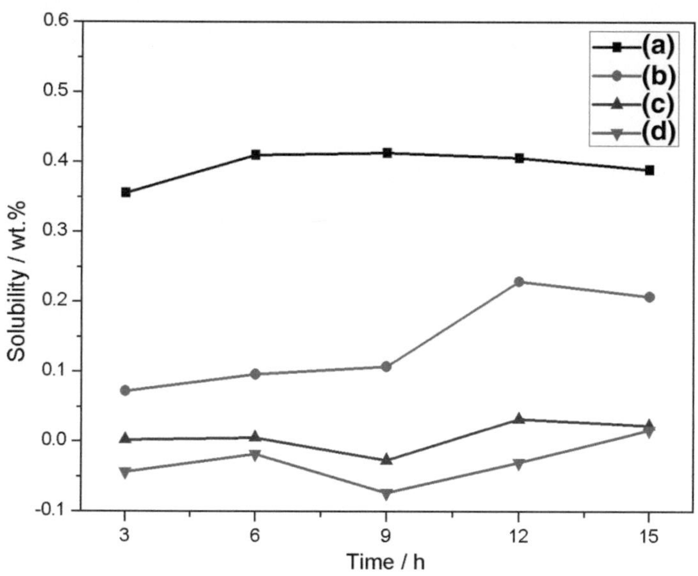

Fig. 9 Solubility of MoS_2 in **a** CsCl–KCl, **b** NaCl–KCl, **c** LiCl–KCl, **d** $CaCl_2$–KCl, measured by the mass loss

Among the chloride eutectic salt chosen in this study, CsCl–KCl possesses the highest solubility of MoS_2 at around 4.1 wt%, while MoS_2 in LiCl–KCl and $CaCl_2$–KCl does not dissolve much.

Conclusion

In this study, the stability of MoS_2 in several eutectic salts was investigated. Results show that MoS_2 has high stability in both physical and chemical processes in the chosen molten salts. Results from the solubility investigating reveal that the solubility of MoS_2 in LiCl–KCl and $CaCl_2$–KCl molten salts was little and may be considered close to zero. The solubility is 2.1 and 4.1 wt% in molten NaCl–KCl and CsCl–KCl, respectively. Though the absorption of salt is obvious, no notable corrosion occurs throughout the immersing process.

References

1. Zhang W, Liu Y (2013) The recent development of metallurgical technology for molybdenum. China Molybdenum Ind 37(3):1–5
2. Chen GZ, Fray DJ, Farthing TW (2000) Direct electrochemical reduction of titanium dioxide to titanium in molten calcium chloride. Nature 407(6802):361–364
3. Li G, Wang D, Jin X, Chen GZ (2007) Electrolysis of solid MoS_2 in molten $CaCl_2$ for Mo extraction without CO_2 emission. Electrochem Commun 9(8):1951–1957
4. Song J, Huang X, Wu J, Zhang X (2017) Electrochemical behaviors of Ti (III) in molten NaCl–KCl under various contents of fluoride. Electrochim Acta 256:252–258
5. Song J, Xiao J, Zhu H (2017) Electrochemical behavior of titanium ions in various molten alkali chlorides. J Electrochem Soc 164(12):E321–E325
6. Song J, Zhang X, Mukherjee A (2016) Electrochemical behaviors of Ce (III) in molten $AlCl_3$–NaCl under various contents of fluoride. J Electrochem Soc 163(14):D757–D763
7. Tan M, He R, Yuan Y, Wang Z, Jin X (2016) Electrochemical sulfur removal from chalcopyrite in molten NaCl–KCl. Electrochim Acta 213:148–154
8. Jiang K, Hu X, Ma M, Wang D, Qiu G, Jin X, Chen GZ (2006) Perovskitization-assisted electrochemical reduction of solid TiO_2 in molten $CaCl_2$. Angew Chem 45(3):428–432

Printed Circuit Board Leached Residue as a Substitute Reducing Agent in Pyrometallurgical Processes

Desmond Attah-Kyei, Guven Akdogan, Daniel K. Lindberg and Christie Dorfling

Abstract The proliferation of technology has resulted in the rise of electronic waste (e-waste) generated. The main focus of recycling e-waste has been to recover the metallic fractions from printed circuit boards (PCBs) due to the inherent high value of metals present such as gold. Hydrometallurgical route, often the most preferred option for recovering the metals, does not address the issue posed by the non-metallic part. In this study, the use of leach residue of PCB as reducing agent in hematite reduction was investigated. The analysis on the leached boards showed that PCB is highly amorphous and has carbon content of 30.10%, oxygen content of 20.1%, and ash and volatile matter of 40.1% and 44.8%, respectively. Thermodynamic modelling and laboratory-scale experiments that simulate solid-state reduction of hematite were performed using FactSage™ and single particle reactor. The results revealed that PCB can be used to partially replace conventional reducing agents.

Keywords Electronic waste · Printed circuit board · Hydrometallurgy · Reductant · Pyrometallurgy

Introduction

The proliferation of technology and remarkable market growth has led to a reduced lifespan of electrical and electronic products resulting in an increase in the electronic waste. It is reported that the world generates between 27.2 and 45.4 million tonnes of e-waste annually and this is expected to increase by about 3–5% every year [1, 2].

D. Attah-Kyei (✉) · G. Akdogan · C. Dorfling
Department of Process Engineering, Stellenbosch University, Banghoek Road, Stellenbosch 7599, South Africa
e-mail: desmond.attah-kyei@aalto.fi

D. Attah-Kyei · D. K. Lindberg
Department of Chemical and Metallurgical Engineering, Aalto University, Kemistintie 1, 02150 Espoo, Finland

D. K. Lindberg
Johan Gadolin Process Chemistry Centre, Åbo Akademi University, Biskopsgatan 8, 20500 Åbo, Finland

J. Lee et al. (eds.), *Materials Processing Fundamentals 2020*, The Minerals, Metals & Materials Series, https://doi.org/10.1007/978-3-030-36556-1_12

Due to the presence of high content of heavy metals and brominated flame retardants (BFR), disposal of this waste by landfilling is harmful to the environment. Printed circuit board (PCB) is the main focus of e-waste recycling because of the inherently high value of contained metals such as gold, silver, and copper. According to Cui and Anderson [3], PCBs contain more base and precious metals than their respective ore and that the gold content in PCBs is 35–50 times higher than gold ore.

The most preferred way of recovering the metals is through hydrometallurgical route through a series of leaching stages [3–5]. The shortcoming of this process, however, is that it does not take into account the non-metallic fractions which form about 60% of the PCB [6]. From an environmental management perspective, a zero-waste approach of recycling should be developed to gain value from and reduce the environmental impact of both the metallic and non-metallic fractions of the PCB waste [6, 7].

Several options for treatment of the non-metallic fraction including material recycling, where the residue may be used as inclusions in concrete or asphalt materials with minimal processing or chemical recycling, where chemicals and fuels are produced from the residue using techniques such as pyrolysis exist. Due to the complex composition of PCB leach residue, recovery by thermal treatment is likely to be the most feasible process route from technical and economical perspectives [8–10].

In this study, the utilization of the non-metallic PCB waste fraction as reductant in pyrometallurgical operations was investigated. Several authors have investigated the recycling of plastics as feedstock for reductive smelting operations [11–13]. One of the major applications in this field involves the use of polymer waste in blast furnaces for steelmaking, where plastics are used as replacement for coke, coal, or oil used for ore reduction and heating. NKK Keihnn Works in Japan first implemented this technology after it was developed by Bremen Steelworks in Germany [14]. The use of polymer waste as substitute for conventional reducing agents provides a number of advantages. The coal resources are conserved since there is a lower consumption of both coke and pulverized coal and there is a reduction in polymer waste being landfilled or incinerated. Moreover, energy resources are saved when plastics are used as reductants. This is because plastics have higher hydrogen-to-carbon ratio than coal [13].

Experiment

Waste printed circuit board obtained from computers was manually dismantled by removing components like batteries and heat sinks. The partially dismantled boards were de-soldered by submerging in 2 M nitric acid. They were then cut into pieces of sizes of about 2 by 2 cm and subsequently crushed using a cutting mill to size less than 2 mm. The crushed boards were leached using sulphuric acid and then with aqua regia. The leach residue was washed with water and air dried.

PCB leach residue was characterised using different techniques including proximate, ultimate, scanning electron microscope (SEM), and X-ray fluorescence (XRF) analyses. Elemental and proximate analyses were performed using Vario EL Cube Elemental analyser and LECO CS 230, respectively. The morphology of the residue was observed using Zeiss Merlin field emission scanning electron microscope, with energy dispersive X-ray spectrometer. Beam conditions were 11 nA current, 20 kV accelerating voltage, and a distance of 9.5 mm. PANanalytical Axios wavelength dispersive spectrometer was used for XRF analysis (Fig 1; Tables 1 and 2).

Hematite reduction tests were performed using PCB, pure carbon (graphite with purity >99%), and blends of PCB and graphite; 1 g of chemical grade hematite (>99%) was reduced with 0.532 g of reducing agent. Reduction tests were carried out in a single particle reactor (SPR) to study reduction of hematite occurring under isothermal conditions. It consists of a quartz glass reactor which is heated to the desired temperature, a nitrogen purge system, off-gas analyser, and video camera (Fig. 2). About 0.5 g of hematite-reduction sample was pelletized using a hydraulic press at 100 bar for 1 min. The pellet was then placed on a sample holder and hanged in the insertion tube. After the reactor has attained the desired temperature, the sample holder is pushed into the hot zone.

FactSage 7.3™ was used to perform thermodynamic simulation of hematite reduction. The properties of leached residue of PCB together with that of hematite and

Fig. 1 SEM image of PCB leached residue

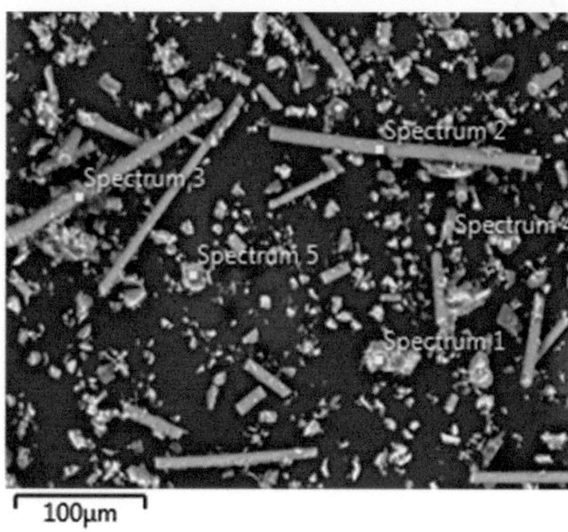

Table 1 Ultimate and proximate analysis of PCB leach residue (wt%)

Carbon	Hydrogen	Nitrogen	Sulphur	Oxygen
30.43	3.10	1.42	0.63	20.72
Moisture content	**Volatile matter**	**Fixed carbon**	**Ash content**	
3.60	44.80	11.50	40.10	

Table 2 XRF results of PCB leach residue (wt%)

Al_2O_3	CaO	SiO_2	Fe_2O_3	MgO	TiO_2	K_2O	P_2O_5	Na_2O
2.62	2.66	31.86	0.31	0.25	0.31	0.02	0.02	0.03

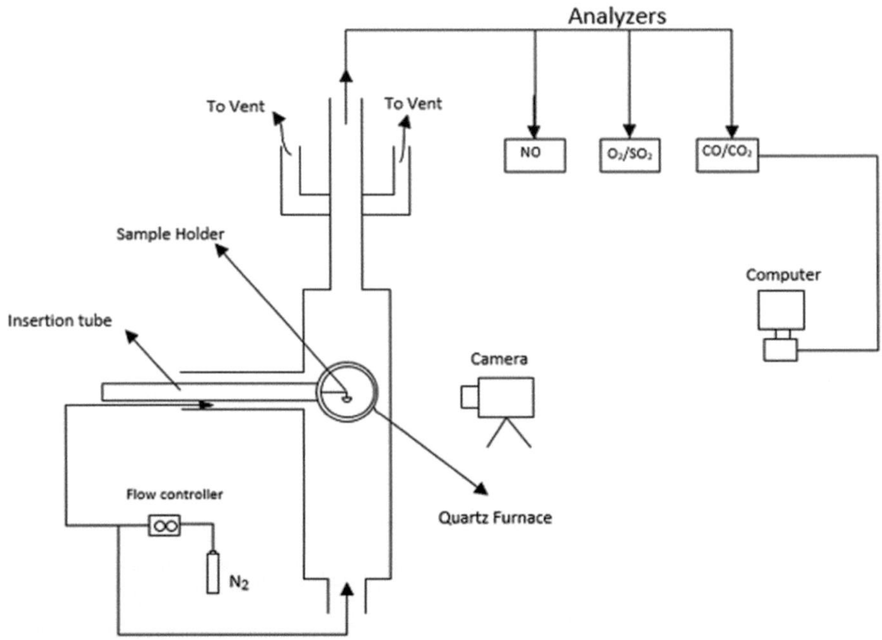

Fig. 2 Schematic diagram of SPR used for the reduction of hematite [15]

graphite were used as input in FactSage™. The FactPS, FToxid, FTmisc, and FSstel databases were selected for the calculation from 500 to 1600 °C using 20 °C intervals.

Results and Discussion

Reduction Tests in SPR

Isothermal reduction tests were carried out in SPR at 900 and 1000 °C under inert conditions using 1 l/min at 25 °C nitrogen. The volume percent of CO and CO_2 were measured and recorded.

Reduction in SPR at 900 °C

Figure 3 shows the volume percent of CO and CO_2 in the off-gas during the reduction tests at 900 °C. It can be observed in the hematite-graphite reduction that CO_2 was seen with no CO peaks. It is believed that hematite is converted to magnetite at 900 °C, due to the release of CO_2. Jung and Yi [16] attributed the initial formation of CO_2 as a combination of Eqs. 1 and 2. They stated that the gas-solid reaction (Eq. 2) is faster than the solid-solid reaction (Eq. 1). CO produced quickly reacts with hematite and releases CO_2. Thus, only CO_2 is observed by the gas analyser. Moreover, the XRD analysis on the hematite-graphite test showed the presence of wustite. This indicates that part of the magnetite reacts with CO and is converted to wustite as shown in Eq. 3.

$$3Fe_2O_{3(s)} + C(s) = 2Fe_3O_{4(s)} + CO_{(g)} \tag{1}$$

$$CO_{(g)} + 3Fe_2O_{3(s)} = 2Fe_3O_{4(s)} + CO_{2(g)} \tag{2}$$

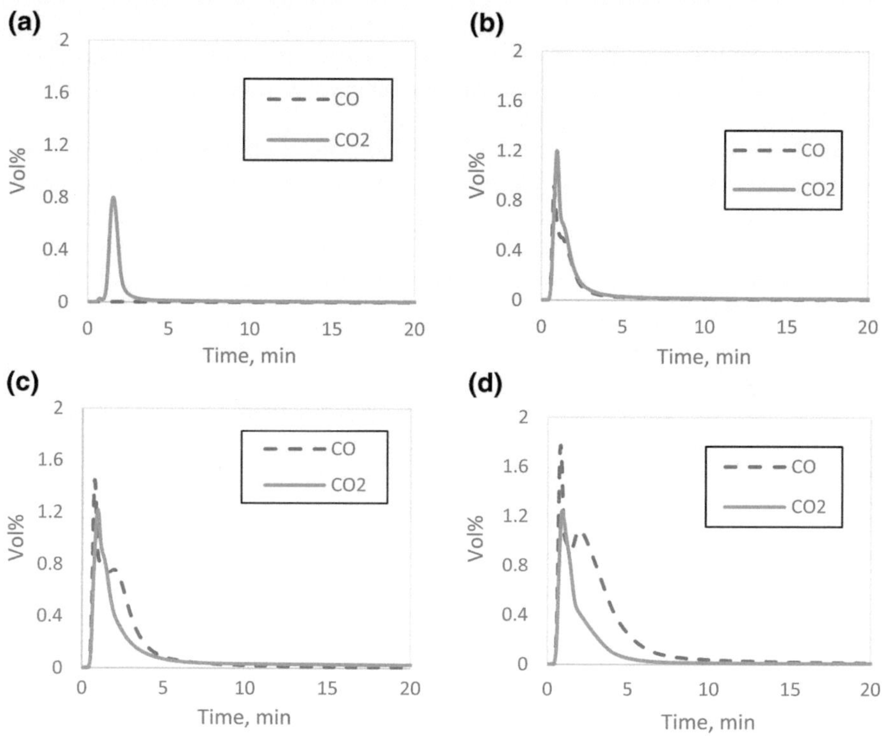

Fig. 3 Volume percent of CO and CO_2 in off-gas during reduction of hematite with **a** Graphite, **b** 40% PCB, **c** 80% PCB, **d** 100% PCB at 900 °C

$$Fe_3O_{4(s)} + CO_{(g)} = 3FeO_{(s)} + CO_{2(g)} \quad (3)$$

$$CO_{2(g)} + C_{(s)} = 2CO_{(g)} \quad (4)$$

$$4CH_{4(g)} + 27Fe_2O_{3(s)} = 18Fe_3O_{4(s)} + 2CO_{2(g)} + 2CO_{(g)} + 3H_2O_{(g)} + 5H_{2(g)} \quad (5)$$

$$3CH_{4(g)} + 8Fe_3O_{4(s)} = 24FeO_{(s)} + 2CO_{2(g)} + CO_{(g)} + 3H_2O_{(g)} + 3H_{2(g)} \quad (6)$$

$$H_2O_{(g)} + CO_{(g)} \rightleftharpoons CO_{2(g)} + H_{2(g)} \quad (7)$$

$$C_nH_m + n/2\,O_2 = nCO + m/2\,H_2 \quad (8)$$

It is observed that when PCB or blends of PCB-graphite were used as reducing agent at 900 °C, both CO and CO_2 are produced in the evolved gas. This may be ascribed to the presence of hydrocarbons and oxygen in PCB. In their study of the kinetics of the reduction of hematite with CH_4 in a thermogravimetric analyser (TGA), Monazam et al. [17] summarized the reduction of hematite to magnetite as Eq. 5 and subsequent formation of wustite from magnetite as Eq. 6. Both equations reveal the presence of CO and CO_2 in the product gas. CO produced is likely to react with H_2O according to Eq. 7 (Table 3).

Rao [18] reported that the CO available governs the hematite reduction process. The presence of CO peak during the reduction of hematite with PCB and blends of PCB at 900 °C shows that PCB acts as a better reductant at lower temperatures compared to graphite. Additionally, Carpenter [13] stated that H_2 and H_2O have higher ability to diffuse into and out of individual pellet and sinter is significantly higher than CO and CO_2. Higher diffusivity promotes faster reduction rates, particularly at lower temperatures. Equations 4 and 5 show that H_2 gas is released during the reduction of hematite using PCB or PCB-graphite blends. The gas analysers, however, are limited to detect only CO and CO_2 but not H_2 which is believed to take part in the reduction process. XRD results show the formation of iron and lower forms of iron oxide when PCB and blends of PCB were used to reduce hematite at 900 °C. This

Table 3 XRD results of products from reduction of hematite in SPR at 900 °C

Phase	Reduction of hematite with			
	Graphite	40% PCB	80% PCB	100% PCB
Fe_3O_4	70%	–	–	–
FeO	30%	100%	35%	–
Fe	–	–	–	8%
Fe_2SiO_4	–	–	65%	36%
SiO_2	–	–	–	55%

confirms that the hydrocarbons in the PCB take part in the reduction of hematite resulting in PCB being a better reductant at lower temperatures.

Reduction in SPR at 1000 °C

Reduction of hematite tests were carried out at 1000 °C. Figure 4 shows the volume percent of CO and CO_2 in the evolved gas during the reduction. Two peaks of CO_2 were observed when pure graphite was used as a reducing agent. The first CO_2 peak observed about 1 min from the start of the experiment may be attributed to the conversion of hematite to magnetite as discussed previously. The second peak may be due to the conversion of magnetite to wustite and subsequently to iron. The formation of metallic iron is accompanied by the CO peak around 8 min. Jung and Yi [16] performed isothermal reduction of hematite using graphite at 1000 °C and observed a similar pattern. They reported that when pure carbon is used in reducing hematite, magnetite to wustite takes relatively longer than that of wustite to metallic iron. Moreover, Rao [18] found that formation of iron phase from the reduction of wustite acts as a catalyst for carbon gasification. Additionally, the CO peak may be ascribed to a dominant Boudouard reaction (Eq. 4).

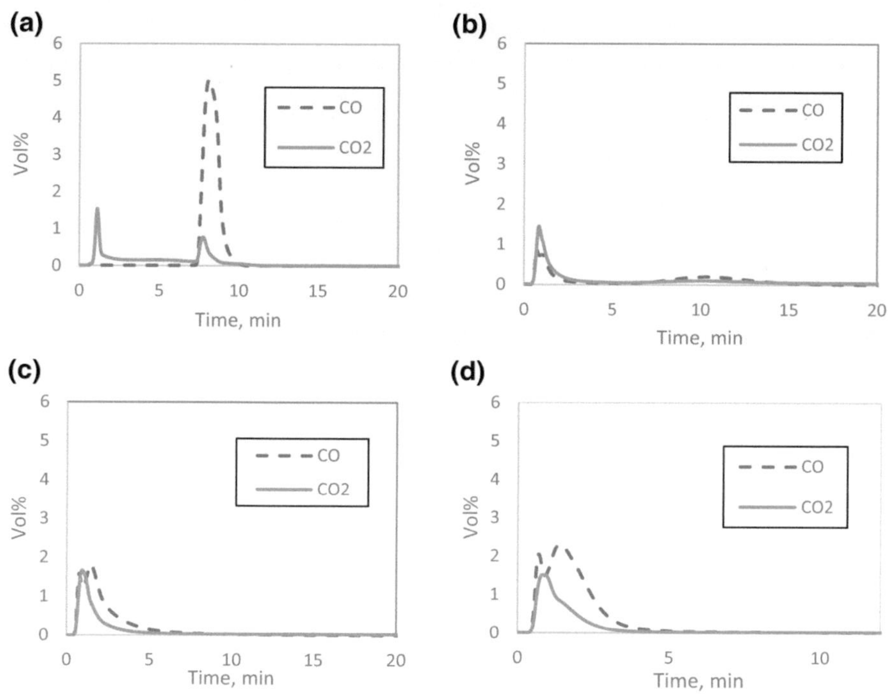

Fig. 4 Volume percent of CO and CO_2 in off-gas during reduction of hematite with **a** Graphite, **b** 40% PCB, **c** 80% PCB, **d** 100% PCB at 1000 °C

Table 4 XRD results of products from reduction of hematite in SPR at 1000 °C

	Reduction of hematite with			
	Graphite	40% PCB	80% PCB	100% PCB
Fe_3O_4	–	–	–	–
FeO	–	44%	–	–
Fe	100%	56%	30%	25%
Fe_2SiO_4	–	–	70%	75%
SiO_2	–	–	–	–

In the reduction of hematite using 40%PCB, two distinct CO peaks were observed at 1 and 10 min, respectively, (Fig. 4b). The first peak that occurred in less than a minute is due to the reduction of hematite by PCB while the second peak is as a result of the reduction by graphite. As discussed earlier when graphite is used during the reduction of hematite, the conversion of magnetite to wustite takes relatively longer than that of wustite to metallic iron. The length of time between the two CO peaks suggests that during hematite-40% PCB reduction test, PCB reduces hematite to magnetite, while wustite and metallic iron is formed due to the reduction of magnetite by graphite.

During the reduction of hematite with 80% PCB and 100% PCB, two CO peaks were observed in less than 5 min of the reduction test. The two CO peaks appear to overlap. The first CO peak may be ascribed to Eq. 8; the second peak may be an indication of the reduction of hematite. XRD results show the presence of metallic iron for all the reduction tests. Metallic iron present in the product of hematite-80% PCB and hematite-100% PCB shows that H_2 is released from PCB and takes part in the reduction of hematite to iron.

XRD also shows that fayalite is produced when PCB is used as a reducing agent which confirms results obtained from FactSage™ (Table 4).

FactSage™ Simulations of Hematite Reduction

Reduction of hematite (Fe_2O_3) was simulated using FactSage 7.3™. This was done to predict the products expected during the reduction tests. Figures 5 and 6 show the mass fraction of the products that were predicted by FactSage 7.3™.

It can be seen that the formation of solid iron starts around 700 °C for the reduction tests. Liquid iron forms at temperatures above 1150 °C for hematite-graphite reduction. The liquid iron phase predicted by FactSage in hematite-graphite reduction consists of iron and carbon whiles that of hematite-PCB is C, Fe, S, and trace amount of N and O. The presence of H_2O in the products of PCB containing reductants indicate that hydrogen present takes part in the reduction of hematite.

It can be seen that the recovery of iron decreases as PCB in the blend increases. This may be ascribed to the lower carbon content in PCB relative to graphite. The CO_2 content in the products is observed to increase as PCB in the blend increase. Dankwah et al. [11] investigated the use of mixed plastics as a reductant in iron

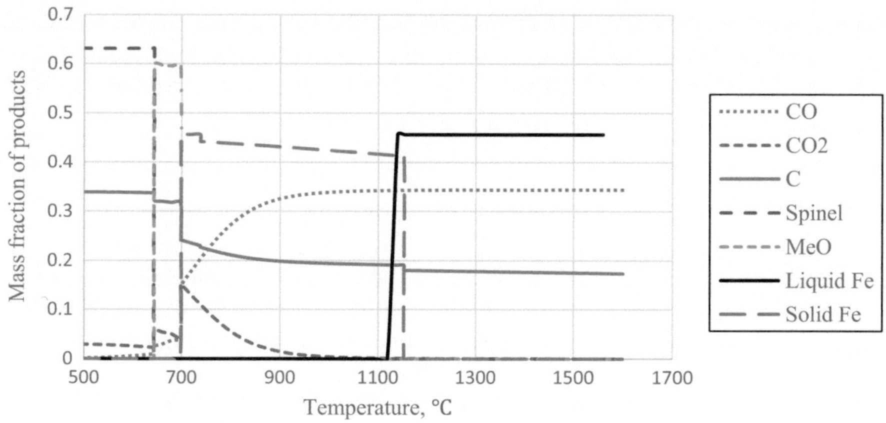

Fig. 5 Mass fraction of products obtained from FactSage™ calculation during the reduction of hematite with graphite

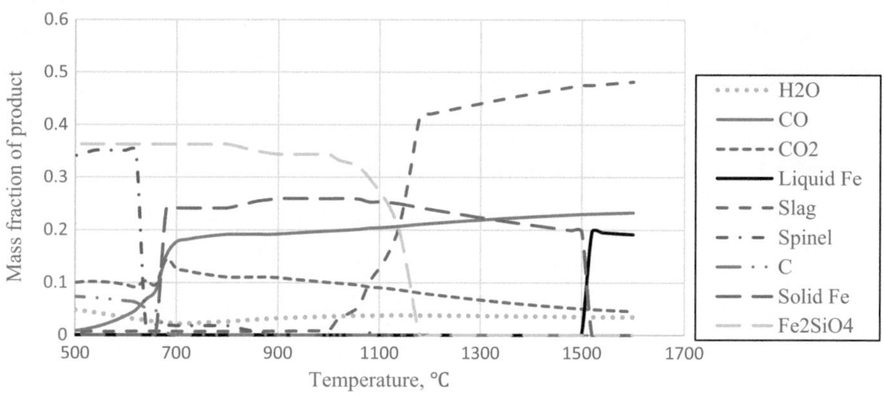

Fig. 6 Mass fraction of products obtained from FactSage™ calculation during the reduction of hematite with 100% PCB

making. They found a high CO_2 emission when polyethylene terephthalate (PET) was used as a reductant and attributed it to high oxygen content. It can be inferred from their observation that the high amount of oxygen present in PCB is responsible for the higher amount of CO_2 during reduction.

When PCB is used as reductant, FactSage™ predicts the formation slag and fayalite (Fe_2SiO_4). This may be attributed to the presence of SiO_2 in the boards. Moreover, it is reported that blend of waste plastics and coke can be used to efficiently increase slag foaming in electric arc furnace steelmaking [12]. The slag predicted by FactSage™ consists mainly of SiO_2, CaO, and Al_2O_3.

At a temperature close to 700 °C, there is a sharp increase in the mass of CO and a decrease in the CO_2 produced. This indicates the gasification of carbon. At temperatures less than 650 °C, FactSage predicts the formation of spinel for all the reduction tests. The spinel consists mainly of Fe_3O_4, with very small quantities of $FeAl_2O_4$ also present when PCB and blends of PCB were used as reducing agents.

Conclusions

The feasibility of using printed circuit board residue after hydrometallurgical treatment as reducing agent was studied. Reduction of hematite tests was performed in single particle reactor, and thermodynamic simulations were carried out with FactSage 7.3. The investigation showed that PCB or blends of PCB-carbon may be used to partially replace conventional reductants. The high amount of SiO_2 in PCB acts as flux which results in increase the slag formation. Hydrogen present in PCB also takes part in reduction and is believed to be a better reductant than carbon at temperatures lower than 1000 °C.

References

1. Tesfaye F, Lindberg D, Hamuyuni J, Taskinen P, Hupa L (2017) Improving urban mining practices for optimal recovery of resources from e-waste. Miner Eng 111:209–221. https://doi.org/10.1016/j.mineng.2017.06.018
2. Cucchiella F, D'Adamo I, Lenny Koh S, Rosa P (2015) Recycling of WEEEs: an economic assessment of present and future e-waste streams. Renew Sustain Energy Rev 51:263–272. https://doi.org/10.1016/j.rser.2015.06.010
3. Cui H, Anderson CG (2016) Literature review of hydrometallurgical recycling of printed circuit boards (PCBs). J Adv Chem Eng 6:1–11. https://doi.org/10.4172/2090-4568.1000142
4. Diaz LA, Lister TE, Parkman JA, Clark GG (2016) Comprehensive process for the recovery of value and critical materials from electronic waste. J Clean Prod 125:236–244. https://doi.org/10.1016/j.jclepro.2016.03.061
5. Sohaili J, Muniyandi SK, Mohamad SS (2012) A review on printed circuit board recycling technology. J Emerg Trends Eng Appl Sci 3:12–18
6. Ogunniyi IO, Vermaak MKG, Groot DR (2009) Chemical composition and liberation characterization of printed circuit board comminution fines for beneficiation investigations. Waste Manag 29:2140–2146. https://doi.org/10.1016/j.wasman.2009.03.004
7. Shuey SA, Taylor P (2005) Review of pyrometallurgical treatment of electronic scrap. In: SME Annual Meeting, pp 1–4. https://doi.org/10.1007/s11837-004-0237-9
8. Bazargan A, Bwegendaho D, Barford J, McKay G (2014) Printed circuit board waste as a source for high purity porous silica. Sep Purif Technol 136:88–93. https://doi.org/10.1016/j.seppur.2014.08.02
9. Fink JK (1999) Pyrolysis and combustion of polymer wastes in combination with metallurgical processes and the cement industry. J Anal Appl Pyrolysis 51:239–252. https://doi.org/10.1016/S0165-2370(99)00019-4
10. Fisher MM, Mark FE, Kingsbury T, Vehlow J, Yamawaki T (2005) Energy recovery in the sustainable recycling of plastics from end-of-life electrical and electronic products. Electron Environ Proc 2005 IEEE Int Symp 2005:83–92. https://doi.org/10.1109/ISEE.2005.1436999
11. Dankwah JR, Amoah T, Dankwah J, Fosu A (2016) Recycling mixed plastics waste as reductant in ironmaking. Ghana Min J 15:73–80

12. Sahajwalla V, Zaharia M, Kongkarat S, Khanna R, Rahman M, Saha-Chaudhury N, O'Kane P, Dicker J, Skidmore C, Knights D (2012) Recycling end-of-life polymers in an electric arc furnace steelmaking process: fundamentals of polymer reactions with slag and metal. Energy Fuels 26:58–66. https://doi.org/10.1021/ef201175t
13. Carpenter AM (2010) Injection of coal and waste plastics in blast furnaces. Report CCC/166
14. Zie A, Stanek W (2001) Forecasting of the energy effects of injecting plastic wastes into the blast furnace in comparison with other auxiliary fuels. Energy 26:1159–1173
15. Karlström O, Hupa L (2019) Energy conversion of biomass char: oxidation rates in mixtures of $O_2/CO_2/H_2O$. Energy 181:615–624. https://doi.org/10.1016/j.energy.2019.05.192
16. Jung SM, Yi SH (2013) A kinetic study on carbothermic reduction of hematite with graphite employing thermogravimetry and quadruple mass spectrometry. Steel Res Int 84:908–916. https://doi.org/10.1002/srin.201200031
17. Monazam ER, Breault RW, Siriwardane R, Richards G, Virginia W, Virginia W (2013) Kinetics of the reduction of hematite (Fe_2O_3) by methane (CH_4) during chemical looping combustion: a global mechanism. Chem Eng J 2013:1–28
18. Rao YK (1971) The kinetics of reduction of hematite by carbon. Metall Trans 2:1439–1447. https://doi.org/10.1007/BF02913373

Part IV
Steelmaking Process Modeling and Composites

Numerical Simulation of Heat Transfer Between Roller and Slab During Medium Thickness Slab Continuous Casting

Shuang Liu, Mujun Long, Pei Xu, Pingmei Tang, Dengfu Chen and Huamei Duan

Abstract In the medium thickness slab continuous casting, the roller will bear higher temperature, which poses a severe challenge to the roller service life. It has significance to study the influence factors on the roller temperature distribution. A three-dimensional model for transient heat transfer between the rotating roller and the moving medium thickness slab at the straightening zone was established. The model considers the coupling process of several physical fields, such as conduction heat transfer, radiation heat transfer, turbulent flow of the cooling water inside the roller, convective heat transfer, and roller rotation. Temperature variation in the roller at the casting speed of 0.9, 1.0, and 1.1 m/min was simulated with the slab surface temperature of 1273, 1323, and 1373 K. Results showed that the maximum temperature on the circumferential roller surface increased with the increase of the slab temperature, but decreased with the increase of casting speed while the slab surface temperature remaining unchanged.

Keywords Heat transfer · Roller cooling · Numerical simulation

Introduction

During medium thickness slab continuous casting, rollers play an indispensable role in supporting, transferring, and straightening the slab. When the roll contacts with the slab, heat transfers from the slab surface to roll surface and results that the roll surface temperature increases. After that, the heat transfers from the roll surface to the cooling water inside the roll, and then the heat is carried out from the roll by cooling water. Unreasonable if the design of the cooling water channel inside the

S. Liu · M. Long (✉) · P. Xu · P. Tang · D. Chen · H. Duan
College of Materials Science and Engineering, Chongqing University, Chongqing 400044, China
e-mail: longmujun@cqu.edu.cn

S. Liu
e-mail: liushuang666@cqu.edu.cn

Chongqing Key Laboratory of Vanadium-Titanium Metallurgy and New Materials, Chongqing University, Chongqing 400044, China

J. Lee et al. (eds.), *Materials Processing Fundamentals 2020*, The Minerals, Metals & Materials Series, https://doi.org/10.1007/978-3-030-36556-1_13

roll, too much heat will accumulate in the roll, and the roll will be in unreasonable high temperature and reduces the roller service life. In the medium thickness slab continuous casting, the conditions of heat transfer near the straightening area between the roller and the slab are more complex than other areas, so it is necessary to study the heat transfer between the roller and the slab and the influence of water channel structure on the roller temperature and yield strength distribution.

At present, some scholars have studied the heat transfer between the roller and the slab. However, there is no in-depth study of the influence factors such as casting speed, slab temperature, and cooling water channel diameter, especially on the heat transfer between the roller and the slab and its internal water channel structure. Pennerstorfer et al. [1] studied the heat transfer between the roller and the slab under dry cooling condition and found that there was a certain relationship between the roller surface temperature and slab surface quality. If the roller surface temperature was too low, the slab surface quality would deteriorate. Javurek et al. [2] compared and analyzed the heat transfer models with multiple dimensions between the roller and the slab and summarized the calculation effects of several calculation models.

Before the straightening zone, the slab temperature is relatively high, which has a great influence on the roller temperature and yield strength distribution. Therefore, the heat transfer between the roller and the slab in the area before straightening is simulated in this paper. Considering the conduction and radiation heat transfer between the roller and the slab, the transient flow of cooling water inside the roller, and the convective heat transfer of air, the temperature distribution of the roller at the casting speed 0.9, 1.0, and 1.1 m/min was compared and analyzed. Then the temperature and yield strength distribution with different diameters of cooling water channels are compared and analyzed, which provides a basis for effectively controlling the temperature distribution and extending the roller service life.

Model Establishment

Mathematic Equations

The heat transfer between the roller and the slab mainly involves conduction heat transfer, radiation heat transfer, and convection heat transfer of cooling water inside the roller. In order to control the slab surface temperature in a reasonable range, the slab after the secondary cooling zone still needs to be sprayed with cooling water. Since there is less cooling water than the secondary cooling zone and it is mainly sprayed on the slab surface, the spray cooling water has less influence on the heat transfer between the roller and the slab after the secondary cooling. Therefore, the influence of spray water on the heat transfer between the roller and the slab is ignored in this model. The cooling water inside the roller can be regarded as a three-dimensional steady-state incompressible fluid, and the momentum equation is solved by the standard equation k-ε equation based on Reynolds average N–S equations [3, 4]. The heat conduction can be described by the three-dimensional unsteady heat conduction differential equation [5, 6]. The equation of energy propagation can be expressed as:

$$\frac{dI(\vec{r},\vec{s})}{ds} + (a+\sigma_s)I(\vec{r},\vec{s}) = an^2\frac{\sigma T^4}{\pi} + \frac{\sigma_s}{4\pi}\int_0^{4\pi} I(\vec{r},\vec{s}')\phi(\vec{s},\vec{s}')d\Omega' \tag{1}$$

where $\vec{r}$, $\vec{s}$, $\vec{s}'$, and s are position vectors, direction vector, scattering direction, and stroke length, respectively. a, n, σ_s, and σ are absorption coefficient, refractive direction, scattering coefficient, and Steven–Boltzmann constant (5.672 $\times$ $10^{-8} w\ m^{-2}\ k^{-4}$), respectively. I, T, φ, and Ω' are radiation intensity, environment temperature, phase function, and solid angle for the space, respectively.

Model Assumption and Material Parameters

To simplify the model and improve the calculation speed, the bearing connections between segmented rollers are ignored, and the number of nodes and units after partition is 304,128 and 1543,373, respectively. The real roller structure and grid division in the model are shown in Fig. 1. The steel grade of the roller is 15CrMo, and Table 1 is the chemical composition of roller steel. The roller is 250 mm in diameter and 1650 mm in length, and the diameter of the central water channel is 33 mm. The emissivity is 0.85, and the thermal conductivity between the roller and the slab is 5243 W/(m^2 K) [7]. The contact angle is between 25° and 30°, where 28° is taken. The pressure at the inlet of cooling water is 0.5 MPa.

In this paper, Fluent software [8, 9] is adopted to calculate and solve the temperature variation and the heat transfer between the roller and the slab near the straightening area of medium thickness slab continuous casting. The discrete ordinates model was selected as the radiation model [10].

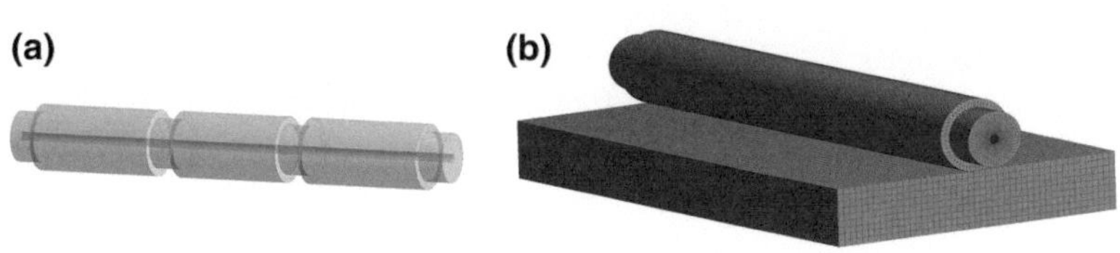

Fig. 1 Structure and meshing details of roller: **a** the real roller structure, **b** model mesh details

Table 1 Chemical composition of roller steel (wt, %)

Name	C	Si	Mn	Mo	Cr
15CrMo	0.15	0.30	0.45	0.45	1.0

Research Scheme

This paper studies the central water-cooling channel roller after the secondary cooling zone. Firstly, the heat transfer between the roller and the slab under different casting speed (0.9, 1.0, 1.1 m/s) and different slab temperatures (1273, 1323, 1373 K) was simulated. Then, the heat transfer between the roller and the slab was simulated under the conditions of different water channel diameters (43, 53 mm) when the casting speed was 0.9 m/s. The experimental scheme is shown in Table 2. In order to quantitatively analyze the influence of different factors on the roller temperature change, the corresponding circumferential temperature is extracted from the roller surface along the counterclockwise direction ($0 \rightarrow \pi/2 \rightarrow \pi \rightarrow 3\pi/2 \rightarrow 2\pi$), as shown in Fig. 2.

According to the calculation of temperature distribution in the roll, the yield strength of the roller was analyzed under different conditions. JMat Pro program was used to calculate the variation of yield strength with temperature for 15CrMo steel.

Table 2 Research cases

Number	Slab surface temperature (K)	Casting speed (m/min)	Channel diameter (mm)
1	1273	0.9	33
2	1273	1.0	33
3	1273	1.1	33
4	1323	0.9	33
5	1323	1.0	33
6	1323	1.1	33
7	1373	0.9	33
8	1373	1.0	33
9	1373	1.1	33
10	1273	0.9	43
11	1273	0.9	53

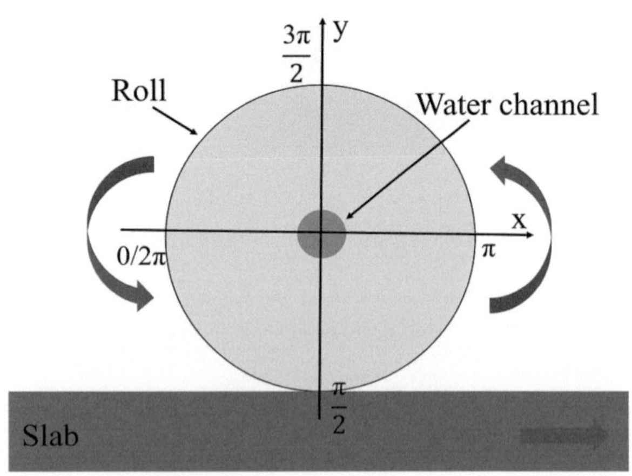

Fig. 2 Surface temperature extraction method of roller

Results

(1) Temperature distribution under different conditions

The roller temperature increases with the increase of slab temperature and decreases with the increase of casting speed. Figure 3 shows the temperature variation under different conditions. The roller temperature decreases gradually from the surface to the inside, and the temperature of the contact area with the slab is higher than other areas. Due to the cooling effect of water channel inside the roller, the temperature in the area close to the cooling water channel is relatively low, and the temperature in the area where directly contacts with the slab is relatively high, which will lead to a large temperature gradient inside the roller, indicating that the cooling water channel of this structure cannot effectively bring heat out, which is not helpful to the extension of the roller service life. The temperature is not obviously increasing at the contact area between the roller and the cooling water channel. It indicates that

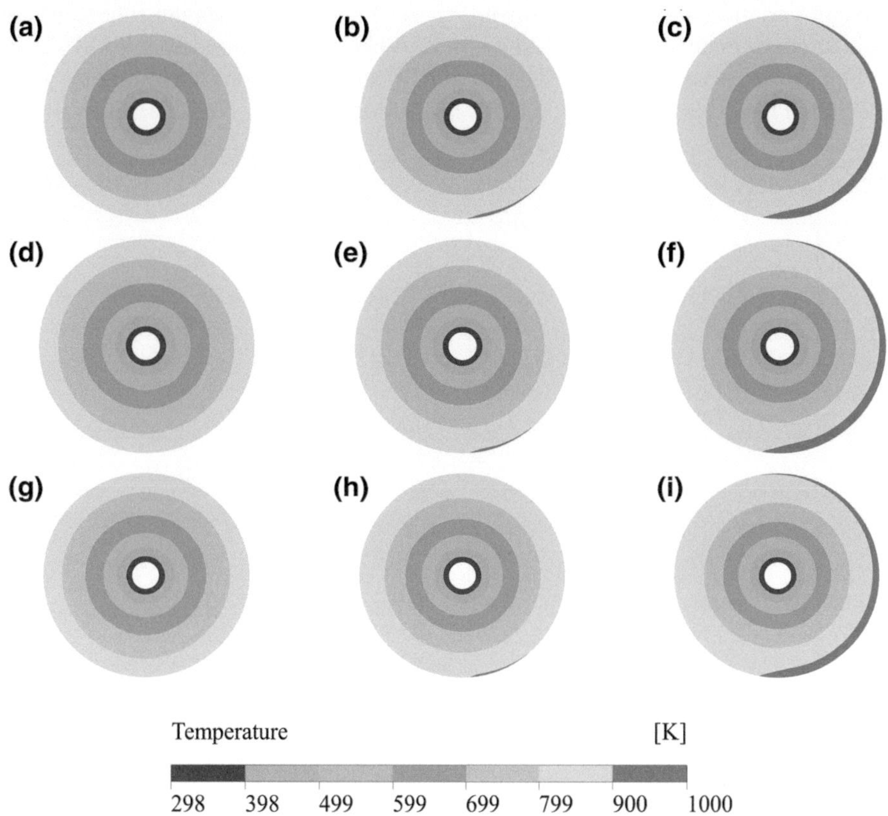

Fig. 3 Roller temperature under different conditions: **a** 0.9 m/min 1273 K, **b** 0.9 m/min 1323 K, **c** 0.9 m/min 1373 K, **d** 1.0 m/min 1273 K, **e** 1.0 m/min 1323 K, **f** 1.0 m/min 1373 K, **g** 1.1 m/min 1273 K, **h** 1.1 m/min 1323 K, **i** 1.1 m/min 1373 K

the heat transfer from the roller to the cooling water channel does not rapidly heat up the cooling water, and under the inlet pressure 0.5 MPa, the heat will not accumulate inside the cooling water, and the heat transferred to the cooling water can be taken out through the flow of cooling water.

(2) Temperature and strength variation under different casting speed

When the slab temperature is 1273 K and the casting speed is 0.9 m/min, 1.0 m/min, and 1.1 m/min, the maximum temperature of roller surface is 885 K, 882 K, and 880 K, respectively. When the slab temperature is 1323 K and the casting speed is 0.9 m/min, 1.0 m/min, and 1.1 m/min, the maximum temperature of roller surface is 912 K, 909 K, and 907 K, respectively. When the slab temperature is 1373 K and the casting speed is 0.9 m/min, 1.0 m/min, and 1.1 m/min, the maximum surface temperature of roller surface is 939 K, 936 K, and 934 K, respectively. When the roller is in the area of $0 \rightarrow \pi/2$, with the rotation of the roller, the distance between the roller surface and slab decreases gradually, and the radiative heat transfer and its temperature on the roller surface increase gradually. Figure 4 shows the roller temperature and yield strength variation with different casting speed. Figure 4d shows the roller surface yield strength under different casting speed. When the slab temperature is 1323 K and the casting speed is 0.9 m/min, 1.0 m/min, and 1.1 m/min, the minimum yield strength of roller surface is 142.96 MPa, 148.91 MPa, and 151.98 MPa, respectively.

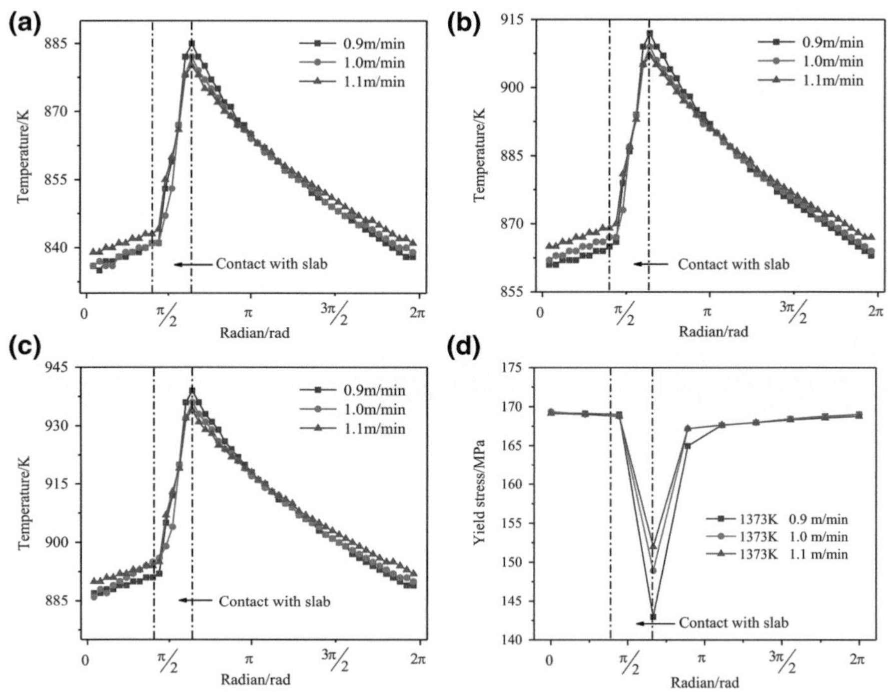

Fig. 4 Roller surface temperature and yield strength under different casting speed: **a** 1273 K, **b** 1323 K, **c** 1373 K, **d** yield strength under different casting speed

(3) Temperature and strength variation under different slab temperatures

When the casting speed is constant, the roller surface temperature increases with the increase of the slab temperature. When the casting speed is 0.9 m/min and the slab temperature is 1273 K, 1323 K, and 1373 K, the maximum temperature of roller surface is 885 K, 912 K, and 939 K, respectively. When the casting speed is 1.0 m/min and the slab temperature is 1273 K, 1323 K, and 1373 K, the maximum temperature of roller surface is 882 K, 909 K, and 936 K, respectively. When the casting speed is 1.1 m/min and the slab temperature is 1273 K, 1323 K, and 1373 K, the maximum surface temperature of roller surface is 880 K, 907 K, and 934 K, respectively. Figure 5 shows the roller surface temperature variation with different slab temperatures. Figure 5d shows the roller surface yield strength under different slab temperatures. When the casting speed is 0.9 m/min and the slab temperature is 1273 K, 1323 K, and 1373 K, the minimum yield strength of roller surface is 169.35 MPa, 167.85 MPa, and 142.96 MPa, respectively.

(4) Temperature and yield strength variation under different cooling water channels

With the increase of cooling water channel diameter, the roller temperature decreases gradually. Figure 6 shows the roller temperature distribution under different water

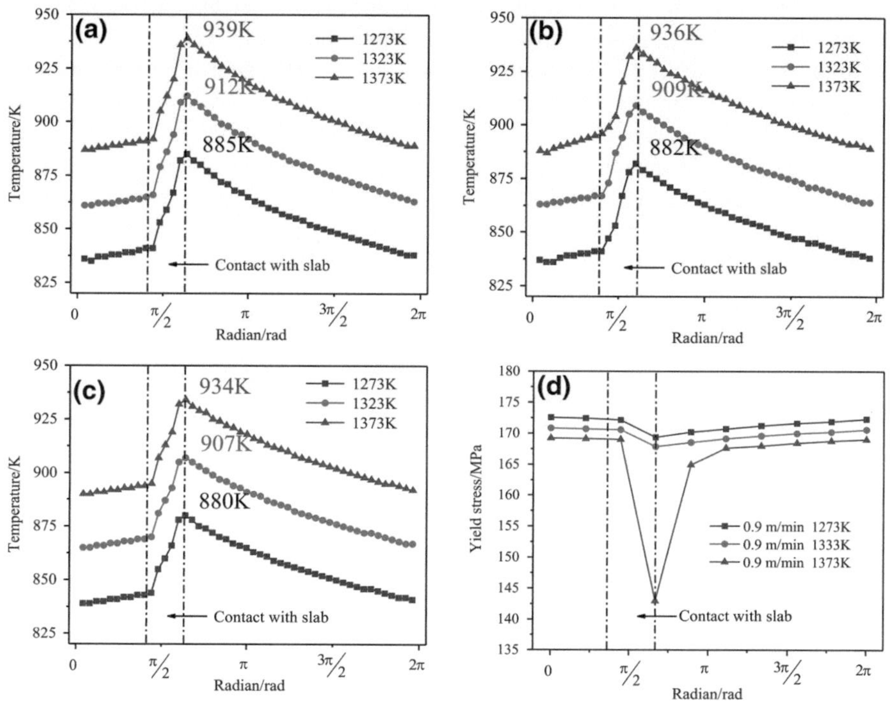

Fig. 5 Roller temperature and yield strength under different slab temperatures: **a** 0.9 m/min, **b** 1.0 m/min, **c** 1.1 m/min, **d** yield strength under different slab temperatures

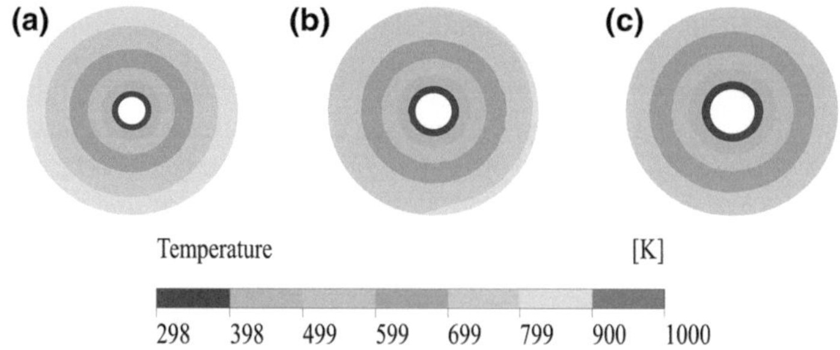

Fig. 6 Roller temperature distribution under different cooling water channel diameters: **a** 33 mm, **b** 43 mm, **c** 53 mm

channel diameters. Figure 7a shows the roller surface temperature variation under different water channel diameters (water channel diameters are 33 mm, 43 mm, and 53 mm, respectively) when the casting speed is 0.9 m/min and the slab temperature is 1273 K. When the water channel diameter is 33 mm, 43 mm, and 53 mm, the maximum roller surface temperature is 875 K, 818 K, and 786 K, respectively. It can be seen that the roller surface temperature decreases with the increase of water channel diameter. Figure 7b shows the variation of yield strength on the roller surface with different diameters of cooling water channel at the casting speed of 0.9 m/min and slab temperature of 1273 K. When the casting speed is 0.9 m/min, the slab temperature is 1273 K, and the water channel diameter is 33 mm, 43 mm, and 53 mm, the minimum roller surface yield strength is 169.35 MPa, 173.38 MPa, and 178.37 MPa, respectively. It can be seen that the roller surface strength increases with the increase of water channel diameter. Therefore, when other conditions remain unchanged, the roller surface temperature can be effectively reduced by appropriately increasing the water channel diameter inside the roller, and the roller service life can be further improved.

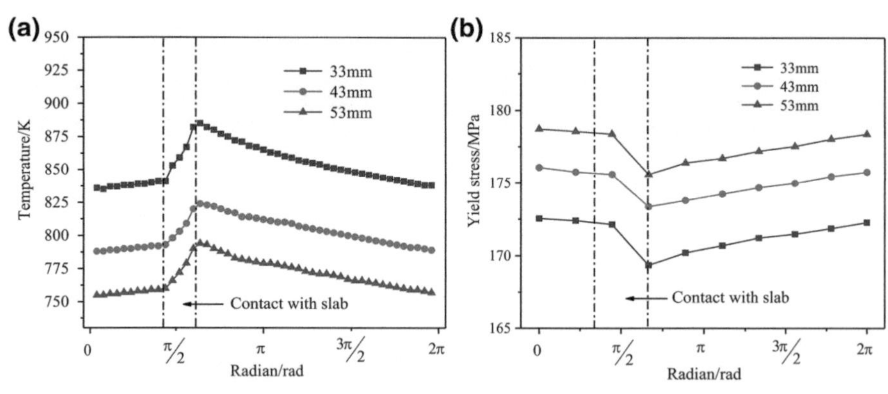

Fig. 7 Roller surface temperature and strength variation under different water channel diameters: **a** roller surface temperature variation, **b** roller surface strength variation

Discussion

The Influence of Casting Speed and Slab Temperature on the Roller Temperature Distribution

According to Fig. 4, when the slab temperature between the roller and the slab is constant, the roller surface temperature increases with the decrease of the casting speed. This is because when the casting speed is low, the contact time between the roller and the slab will be prolonged, and the heat transfer between the roller and the slab will increase. In the contact area between the roller and the slab, the roller surface temperature rises faster. When the roller is out of contact with the slab, the roller only receives radiation heat transfer from the slab, and the heat received from the slab surface is less than conduction heat transfer, and the surface temperature in the corresponding area of the roller decreases gradually with the increased distance between the roller surface and slab. According to Fig. 5, when the casting speed is constant, the roller surface temperature increases with the increase of slab temperature, which is caused by the increased slab temperature. When the roller is in the area of $0 \rightarrow \pi/2$, with the rotation of the roller, the distance between the roller surface and slab decreases gradually, and the radiative heat transfer and its temperature on the roller surface increase gradually. Similarly, in the contact area between the roller and the slab, the roller surface temperature increases rapidly. When the roller is out of contact with the slab, the roller only receives radiation heat transfer from the slab, the heat received from the slab surface is less than the conduction heat transfer, and the roller surface temperature in the corresponding area decreases gradually with the increased distance between the roller surface and slab.

The Influence of Water Channel Diameter on the Roller Temperature

The roller temperature distribution is similar to the above. The temperature of the direct contact area between the roller surface and slab is the highest. With the roller rotation, its surface temperature gradually decreases. The temperature inside the roller decreases with the decrease distance from the central water channel. With the increase of the water channel diameter inside the roller, the roller temperature decreases gradually, and the temperature gradient between its surface and interior decreases gradually. This shows that appropriately increasing the diameter of water channel inside the roller is conducive to improving the cooling effect, which further verifies the feasibility of reducing the roller temperature by increasing the diameter of waterway. This provides a direction for optimizing the structural parameters of water channel and improving the roller high temperature service life during continuous casting.

Analysis on the Yield Strength of the Roller

As shown in the calculation by JMat Pro program, the roller surface strength decreases as the temperature increases. Figures 4d and 5d show the roller surface yield strength under different casting speed and slab temperature. The roller surface yield strength increases with the increase of casting speed and decreases as the slab temperature increases. Figure 7b shows the variation of yield strength on the roller surface with different diameters of cooling water channel at the casting speed of 0.9 m/min and slab temperature of 1273 K, indicating that the roller surface strength increases with the increase of water channel diameter.

Under the condition of slab temperature at 1373 K, casting speed at 0.9 m/min, and water channel diameters of 33 mm, the roller surface temperature reaches the maximum 939 K, while the yield strength of roll is the lowest at 142.96 MPa. Usually, the working pressure on the roll is ~70 MPa, which is much lower than the lowest yield strength of the roll, indicating that the roller is safely working under the current conditions.

Conclusions

A three-dimensional transient heat transfer model between the roller and the slab is established, and the influence of casting speed, slab temperature, and water channel diameter on the temperature and strength in the roller is analyzed. The conclusions are as follows:

(1) The maximum roller surface temperature increases with the increase of slab temperature. Under the slab temperature of 1273 K, 1323 K, and 1373 K at casting speed 0.9 m/min, the maximum roller surface temperature is 885 K, 912 K, and 939 K, respectively. With fixed slab temperature, the maximum roller surface temperature will decrease as the casting speed increases.

(2) Appropriately increasing the water channel diameter can effectively reduce the roller surface temperature. Under the casting speed at 0.9 m/min and slab temperature at 1273 K, the maximum roller surface temperature is 875 K, 818 K, and 786 K when the water channel diameter is 33 mm, 43 mm, and 53 mm, respectively.

(3) The roller surface strength decreases as the temperature increases. The minimum yield strength of the roller under the calculated conditions is 142.96 MPa, which is higher than the working pressure ~70 MPa, indicating that that roller is working in safe condition.

Acknowledgements The work is financially supported by the National Natural Science Foundation of China (NSFC, Project No. 51874059, 51874060 and U1960113). The authors would like to thank the support by the Natural Science Foundation of Chongqing (Project No. cstc2018jcyjAX0647, cstc2018jszx-cyzdX0076).

References

1. Pennerstorfer P, Watzinger J, Enzinger CH (2015) It's all about temperature—dry casting for optimum surface quality. Met Mag 1:301–305
2. Javurek M, Ladner P, Watzinger J (2015) Secondary cooling: roll heat transfer during dry casting. In: Paper presented at the METEC and 2nd European steel technology and application days (ESTAD) conference, Germany, 15–19 June 2015
3. Seyedein SH, Hasan M (1997) A three-dimensional simulation of coupled turbulent flow and macroscopic solidification heat transfer for continuous slab casters. Int J Heat Mass Trans 40(18):4405–4423
4. Liang CZ, Mao QZ, Xi HL (2012) Study on fluid flow in a large round bloom continuous casting mold. Adv Mater Res 402:196–201
5. Jian Z, Chen D, Wang S, Long M (2015) Compensation control model of superheat and cooling water temperature for secondary cooling of continuous casting. Steel Res Int 82(3):213–221
6. Wang Z, Man Y, Wang X, Zhang X, Yang L, Lu H, Xiong W (2014) Inverse problem-coupled heat transfer model for steel continuous casting. J Mater Process Tech 214(1):44–49
7. Sun JQ, Li HJ, Zhang XZ (1997) Study on heat transfer between roller and slab in continuous casting machine. J Iron Steel Res 1(3):10–14
8. Zhang LL, Chen DF, Chen HB, Long MJ, Xie X (2016) Study on transport phenomena in the beam blank continuous casting mould coupling copper mould with molten steel. Ironmaking Steelmaking 44(3):1–9
9. Thomas BG, Najjar FM (1991) Finite element modelling of turbulent fluid flow and heat transfer in continuous casting. Appl Math Model 15(5):226–243
10. Liu Y, Su FY, Wen Z, Li Z, Yong HQ, Feng XH (2014) CFD modeling of flow, temperature, and concentration fields in a pilot-scale rotary hearth furnace. Metall Mater Trans B 45(1):251–261

Mathematical Simulation on the Influence of Melting Rate and Melting Current on Droplet Behavior During Electroslag Remelting Process

Tianjie Wen, Xiujie Li, Anjun Xu and Lifeng Zhang

Abstract A transient three-dimensional model was established and calculated by the commercial software ANSYS Fluent 14.0 to simulate the electroslag remelting process. Based on this model, a quantitative analysis of droplet behavior during dripping process was conducted by monitoring the mass flow rate of steel phase on monitor surfaces to investigate the influence of melting rate and current on the equivalent diameter, dripping frequency, and dripping time of droplets. Results showed that the droplet size was proportional to the melting rate and the dripping frequency was proportional to the cubic root of the melting rate, while the droplet size was proportional to the melting current and the dripping frequency was inversely proportional to the melting current.

Keywords Electroslag remelting · Droplet formation behavior · Monitor surface · Mathematical model

Introduction

Electroslag remelting (ESR) is a secondary refining process which utilizes the Joule heat generated in slag to form the droplets at the tip of electrode. Droplets dramatically change the distribution of current density, which further influenced the temperature and fluid flow field [1]. Besides, droplets have large specific surface area, which is beneficial to remove nonmetallic inclusions in ESR process. So it is necessary to study droplet behavior in slag pool.

Due to the opaque mold and high temperature, the formation and dripping of droplets cannot be observed directly, so several investigations including physical simulation and mathematical simulation were proposed to simulate the droplet behavior. Campbell [2] designed a transparent apparatus which utilized aluminum as electrode, LiCl-KCl eutectic as slag, and Pyrex or silica as mold to visually observe phenomena occurring in the slag and discussed the relationship between the droplet sizes with

T. Wen · X. Li · A. Xu · L. Zhang (✉)
School of Ecological and Metallurgical Engineering, University of Science and Technology Beijing, Beijing 100083, China
e-mail: zhanglifeng@ustb.edu.cn

J. Lee et al. (eds.), *Materials Processing Fundamentals 2020*, The Minerals, Metals & Materials Series, https://doi.org/10.1007/978-3-030-36556-1_14

parameters including interfacial tension, electrode size, and melting rate. Based on this physical model, Cao et al. [3] proposed that the droplet size showed a parabolic distribution with the change of filling ratio and the maximum value was reached at the filling ratio of 0.60. Kharicha et al. [4, 5] established two-dimensional model to numerically simulate the dripping process of droplets and discussed the influence of interfacial tension on droplet behaviour.

In the current study, based on a transient three-dimensional (3D) mathematical model, the mass flow rate of steel phase was monitored by monitor surface in the slag area to track droplets. Therefore, the influence of melting rate and melting current on droplet size and dripping frequency was discussed in this work.

Mathematical Model

In order to investigate the formation and dripping process of droplets during ESR process, a transient 3D mathematical model was established. Figure 1 shows the mesh of the model, which assumed the electrode tip to be flat while ignored the

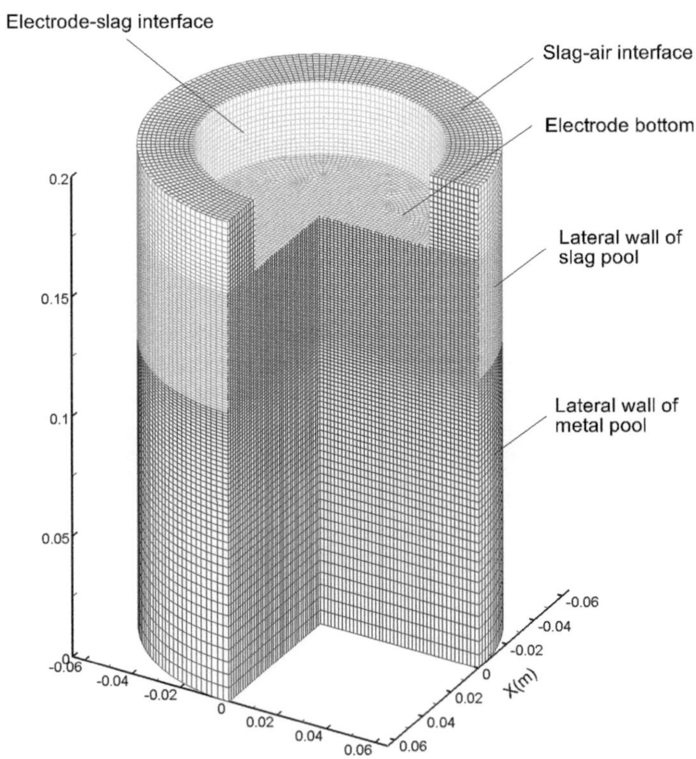

Fig. 1 Mesh model of ESR process

electrode domain. The electrode with diameter of 90 mm was used to produce a φ 130 mm ingot, and the height of slag pool was 65 mm, in which the immersion depth of the electrode was 30 mm.

The slag domain especially the center part was locally refined to guarantee the accuracy of the droplet simulation. The continuum mixture conservation equations for continuity, momentum, energy [6], and electromagnetism [7–9] were as follows:

$$\frac{\partial \rho}{\partial t} + \nabla \cdot (\rho \vec{\mathbf{u}}) = 0 \tag{1}$$

$$\frac{\partial (\rho \vec{\mathbf{u}})}{\partial t} + \nabla \cdot (\rho \vec{\mathbf{u}} \times \vec{\mathbf{u}}) = -\nabla \overrightarrow{P} + \mu_{\text{eff}} \nabla^2 \vec{\mathbf{u}} + \rho g + \vec{\mathbf{F}}_b + \vec{\mathbf{F}}_e \tag{2}$$

$$\frac{\partial}{\partial t}(\rho H) + \nabla \cdot (\rho \vec{\mathbf{u}} H) = \nabla \cdot (k_{\text{eff}} \nabla T) + Q_{\text{Joule}} \tag{3}$$

$$\nabla \cdot \vec{\mathbf{J}} = \nabla \cdot (\sigma \nabla \varphi) = 0 \tag{4}$$

$$\nabla \cdot \vec{\mathbf{B}} = 0 \tag{5}$$

where ρ is fluid density (kg m^{-3}); t is time (s); $\vec{\mathbf{u}}$ is fluid velocity (m s^{-1}); $\vec{\mathbf{P}}$ is pressure (Pa); u_{eff} is effective viscosity (Pa s); g is gravity acceleration (m s^{-2}); H is enthalpy or heat content (J kg^{-1}); k_{eff} is temperature-dependent effective thermal conductivity (W m^{-1} K^{-1}); $\vec{\mathbf{J}}$ is current density (A m^{-2}); $\vec{\mathbf{B}}$ is magnetic flux intensity (T); $\vec{\mathbf{F}}_b$ is thermal buoyancy force (N m^{-3}), which is determined by Boussinesq approximation; $\vec{\mathbf{F}}_e$ is the Lorentz force (N m^{-3}); and Q_{Joule} is the source term of the Joule heat (W m^{-3}).

$$F_b = \rho \beta g (T - T_{\text{ref}}) \tag{6}$$

$$F_e = \frac{1}{2}(J_r B_r + J_i B_i) \tag{7}$$

$$Q_{\text{Joule}} = \frac{1}{2\sigma}(J_r J_r + J_i J_i) \tag{8}$$

where J_r and J_i are real and imaginary parts of current density, respectively.

A horizontal monitor surface was set up in the slag area (40 mm below the electrode tip) to get the mass flow rate of steel phase at each time. The monitoring data of one case were shown in Fig. 2 as an example, from which the equivalent diameter and the dripping frequency of droplets could be analyzed. The mass flow rate is a periodic wave along time recording the process that separates droplets drip across the monitor surfaces. The mass of one droplet could be calculated by the integral of the mass flow rate within one period:

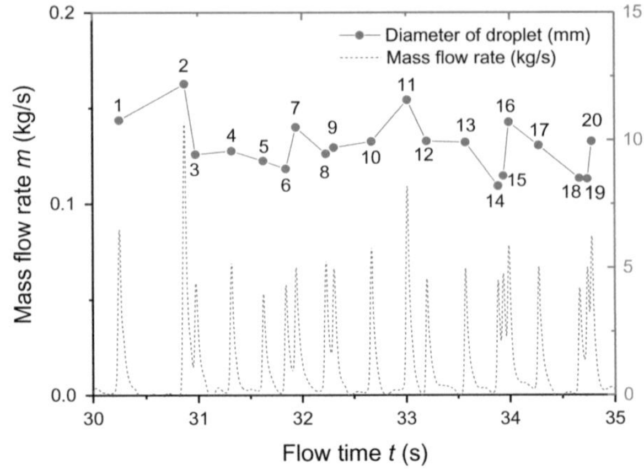

Fig. 2 A typical pattern for mass flow rate

$$M_i = \int_{t_{i0}}^{t_{i1}} m(t)\mathrm{d}t = \frac{4}{3}\pi \cdot \rho_{\text{steel}}\left(\frac{D_i}{2}\right)^3, \quad i = 1, 2, \ldots, n \tag{9}$$

where M_i is the mass of droplet i (kg); t_{i0} and t_{i1} are the start time and the end time of one period (s); $m(t)$ is the mass flow rate at time t (kg s^{-1}); ρ_{steel} is the density of liquid steel (kg m^{-3}); and D_i is the equivalent diameter of droplet i (m).

Within one droplet period, the peak time is used to represent the time when the droplet passes through the monitor surface. Hence, the interval time T can be simply defined as the time between two adjacent peak time.

$$\bar{T} = \frac{1}{n}\sum_{1}^{n}(t_{i+1} - t_i) = f^{-1}, \quad i = 1, 2, \ldots, n-1 \tag{10}$$

where $\bar{T}$ is the average interval time (s); t_i is the peak time of droplet i (s); and f is the dripping frequency (Hz).

In order to investigate the influence of melting rate and melting current, the cases of different melting rates (0.01, 0.02, 0.03, 0.04, 0.05 kg) and melting current (1, 2, 3, 4, 5 kA) were simulated. The frequency of AC current used in remelting was 50 Hz, and the interfacial tension of slag-steel interface was 1.0 N m^{-1}. Other material properties and boundary conditions can be found in previous published work [10].

Influence of Melting Rate

Figure 3 showed the mass flow rate of different melting rates with the melting current of 3 kA. With the increase of melting rate, the number of droplets within the same time and the peak value of mass flow rate, which indicated the dripping frequency and

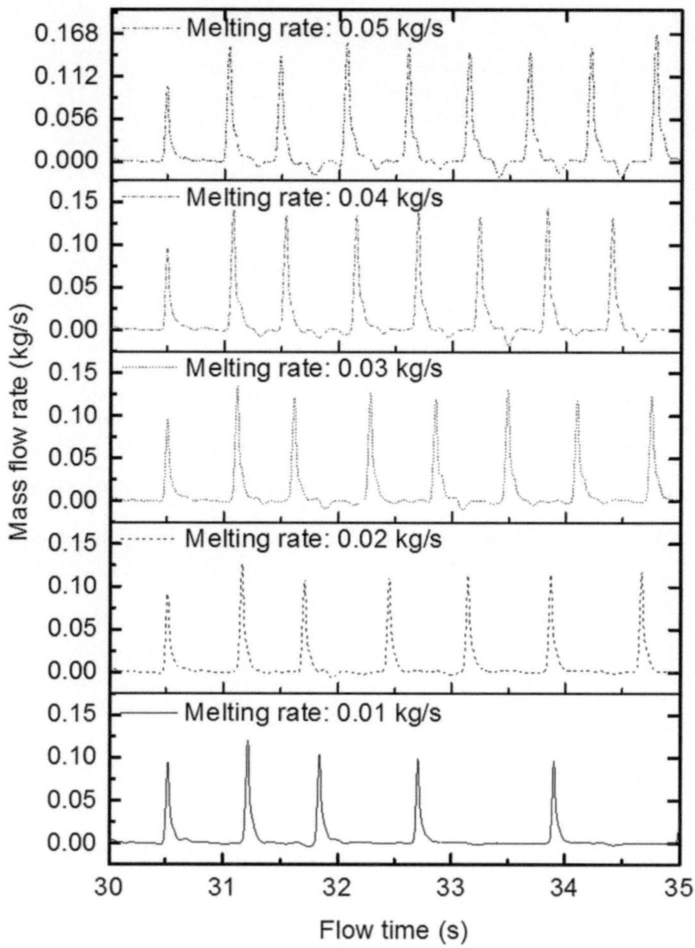

Fig. 3 Mass flow rate on monitor surface with different melting rates

droplet size, respectively, increased simultaneously. Increasing melting rate resulted in more liquid steel to form a droplet at the same time. Specifically, the droplet grew larger, which will fortify the gravity and the tension of droplets. As the gravity is a body force while the tension is a face force, the increase of gravity will exceed the increase of tension, leading to easier detachment of droplets. Hence, the dripping frequency was also increased.

The quantitative analysis of the relationship between melting rate with equivalent diameter and dripping frequency has been shown in Fig. 4. Results showed that the droplet size was proportional to the melting rate and the dripping frequency was proportional to the cubic root of the melting rate, which was in accordance with the explanation above.

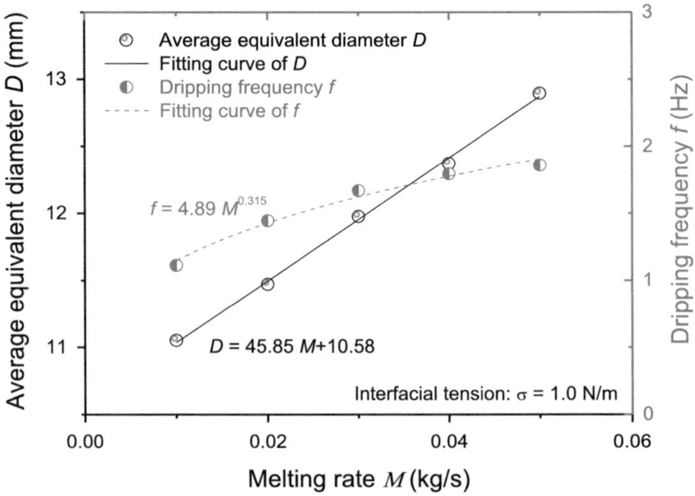

Fig. 4 Relationship between melting rate and droplet size as well as dripping frequency

According to mass conservation, within one dripping period, the mass of melted electrode equals the droplets formed. Therefore, the equivalent diameter (radius) as a function of melting rate and dripping frequency could be conducted as:

$$r = \left(\frac{3}{4\pi}\right)^{1/3}\left(\frac{\dot{M}}{f\rho_{\text{steel}}}\right)^{1/3} = 0.62\left(\frac{\dot{M}}{f\rho_{\text{steel}}}\right)^{1/3} \tag{14}$$

where r was the equivalent radius of droplets (m), $\dot{M}$ was the melting rate of electrode(kg s^{-1}), f was the dripping frequency of droplets (Hz), and ρ_{steel} was density of steel phase(kg m^{-3}). This equation was the same with the equation derived from dynamics analysis by Campbell [2]. Furthermore, the equation was transformed to get the function of theoretical melting rate:

$$\dot{M}_t = f\rho_{\text{steel}}\left(\frac{d}{1.24}\right)^3 \tag{15}$$

The comparison of theoretical melting rate $\dot{M}_t$ and practical melting rate in simulation $\dot{M}_s$ with interfacial tension of 0.5 and 1.0 N m^{-1} is showed in Fig. 5, illustrating an obvious positive proportion between these two melting rates. Since a larger interfacial tension caused a smaller scale factor, it could be concluded that the interfacial tension has an influence on this equation, which needs further investigation.

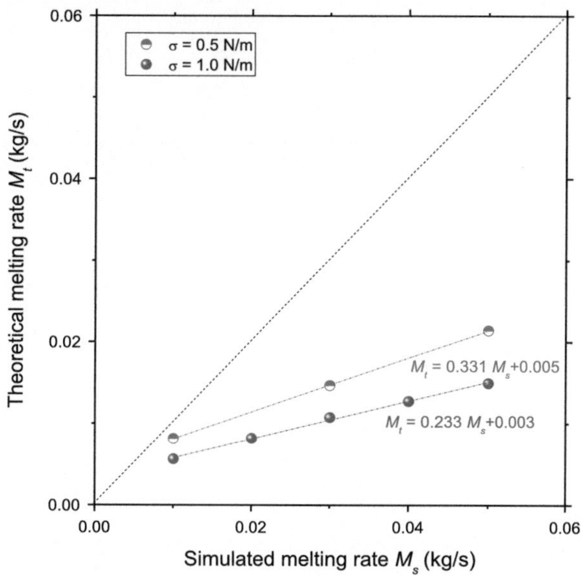

Fig. 5 Comparison of theoretical melting rate and simulated melting rate

Influence of Melting Current

Melting current, as a vital energy source, has a great influence on temperature distribution, fluid flow field, and the electromagnetic field. Noticeably, as the melting rate of electrode was set as inlet velocity rather than the real melting process according to the heat condition, this model ignored the change of melting rate while melting current increased. Figure 6 showed the mass flow rate of different melting current with the melting rate of 0.03 kg s^{-1}. Similar to the effect of melting rate, with the increase of melting current, the number of droplets within the same time and the peak value of mass flow rate also increased simultaneously. However, quantitative analysis (Fig. 7) showed that despite the minor rise of peak value, the equivalent diameter of droplets decreased with the ascend of the melting current.

Figure 7 indicated that the droplet size was proportional to the melting current and the dripping frequency was inversely proportional to the melting current. This was resulted by the change of fluid velocity and Lorentz force as shown in Fig. 8. With the increase of melting current, both the fluid velocity and the Lorentz force near the droplet turned intenser. On the one hand, the fluid flow near the droplet moved downward, driving the droplet to detach from the electrode tip; on the other hand, the Lorentz force of droplet pointed to the center, which will accelerate the neck shrinkage of the droplet. Both of these changes made the droplet easier to detach, which increased the dripping frequency and decreased the size of droplets.

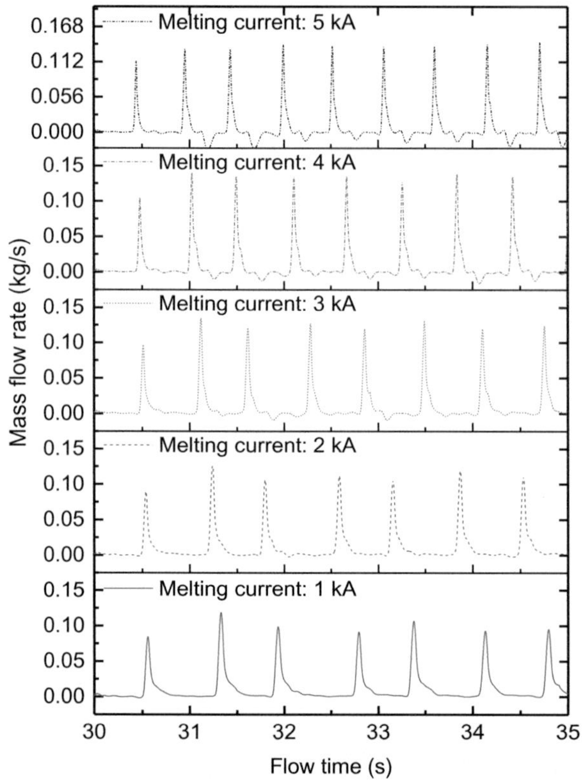

Fig. 6 Mass flow rate on monitor surface with different melting current

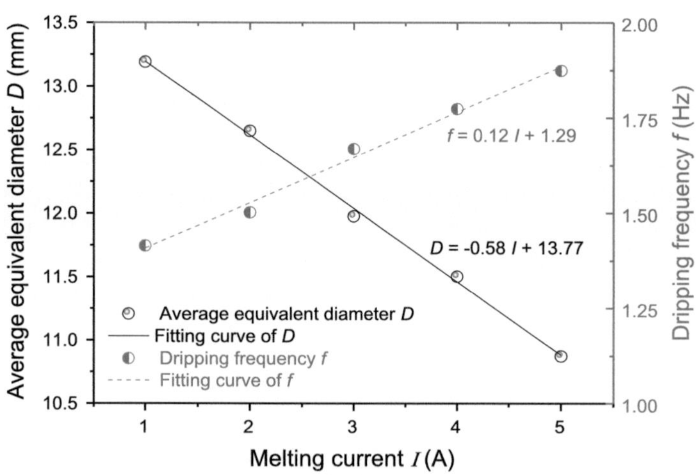

Fig. 7 Relationship between melting current and droplet size as well as dripping frequency

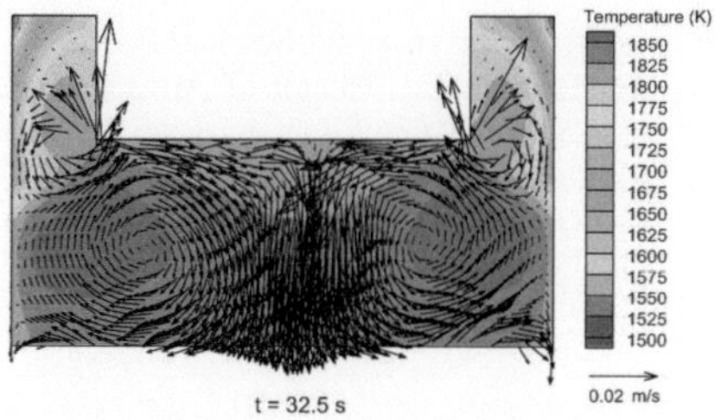

(a) Temperature distribution(counter) and fluid velocity(vector), 1 kA

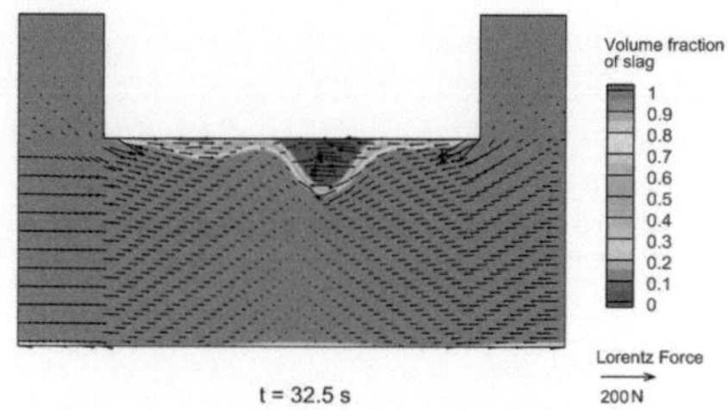

(b) Volume fraction of slag(counter) and Lorentz force (vector), 1 kA

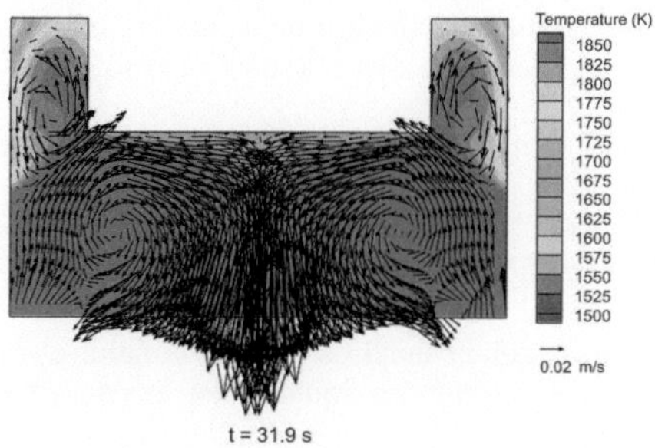

(c) Temperature distribution(counter) and fluid velocity(vector), 5 kA

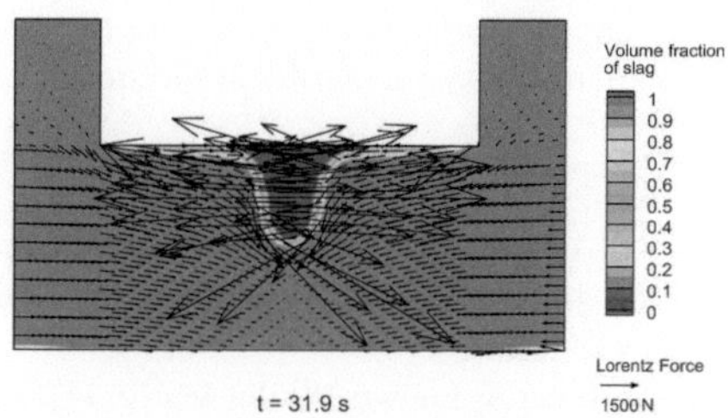

(d) Volume fraction of slag(counter) and Lorentz force (vector), 5 kA

Fig. 8 Influence of melting current of fluid velocity and Lorentz force

Conclusions

In the current paper, based on a transient 3D mathematical model of ESR process, the droplet size and dripping process were investigated by the monitor surface of mass flow rate. The following conclusions were obtained:

(1) The mass flow rate is a periodic wave along time recording the process that separates droplets drip across the monitor surfaces. The mass of one droplet could be calculated by the integral of the mass flow rate within one period, and the interval time can be defined as the time between two adjacent peak time.
(2) Increasing melting rate made droplets detach more easily, resulting in the increase of droplet size and dripping frequency. With the same melting current, the droplet size was proportional to the melting rate and the dripping frequency was proportional to the cubic root of the melting rate.
(3) With the increase of melting current, both the fluid velocity and the Lorentz force near the droplet turned intenser, which drove the droplet to detach from the electrode tip and accelerate the neck shrinkage of the droplet. With the same melting rate, the droplet size was proportional to the melting current and the dripping frequency was inversely proportional to the melting current.

Acknowledgements The authors are grateful for support from the National Science Foundation China (Grant No. U1860206 and No. 51725402), Beijing International Center of Advanced and Intelligent Manufacturing of High Quality Steel Materials (ICSM), Beijing Key Laboratory of Green Recycling and Extraction of Metals (GREM), and the High Quality Steel Consortium (HQSC) at the School of Metallurgical and Ecological Engineering at University of Science and Technology Beijing (USTB), China.

References

1. Wang Q, Wang F, Li G, Gao Y, Li B (2017) Simulation and experimental studies of effect of current on oxygen transfer in Electroslag remelting process. Int J Heat Mass Transf 113:1021–1030
2. Campbell J (1970) Fluid flow and droplet formation in the Electroslag remelting process. JOM 22(7):23–35
3. Cao Y, Dong Y, Jiang Z, Cao H, Hou D, Feng Q (2016) Research on droplet formation and dripping behavior during the Electroslag remelting process. Int J Miner Metall Mater 23(4):399–407
4. Kharicha A, Ludwig A, Wu M (2011). Droplet formation in small Electroslag remelting processes. In: Proceedings of the 2011 international symposium on liquid metal processing & casting
5. Kharicha A, Wu M, Ludwig A, Karimi-Sibaki E (2016) Simulation of the electric signal during the formation and departure of droplets in the Electroslag remelting process. Metall Mater Trans B 47(2):1427–1434
6. Thomas BG, Zhang LF (2001) Mathematical modeling of fluid flow in continuous casting. ISIJ Int 41(10):1181–1193
7. Wang Z, Quan Y, Lu B (2011) Engineering electromagnetic field. Tsinghua University Press, Beijing
8. Mostaghimi J, Boulos M (1989) Two-dimensional electromagnetic field effects in induction plasma modelling. Plasma Chem Plasma Process 9(1):25–44
9. Xue S, Proulx P, Boulos M (2003) Effect of the coil angle in an inductively coupled plasma torch: a novel two-dimensional model. Plasma Chem Plasma Process 23(2):245–263
10. Wen T, Zhang H, Li X, Le Y, Ren Y, Liu H, Zhang L (2018) Numerical Simulation on the oxidation of lanthanum during the Electroslag remelting process. JOM 70(10):2157–2168

Numerical Simulation on the Multiphase Flow During the KR Process Using the Eulerian–Eulerian Modeling

Yanyu Zhao, Wei Chen and Lifeng Zhang

Abstract A three-dimensional model based on a full-scale ladle and impeller was established to investigate the multiphase flow behavior and surface profile distribution. The hot metal phase, air phase, and argon gas phase were simulated using the Euler–Euler method. The effect of the argon gas injection, the baffles, and the inclination of the ladle bottom on the distribution of the fluid flow and surface level was revealed. The results indicate that the argon injection had no significant effect on the overall flow field distribution, because the effect of argon blowing was not significant compared to the vigorous stirring of the impeller. The inclination of the ladle bottom caused an asymmetrical distribution of the bottom flow field, but had little effect on the upper flow field distribution. However, the presence of the baffles directly reduced the flow field strength and smoothed the surface profile distribution.

Keywords Multiphase flow · Euler–Euler method · Baffle · KR process

Introduction

The knotted reactor (KR) agitation desulfurization method was developed in 1965 by Japan's Nippon Steel Guang Tian institution, and was used in a hot metal pretreatment desulfurization technology. The KR desulfurization method had the advantages of high desulfurization efficiency and stable desulfurization effect, which was widely used by major steel companies to produce low sulfur content steel [1–3]. However, the forced vortex zone in the ladle will decrease the mixing effect during the KR process. Thus, it is of great significance to improve the flow field distribution during the KR process. Yianneskis et al. [4] used the water model to investigate the liquid flow in a baffled stirred reactor vessel driven by a six-blade disk impeller. The results indicated that the formation of ring vortices above and below the impeller depended on the clearance and the unsteady method should be used to predict fluid flow. Gao et al. [5] proposed a two-dimensional volume of fluid (VOF) model to simulate the behavior

Y. Zhao · W. Chen · L. Zhang (✉)
School of Ecological and Metallurgical Engineering, University of Science and Technology Beijing, Beijing 100083, China
e-mail: zhanglifeng@ustb.edu.cn

J. Lee et al. (eds.), *Materials Processing Fundamentals 2020*, The Minerals, Metals & Materials Series, https://doi.org/10.1007/978-3-030-36556-1_15

of bubbles in a physical model of the aluminum degassing ladle. Abreu-Lopez et al. [6, 7] studied the liquid velocity, gas holdup, vortex size, mass transfer coefficient, and kinetics with four different impeller designs using the physical modeling and mathematical simulation. The impeller designed with blades, in particular that with four blades, was found to have superior performance. Using a coupled k-ε model and VOF model, Maniruzzaman and Makhlouf [8, 9] found that the energy dissipation rate was mainly affected by the impeller's rotating speed and the randomly distributed solid particles were agglomerated into relatively large clusters due to the stirring effect. Ji et al. [10] developed a new type of stirring vessel of water model for KR mechanical desulfurization using the CFD method. The dead zones were decreased with the new designed impeller. The previous literature was more focused on the design of the impeller, and there was little research on the influence of ladle geometry. Therefore, in the current study, a three-dimensional model based on a full-scale ladle and impeller was established to investigate the argon gas injection, the baffles, and the inclination of the ladle bottom on the distribution of the fluid flow and surface level using the Euler–Euler method.

Mathematical Formulation

The ladle used in the current study has a capacity of 230 t. The total height of the ladle was 4105 mm, and the top and bottom diameters of the ladle were 3869 mm and 3248 mm, respectively. The detailed model can be seen in Fig. 1a. The shape of the impeller was a cross shape commonly used in factories. The rotational radius

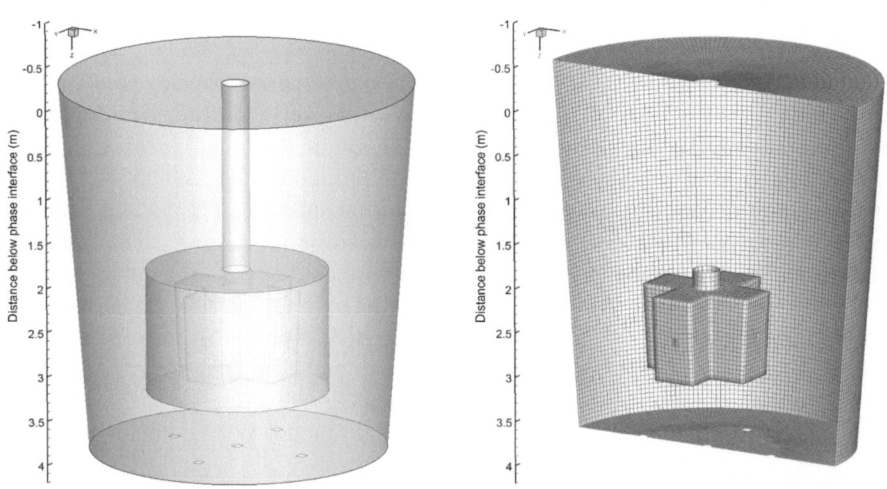

(a) Geometry (b) Mesh

Fig. 1 Calculation domain

Table 1 Parameters used in the current study

Parameter	Value	Parameter	Value
Hot metal density (kg m^{-3})	7200	Argon gas viscosity (kg m^{-1} s^{-1})	8.1×10^{-5}
Hot metal viscosity (kg m^{-1} s^{-1})	0.006486	Rotate speed (r min^{-1})	100
Air density (kg m^{-3})	1.225	Bottom gas flow rate (NL min^{-1})	500
Air viscosity (kg m^{-1} s^{-1})	1.79×10^{-5}	Impeller gas flow rate (NL min^{-1})	500
Argon gas density (kg m^{-3})	0.26	Immersion depth (mm)	1800

was 727 mm. The initial level of the hot metal was set as 3601 mm, and the impeller immersion depth was 1800 mm. The entire calculation domain was divided into approximately 0.53 million structured mesh cells [7, 10]. A fine graded mesh was adopted near the interface. The mesh distribution can be seen in Fig. 1b. Other detailed parameters used in the current study are summarized in Table 1.

The continuity equation and momentum balance for phase q are calculated as Eqs. (1) and (2), respectively.

$$\frac{1}{\rho_q}\left[\frac{\partial}{\partial t}\left(\alpha_q\rho_q\right)+\nabla\cdot\left(\alpha_q\rho_q u_q\right)\right]=0 \tag{1}$$

$$\frac{\partial\left(\alpha_q\rho_q u_q\right)}{\partial t}+\nabla\cdot\left(\alpha_q\rho_q u_q u_q\right)=-\nabla\cdot\left(\alpha_q\tau_q\right)-\alpha_q\nabla P+\alpha_q\rho_q g+F \tag{2}$$

where α_q is the volume fraction, ρ_q is the density, and F is the momentum exchange between the phases due to interface force.

In the standard k-ε model, two additional scalar transport equations, of turbulent kinetic energy k and its dissipation rate ε, are required to model turbulence:

$$\alpha_q\rho_q\left(\frac{\partial k}{\partial t}+u_q\cdot\nabla k\right)=\nabla\left[\left(\alpha_q\frac{\mu_t}{\sigma_k}\right)\nabla k\right]+\alpha_q G_k-\alpha_q\rho_q\varepsilon \tag{3}$$

$$\alpha_q\rho_q\left(\frac{\partial \varepsilon}{\partial t}+u_q\cdot\nabla \varepsilon\right)=\nabla\left[\left(\alpha_q\frac{\mu_t}{\sigma_\varepsilon}\right)\nabla \varepsilon\right]+\alpha_q C_{1\varepsilon}\frac{\varepsilon}{k}G_k-\alpha_q C_{2\varepsilon}\rho_q\frac{\varepsilon^2}{k} \tag{4}$$

where $C_{1\varepsilon}=1.44$, $C_{2\varepsilon}=1.92$, $\sigma_k=1.0$, $\sigma_\varepsilon=1.3$.

The top surface of the ladle was assumed to be flat with zero shear stress. The non-slip condition was adopted at static wall. The MRF conditions were set in the rotating zone, and the rotation speed was set as 100 r/min. Four different cases were calculated, including the normal case without gas injection, the normal case with gas injection, the ladle with the baffle, and the ladle with the slope bottom, as shown in Fig. 2.

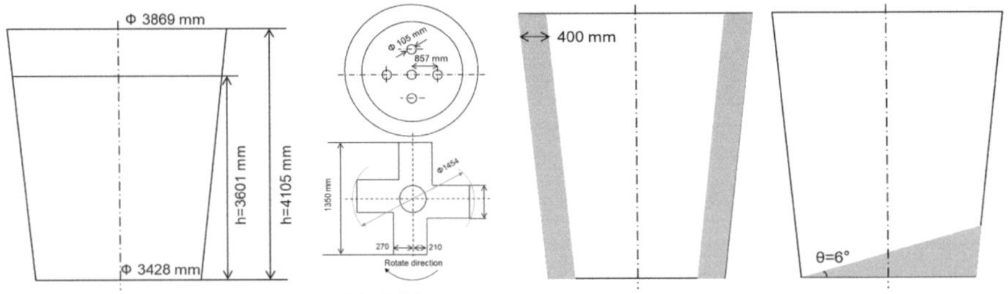

(a) Without gas injection (b) With gas injection (c) With baffle (d) With slope bottom

Fig. 2 Ladle geometry distribution

Multiphase Flow Distribution

The velocity of hot metal in the ladle during the KR desulfurization process was mainly affected by the rotation speed and the immersion depth of the impeller. The argon gas injection seems to reduce the dead zone ratio (the velocity magnitude that is small than 0.15 m/s) under a rotation speed and the immersion depth. Thus, the effect of the argon gas injection on the flow field distribution was investigated, which can be seen in Fig. 3a and b. The injection positions are shown in Fig. 2b. The total gas flow rate in the ladle bottom and impeller was the same and set as 500 NL/min. The comparison results show that the velocity magnitude at the upper region of the ladle was increased slightly after the injection of the argon gas, but it has no significant effect on the overall flow field distribution. The dead zone ratio near the bottom was not reduced. This may be due to the intense agitation that counteracted the effects of the gas injection. The presence of the ladle baffle had a significant impact on the flow field distribution, as shown in Fig. 3c. The baffle made the velocity magnitude distribution significantly reduced, and the circulation structure inside the ladle was changed. The effect of the slope of the ladle bottom on the flow field is shown in Fig. 3d. An asymmetrical flow field distribution was induced due to the inclined distribution at the bottom of the ladle. The dead zone near the bottom of the ladle was also significantly decreased.

The velocity magnitude distribution along the line $X = 1.05$ m and $Z = 3.0$ m under different conditions is compared in Fig. 4. The hot metal near the impeller reached the maximum velocity magnitude, 3.7 m/s. As mentioned before, the argon gas injection hardly affected the velocity distribution in the lower region of ladle. The normal case with gas injection, the ladle with the baffle, and the ladle with the slope bottom can slightly increase the upper flow field of the ladle. The ladle with the baffle had the greatest influence on the flow field. The maximum velocity magnitude was increased to 4.0 m/s, but the overall flow field distribution was decreased.

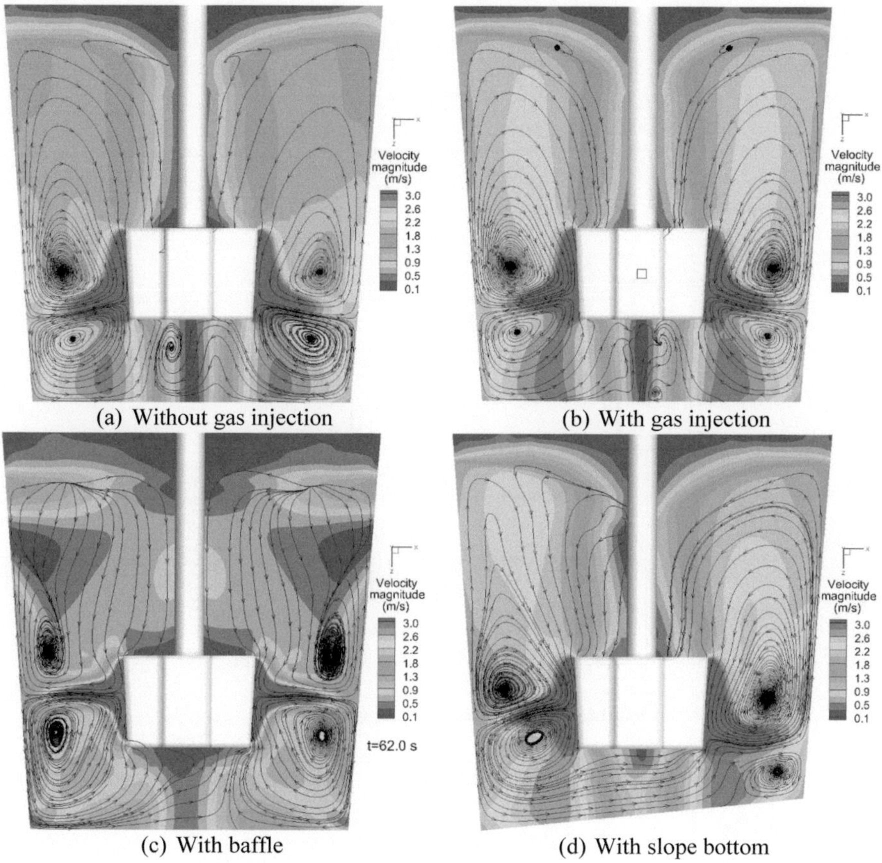

Fig. 3 Velocity and streamline distribution under different conditions

Surface Profile Distribution

The free surface distribution under different conditions is compared in Fig. 5. The results demonstrate that the argon gas injection and the ladle with the slope bottom had little effect on the surface-level distribution compared to the normal case. However, the velocity distribution on the free surface was decreased and increased, respectively. The presence of the ladle baffle made the distribution of the free surface smoother, as shown in Fig. 5c. The velocity magnitude was also reduced accordingly. The results in Fig. 5b correspond to the results of Fig. 3b. The argon gas under the 500 NL/min flow rate was mainly distributed in the lower part of the impeller due to the intense agitation. Detailed surface-level distribution at the plane $Y = 0$ is shown in Fig. 6. The free surface-level difference was about 150 mm with the effect of the baffle. Moreover, the free surface-level difference was more than 600 mm in other cases.

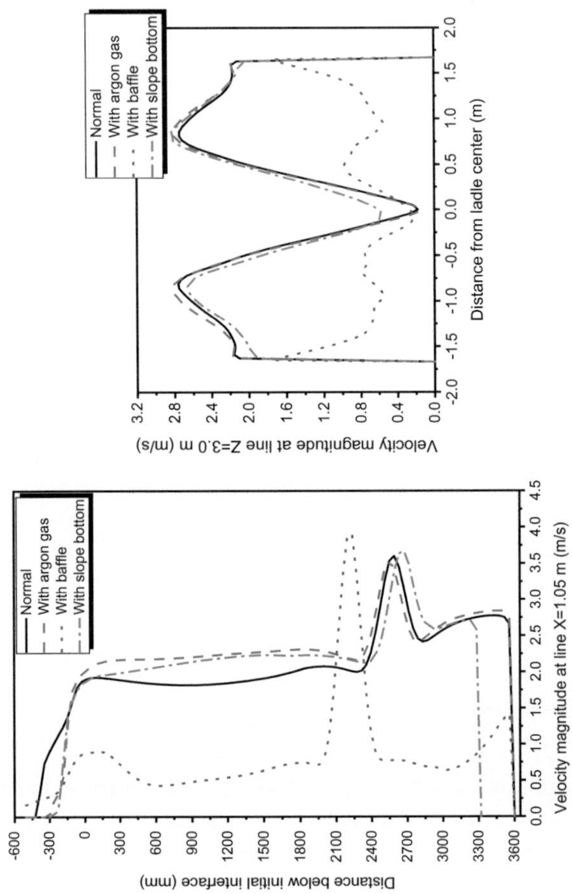

(a) Along line X=1.05 m

(b) Along line Z=3.0 m

Fig. 4 Velocity magnitude distribution

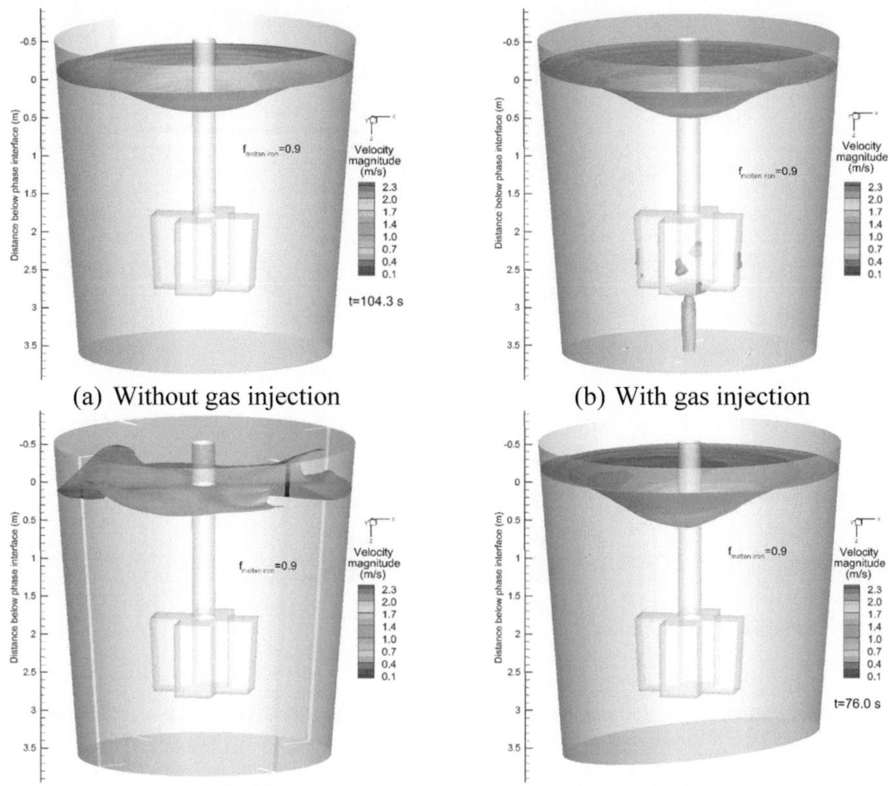

(a) Without gas injection (b) With gas injection

(c) With baffle (d) With slope bottom

Fig. 5 Free surface distribution under different conditions

Conclusions

(1) The velocity magnitude at the upper region of the ladle was increased slightly with a 500 NL/min injection of the argon gas, but it has no significant effect on the overall flow field distribution. The effect of gas injection on the gentling liquid level and reducing the dead zone ratio was small.
(2) The presence of the ladle baffle had an obvious impact on the flow field and free surface distribution. The free surface-level difference was decreased to 150 mm. However, the overall flow velocity in the ladle was correspondingly reduced.
(3) The inclination of the bottom of the ladle caused an asymmetrical flow field distribution. The dead zone near the bottom of the ladle was significantly decreased. It had little effect on the surface-level distribution compared to the normal case. Nevertheless, the velocity distribution on the free surface was increased.

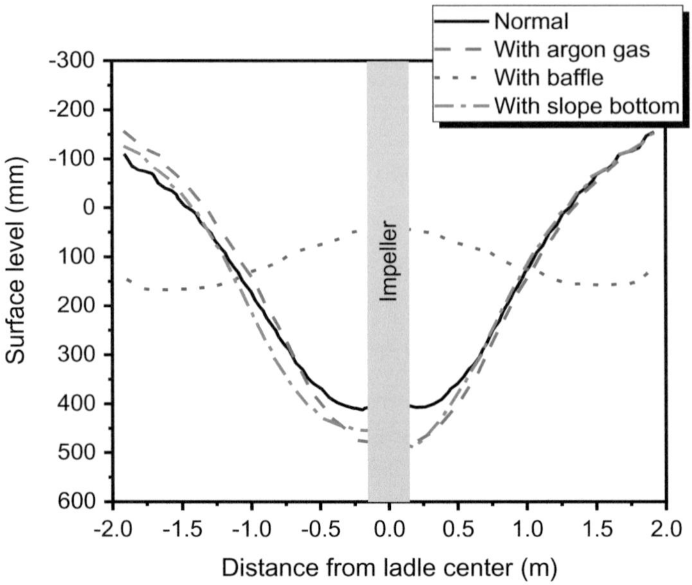

Fig. 6 Surface-level distribution

Acknowledgements The authors are grateful for support from the National Science Foundation China (Grant No. U1860206 and No. 51725402, Beijing International Center of Advanced and Intelligent Manufacturing of High Quality Steel Materials (ICSM), Beijing Key Laboratory of Green Recycling and Extraction of Metals (GREM), and the High Quality Steel Consortium (HQSC) at the School of Metallurgical and Ecological Engineering at University of Science and Technology Beijing (USTB), China.

References

1. Khopkar AR, Kasat GR, Pandit AB, Ranade VV (2006) CFD simulation of mixing in tall gas-liquid stirred vessel: role of local flow patterns. Chem Eng Sci 61(9):2921–2929
2. Chew CM, Ristic RI, Reynolds GK, Ooi RC (2004) Characterisation of impeller driven and oscillatory mixing by spatial and temporal shear rate distributions. Chem Eng Sci 59(7):1557–1568
3. Khopkar AR, Rammohan AR, Ranade VV, Dudukovic MP (2005) Gas-liquid flow generated by a Rushton turbine in stirred vessel: CARPT/CT measurements and CFD simulations. Chem Eng Sci 60(8–9):2215–2229
4. Yianneskis M, Popiolek Z, Whitelaw J (1987) An experimental study of the steady and unsteady flow characteristics of stirred reactors. J Fluid Mech 175:537–555
5. Gao G, Wang M, Shi D, Kang Y (2019) Simulation of bubble behavior in a water physical model of an aluminum degassing ladle unit employing compound technique of rotary blowing and ultrasonic. Metall Mater Trans B Process Metall Mater Process Sci 50(4):1997–2005
6. Abreu-López D, Dutta A, Camacho-Martínez JL, Trápaga-Martínez G, Ramírez-Argáez MA (2018) Mass transfer study of a batch aluminum degassing ladle with multiple designs of rotating impellers. JOM 70(12):2958–2967

7. Abreu-López D, Amaro-Villeda A, Acosta-González FA, González-Rivera C, Ramírez-Argáez MA (2017) Effect of the impeller design on degasification kinetics using the impeller injector technique assisted by mathematical modeling. Metals 7(4):132
8. Maniruzzaman M, Makhlouf M (2002) Mathematical modeling and computer simulation of the rotating impeller particle flotation process: Part I. Fluid flow. Metall Mater Trans B Process Metall Mater Process Sci 33(2):297–303
9. Maniruzzaman M, Makhlouf M (2002) Mathematical modeling and computer simulation of the rotating impeller particle flotation process: Part II. Particle agglomeration and flotation. Metall Mater Trans B Process Metall Mater Process Sci 33(2):305–314
10. Ji J, Liang R, Feng Y, He J (2012) Study on the characteristics of fluid flow in stirring vessel of new type stirring of KR desulphurization. J Iron Steel Res Int s1:171–174

Part V
Molten Metal Processing

The Effect of Side Arcs on Current Distributions in a Submerged Arc Furnace for Silicon Production

Y. A. Tesfahunegn, T. Magnusson, M. Tangstad and G. Saevarsdottir

Abstract Recent electromagnetic modeling efforts for submerged arc furnace give an opportunity to improve understanding of the current distribution, which is critical for proper operation of furnaces for silicon production. This paper presents calculations of electric current distributions inside an industrial smelter. A 3D model has been developed in ANSYS Maxwell using the AC, eddy current solver. The modeled furnace operates with three-phase AC. In each phase, electrode, main arc, crater, crater wall, and side arcs that connect electrode and crater wall are taken into account. In this work, the number of side arcs is varied to study the effect on current distributions in different parts of the furnace, as well as skin and proximity effects in the electrodes. The system resistance, active and reactive power distributions are also investigated.

Keywords Current distribution · Current paths · Power distributions · Submerged arc furnace · Side arcs

Introduction

Numerical modeling has an increasing role in enhancing the understanding of current distribution inside submerged arc furnaces (SAFs), with the potential to improve furnace operation. As the existing technology is not able to measure the current distribution directly, metallurgists operate the smelters based on the analysis of limited data

Y. A. Tesfahunegn (✉) · G. Saevarsdottir
School of Science and Engineering, Reykjavik University, Menntavegur 1, 101 Reykjavik, Iceland
e-mail: yonatant@ru.is

G. Saevarsdottir
e-mail: gudrunsa@ru.is

T. Magnusson
United Silicon, Stakksbraut 9, 230 Reykjanesbæ, Iceland
e-mail: tm@silicon.is

M. Tangstad
Department of Materials Science and Engineering, NTNU, 7491 Trondheim, Norway
e-mail: merete.tangstad@ntnu.no

J. Lee et al. (eds.), *Materials Processing Fundamentals 2020*, The Minerals, Metals & Materials Series, https://doi.org/10.1007/978-3-030-36556-1_16

at hand. Recent dig-outs of industrial furnaces have expanded available information on location-dependent charge properties, which are the required inputs for numerical models. Improved material property data makes the developed mathematical models more reliable in predicting furnace behavior.

The geometry of the zones in a silicon furnace is dependent on the operation history, and hence, a number of different geometries, sizes, and compositions are possible in the various parts of the furnace. A report from excavations of industrial furnaces published by Tranell et al. [1] describes the different zones in a FeSi furnace. Myrhaug [2] reported similar features from a pilot-scale excavation operating around 150 kW. Tangstad et al. [3] published results from excavation of industrial furnaces, where the interior of the furnace is divided into zones depending on the materials and their degree of conversion. Mapping the material distribution gives a basis for electric modelling once the location-dependent physical properties of the charge materials such as electrical conductivity are quantified.

Complete numerical modeling of SAF requires electrical, chemical, thermal, and fluid flow considerations. In this paper, we only consider the electrical aspect, which needs electrical conductivity of the different parts of the furnace. Some previous work has been done to address this issue. Krokstad [4] outlined an experimental method to measure the electrical conductivity of silicon carbide, and Vangskåsen [5] looked in detail at the metal producing mechanisms. Mølnås [6] and Nell [7] have also published data on dig-out samples and material analysis that are relevant. These are some of the essential inputs necessary to set up a reasonably realistic modeling domain with correct physical properties to model the current and power distributions within a furnace, and this opens a unique opportunity to create a model which enables understanding of the current and power distributions in the furnace. These results can be used in the development of furnace control strategies that can enable improved silicon recovery and specific power consumption.

The recent developments of electrical numerical modeling show that several components of the furnace are included. Tesfahunegn et al. [8, 9] developed a 3D numerical furnace model that contains electrodes, main arcs, side arcs, crater wall, crater, and other parts. ANSYS Fluent was the modeling platform, using the electric potential solver. The authors showed that the current distribution effect without taking into account the main arc. As a continuation of their work, they have implemented a vector potential method using a user-defined function in ANSYS Fluent environment to calculate dynamic current distributions [10, 11]. That model only includes the electrodes and predicts skin and proximity effects. The same authors [12] extend their work on the model that has been developed in [8, 9] to study the alternating current and power distributions using eddy current solver (AC). Other researchers have developed different numerical models for SAF based on computational fluid dynamics (CFD) and finite element method (FEM). Herland et al. [13] studied proximity effects in large FeSi and FeMn furnaces using FEM. In their model, they have included different parts of the furnaces. Dhainaut [14] presented computations of electric field in SAF using CFD. The author showed the effect of contact resistance by studying the contact between two coke particles before dealing with a full-scale furnace. The furnace is partitioned in layers to consider different materials, and no assumption has been made on the current path. Bezuidenhout et al. [15] applied CFD on a three-phase electric smelting furnace to investigate the electrical aspects, thermal and flow behaviors. They showed relationships between electrode positions, current distribution, and slag electrical resistivity. Darmana et al. [16] developed a modeling concept applicable for SAFs using CFD that considers various physical phenomena such as thermodynamics, electricity, hydrodynamics, heat radiation, and chemical reactions. Wang et al. [17] investigated the thermal behavior inside three different electric furnaces for MgO production.

This paper studies the effects the number of side arcs has on current distributions in different parts of the furnace. It also presents skin and proximity effects in the electrodes. The system resistance, active and power distributions are well mapped as well. To accomplish the objective, a 3D model has been developed in ANSYS Maxwell [18] using eddy current solver. Electrode, main arc, crater, crater wall, and side arcs that connect the electrode and crater wall are taken into account for each phase. Other furnace parts such as carbon block, steel shell, and aluminum block are also incorporated.

The Process

In the silicon production process, quartz and carbon materials that are called charge are fed into a submerged arc furnace. Three electrodes penetrate the charge from above. Three-phase 50 Hz AC electric current is fed to the respective electrodes and passes through the contents of the furnace between them, heating the contents of the furnace through Ohmic heating in the process.

The overall reaction for producing silicon metal is:

$$SiO_2 + 2C = Si + 2CO(g) \tag{1}$$

This reaction, however, happens through a series of sub-reactions, changing the properties of the charge along the way as intermediary reaction products are formed. The current passes from the electrodes through the raw-material charge and an electric arc burning at the tip of the electrode. The arc, which consists of thermal plasma in the range of 10000–20000 K [19], provides heat for energy-consuming silicon-producing reaction (4), while the SiC-forming reaction and SiO(g) condensation reactions (2) and (3) take place at a lower temperature higher up in the furnace, see Schei et al. [20].

$$SiO(g) + 2C = SiC + CO(g) \tag{2}$$

$$2SiO(g) = Si + SiO_2 \tag{3}$$

$$SiO_2 + SiC = SiO(g) + CO(g) + Si(l) \tag{4}$$

It is essential for the silicon recovery in this process that there is a balance between the high-temperature reactions (4) and the low-temperature reactions (2) and (3). Therefore, it is necessary that sufficient heat is released in the lower arc region to drive reaction (4), while a certain part should be released in the raw-material charge to drive reactions (2) and (3). The stoichiometry of reaction (4) is affected by temperature, and the ratio of P_{SiO}/P_{CO} is decreased at a higher temperature, thus enabling a better silicon recovery. Therefore, sufficient arcing is important for good silicon recovery.

Computational Model

In this section, we describe the mathematical modeling, furnace geometry, material properties, mesh generation, and boundary conditions.

Mathematical Modeling

In this paper, we will focus on the electrical aspects of SAF. The 3D electrical model is developed in ANSYS Maxwell [18] using eddy current solver, which is suitable for low-frequency devices and phenomena. It solves sinusoidally varying magnetic fields in the frequency domain. The solution assumes frequency to be the same throughout the domain. Induced fields such as skin and current proximity effects are also considered. It is a quasi-static solver. To solve for the magnetic field, $\boldsymbol{H}$, the solver computes the values as follows [18]:

$$\nabla \times \left(\frac{1}{\sigma + j\omega\varepsilon} \nabla \times \boldsymbol{H} \right) = -j\omega\mu\boldsymbol{H} \tag{5}$$

where σ, ω, μ, and ε are electrical conductivity, circular frequency, magnetic permeability, and electrical permittivity. The magnetic permeability is typically given by $\mu = \mu_r\mu_0$, where $\mu_0 = 4\pi \times 10^{-7}$ (H/m) is the constant magnetic permeability of vacuum and μ_r (–) is the relative magnetic permeability. Once Eq. (5) is solved, the electric field ($\boldsymbol{E}$) and the electric current density ($\boldsymbol{J}$) are solved using Faraday's and Ampere's laws. Also, $\boldsymbol{J}$ and $\boldsymbol{E}$ are related by Ohm's law. The equation is solved using the finite element method.

Furnace Geometry and Material Properties

The computational domain is based on the actual design of a 32 MW industrial furnace with AC frequency of 50 Hz. A simplified schematic drawing of the furnace is shown in Fig. 1. Due to the proprietary right, the dimensions of the furnace are not indicated in the figure. Hence, the furnace is partitioned into different zones based on the material properties. Included in the modeling are the furnace lining, three electrodes, charge, molten material, three arcs below electrodes, side arcs, and three craters with crater walls made of carbides. We assume that several concentrated side arcs are distributed around the circumference of the electrode near the tip electrodes, and the circular distances between each side arc are held constant.

For brevity, a section of the furnace and one electrode are depicted in Fig. 1. Main and side arcs are included in each phase. The main arc is burning below each electrode, with an arc length of 10 cm and a diameter of 5 cm [21], and the side arcs which are shorter in length are connecting the respective phase crater wall to the side of electrodes. The curvature of the three crater walls is assumed to be a circular section with a diameter of 100 cm [21]. Each of the zones is assumed to have constant electrical conductivity. The conductivity of each zone is taken from various literature sources and summarized in Table 1.

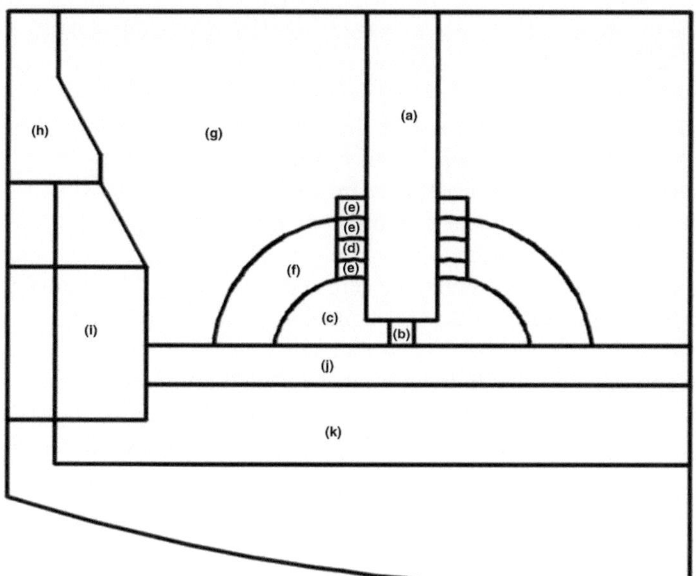

Fig. 1 Schematic of the industrial Silicon SAF with different zones **a** electrode, **b** arc, **c** crater, **d** side arc, **e** gap, **f** carbide, **g** charge, **h** alumina brick, **i** carbon block and carbide, **j** molten material, and **k** carbon block

Table 1 Electrical conductivity of different zones

Zones	Electrical conductivity (S/m)
Electrode [4]	225,000
Arc [21]	7000
Crater	$1e^{-14}$
Carbide [4]	400
Charge	0.15
Molten material [22]	1,388,900
Carbon block [4]	225,000
Alumina brick	$1e^{-14}$
Steel shell [13]	$6.3e^{+10}$

Mesh Generation and Boundary Conditions

Mesh generation is a crucial part of any computational method. It has a significant influence on the runtime and memory use of simulation, as well as the accuracy and stability of the solution. Since the eddy current solver utilizes an adaptive mesh refinement algorithm, the material volumes described in Fig. 1 were meshed according to the method. This type of meshing technique provides automated mesh refinement capability based on reported energy error in simulation.

The model boundary conditions were imposed based on the positions of the surfaces in the model. Two types of boundary conditions are required, i.e., the natural and Neumann. The natural boundary condition is used for interface between objects. It means that the $\boldsymbol{H}$ field is continuous across the boundary. The Neumann boundary condition is applied for an exterior boundary of the solution domain, and the $\boldsymbol{H}$ field is tangential to the boundary and flux cannot cross it. To impose appropriate boundary conditions on the $\boldsymbol{H}$ field, a large far-field around the furnace which is filled with air is created. The top surface of the three electrodes is excited by current with equivalent value of $I_{\text{rms}} = 99$ kA. The phase shift between electrodes is 120°.

Numerical Cases

The furnace described in Sect. 3.2 was used to determine the current and power distribution as well as the resistance of the system. We vary the number of side arcs from 1 to 14 with an increment of 1. Other parameters, such as the charge conductivity and main arcs, are kept constant. Hence, a total of 14 simulation cases have been performed.

For all cases, the simulations were performed by the adaptive meshing algorithm using energy error as a convergence criterion. The energy error set to 2%. For all cases, the initial mesh size is ~0.7e^{+06} elements, and the simulation is converged with the mesh size is of ~1.5e^{+06} elements. The simulation time per case on average is around 3 h.

Since the required results are not directly obtained from the simulation, we need to perform post-processing. The current is calculated from current density by integrating on the surface of interest. The active power density, p (W/m^3), given by $p = |\boldsymbol{J}|^2/2\sigma$, and the reactive power density q (W/m^3), given by $q = (\pi f/\mu)|\boldsymbol{B}|^2$. By integrating the respective power densities over different material domains and the entire furnace, we obtain active power, P (MW), and reactive power, Q (MW). After calculating the active and reactive powers, resistance (R) of the system is determined.

Figure 2 shows the resulting nonuniform current density on the cross section of an electrode for a selected number of side arcs (4, 7, 11, and 14). It can be observed that there are spots with a higher current density. These spots are side arc attachments, and as the number of side arcs increases, the area of the spots decreases. It also shows the skin effect is more pronounced for fewer side arcs, whereas the proximity effect is more noticeable for a higher number of side arcs.

Figure 3 shows the percentage of the total electrode current passing into the side arcs. The current is increasing in a nonlinear manner as the number of side arcs increases. As Fig. 3 shows, 22–37% of the total current pass through the side arcs into the crater wall as the side arc number changes from 4 to 14. It means that the heat generated below the bottom of the electrode reduces in that order.

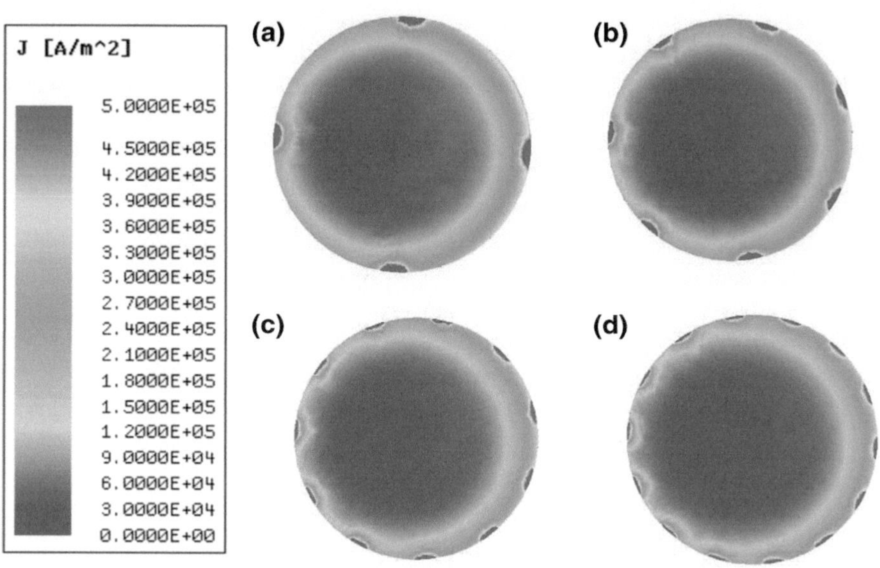

Fig. 2 Current distributions as the number of side arcs change: **a** 4, **b** 7, **c** 11, and **d** 14

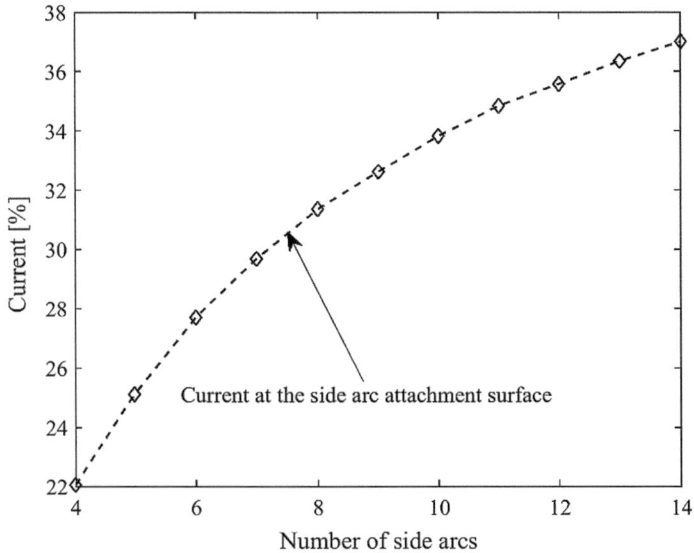

Fig. 3 Current distributions at arc attachment

Table 2 shows the active and reactive power percentage distributions in some selected zones with variations of the number of side arcs. The other zones are less than one percent; hence, they are not included. The majority of the active power is shared between the main arcs and crater wall. Approximately 6–8% is generated in the electrodes and about 2% in the side arcs. The active power decreases in the main arcs as the number side arcs increase and vice versa for the crater wall. The trends for the main arc, crater wall, and side arcs are shown in Fig. 4. The main contributors to the reactive power are the far-field, charge, and electrodes as depicted in Table 2. The other parts have contributions less than 2% and are left out from the table. Since we have not included other parts outside the furnace, such as bus bars and flexibles, the contributions of each zone could be smaller than the reported percentages. In general, the side arc effect on reactive power is negligible.

In Fig. 5, the normalized resistance of the system is designated, assuming an operating system resistance of 1.24 mΩ. Most furnaces are operated at a constant resistance. Overall the trend that can be observed is that increasing the number of side arcs will result in a reduction of the system resistance. Fewer than four side arcs, the resistance increases more than 5%. However, more than five side arcs, the resistance decreases from 1 to 27%.

Table 2 Active and reactive power distributions in percentage for different zones

Zones	Number of side arcs										
	4	5	6	7	8	9	10	11	12	13	14
Active power (%)											
Electrode	5.84	6.11	6.34	6.59	6.77	6.96	7.12	7.27	7.40	7.53	7.63
Main arcs	63.22	58.64	55.04	52.24	50.14	47.87	46.24	44.78	43.61	42.53	41.57
Side arcs	1.52	1.72	1.82	1.88	1.87	1.92	1.91	1.89	1.86	1.83	1.78
Crater wall	28.77	32.96	36.17	38.69	40.58	42.66	44.14	45.47	46.54	47.52	48.41
Reactive power (%)											
Electrode	8.45	8.33	8.33	8.29	8.26	8.24	8.22	8.20	8.18	8.17	8.14
Charge	25.26	25.24	25.45	25.50	25.42	25.59	25.62	25.64	25.67	25.69	25.65
Far-field	60.64	60.58	61.11	61.22	61.02	61.44	61.51	61.58	61.63	61.69	61.60

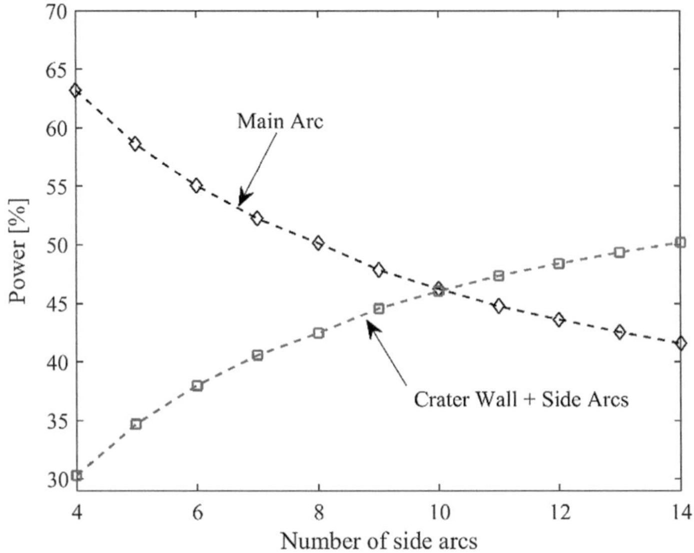

Fig. 4 Active power distributions in selected zones

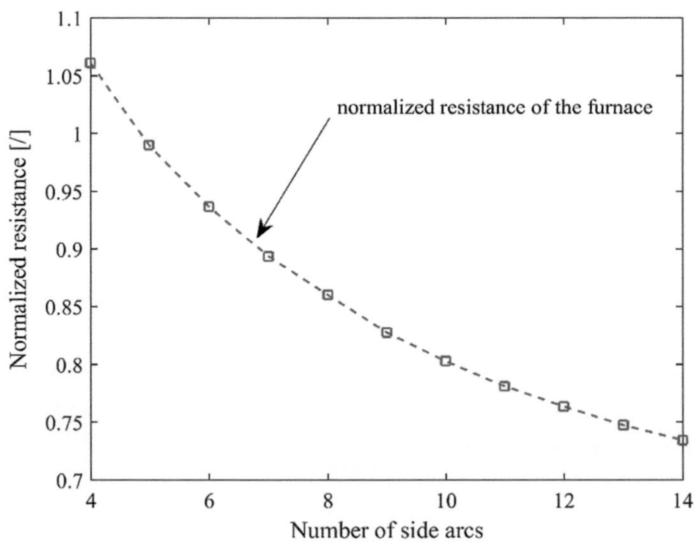

Fig. 5 Normalized resistance of the furnace

Conclusions

This paper presents the effects of number of side arcs on current and power distributions as well as resistance of an industrial submerged arc furnace that produces silicon. A 3D model has been developed in ANSYS Maxwell using eddy current solver. Electrodes, main arcs, crater, crater wall, and side arcs that connect electrode and crater wall are considered for each phase. In this paper, the current distributions at the arc attachment and the power distributions in different parts of the furnace are presented by varying the number of side arcs. The skin and proximity effects are affected by the number of side arcs. More current passed through the crater wall as the number of side arcs increases, and the opposite effect discerned in the main arc. It was observed that the resistance of the furnace decreased as the number of side arcs increased.

Acknowledgements The Icelandic Technology development fund is greatly acknowledged for their funding of this work.

References

1. Tranell G, Andersson M, Ringdalen E, Ostrovski O, Stenmo JJ (2010) Reaction zones in a FeSi75 furnace—results from an industrial excavation. Paper presented at the 12th international Ferro-alloys congress (INFACON XII), Helsinki, Finland, 6–9 June 2010
2. Myrhaug EH (2003) Non-fossil reduction materials in the silicon process -properties and behavior. Ph.D. thesis, NTNU
3. Tangstad M, Ksiazek M, Andersen J E (2014) Zones and materials in the Si furnace. Presented at the 12th silicon for the chemical and solar industry, Trondheim, Norway, 24–27 June 2014
4. Krokstad M (2014) Electrical resistivity of industrial SiC crusts. M.Sc. thesis, NTNU
5. Vangskåsen J (2012) Metal-producing mechanisms in the carbothermic silicon process. M.Sc. thesis, NTNU
6. Mølnås H (2010) Investigation of SiO condensate formation in the silicon process. Project report in TMT 4500, NTNU, Norway
7. Nell J, Joubert C (2013) Phase chemistry of digout samples from a ferrosilicon furnace. Paper presented at the 13th international Ferro-alloys congress (INFACON XIII), Almaty, Kazakhstan, 9–12 June 2013
8. Tesfahunegn YA, Magnusson T, Tangstad M, Saevarsdottir G (2018) Effect of electrode shape on the current distribution in submerged arc furnaces for silicon production—a modelling approach. J South Afr Inst Min Metall 118(6):595–600
9. Tesfahunegn YA, Magnusson T, Tangstad M, Saevarsdottir G (2018) Effect of carbide configuration on the current distribution in submerged arc furnaces for silicon production—a modelling approach. In: Nastac L, Pericleous K, Sabau A, Zhang L, Thomas B (eds) CFD modeling and simulation in materials processing 2018. The minerals, metals & materials series. Springer, Cham, pp 175–185
10. Tesfahunegn YA, Magnusson T, Tangstad M, Saevarsdottir G (2018) Dynamic current distribution in the electrodes of submerged arc furnace using scalar and vector potentials. In: Shi Y et al (eds) Computational science—ICCS 2018. ICCS 2018. Lecture notes in computer science, vol 10861. Springer, Cham, pp 518–527
11. Tesfahunegn YA, Magnusson T, Tangstad M, Saevarsdottir G (2018) The effect of frequency on current distributions inside submerged arc furnace. Paper presented at the IEEE MTT-S

international conference on numerical and electromagnetic and multiphysics modeling and optimization, Reykjavik, Iceland, 08–11 Aug 2018
12. Tesfahunegn YA, Magnusson T, Tangstad M, Saevarsdottir G (2019) Dynamic current and power distributions in a submerged arc furnace. In: Lambotte G, Lee J, Allanore A, Wagstaff S (eds) Materials processing fundamentals 2019. The minerals, metals & materials series. Springer, Cham
13. Herland EV, Sparta M, Halvorsen SA (2018) 3D models of proximity effects in large FeSi and FeMn furnaces. J South Afr Inst Min Metall 118(6):607–618
14. Dhainaut M (2004) Simulation of the electric field in a submerged arc furnace. Paper presented at the 10th international Ferro-alloys congress (INFACON X), Cape Town, South Africa, 1–4 Feb 2007
15. Bezuidenhout JJ, Eksteen JJ, Bardshaw SM (2009) Computational fluid dynamic modelling of an electric furnace used in the smelting of PGM containing concentrates. Miner Eng 22:995–1006. https://doi.org/10.1016/j.mineng.2009.03.009
16. Darmana D, Olsen JE, Tang K, Ringldalen E (2012) Modelling concept for submerged arc furnaces. Paper presented at the ninth international conference on CFD in the minerals and process industries CSIRO, Melbourne, Australia, 10–12 Dec
17. Wang Z, Fu Y, Wang N, Feng L (2014) 3D numerical simulation of electrical arc furnaces for the MgO production. J Mater Process Technol 214:2284–2291. https://doi.org/10.1016/j.jmatprotec.2014.04.033
18. Maxwell, ver. 18.0 (2018) ANSYS Inc., Southpointe, 275 Technology Drive, Canonsburg, PA 15317
19. Saevarsdottir GA, Bakken J, Sevastyanenko V, Liping G (2001) Arc simulation model for three-phase electro metallurgical furnaces. Paper presented at the 9th international Ferro-alloys congress (INFACON IX), Quebec City, Canada, 3–6 June 2001
20. Schei A, Tuset JK, Tveit H (1998) Production of high silicon alloys. Tapir Forlag, Trondheim
21. Sævarsdottir GA (2002) High current ac arcs in silicon and ferrosilicon furnaces. Ph.D. thesis, NTNU
22. Sasaki H, Ikari A, Terashima K, Kimura S (1995) Temperature dependence of the electrical resistivity of molten silicon. Jpn J Appl Phys. https://doi.org/10.1143/JJAP.34.3426

Empirical Study of Laser Cleaning of Rust, Paint, and Mill Scale from Steel Surface

Jean-Michaël Deschênes and Alex Fraser

Abstract Since the last decades, the demand for laser cleaning technology has significantly increased. Laser cleaning is the process by which contaminant or impurities are removed from a material surface by using a highly energetic focalised laser beam. Typical applications of laser cleaning are paint and coating stripping, mold cleaning, and rust removal. As many industrial applications require a short processing time, it is therefore a necessity to make the laser cleaning process as efficient as possible. In this paper, we compared the performance of two types of fiber laser sources for laser cleaning: single mode and multimode. As these two types of lasers offer specific characteristics, we perform an empirical study to compare their performances in terms of cleaning speed for typical cleaning applications. Emphasis is put into the optimization of the optical parameters that maximize the cleaning speed with minimal substrate damage. Advantages and limitations of each system are also discussed.

Keywords Laser cleaning · Rust · Paint · Mill scale

Introduction

In the past few years, laser cleaning has attracted more attention for various industrial applications. Main markets for laser cleaning are transportation and energy. In transportation, and especially automotive, high-volume manufacturing processes could benefit from fixed inline high-speed solutions. On the other hand, the maintenance market for boats, planes, and trains requires solutions that are portable and with long fiber reach. For energy, the need is mainly for maintenance of gas, hydro, or nuclear plants where highly rugged and versatile solutions with long fiber length are also necessary.

Laser cleaning was introduced to the market to solve the problems that conventional methods such as dry ice blasting, media blasting, or cleaning using chemical solvents are presented. Those traditional methods can be highly abrasive to substrate and produce a high amount of waste which can cause safety and environmental issues.

J.-M. Deschênes · A. Fraser (✉)
Laserax Inc., 2811 Avenue Watt, Québec, QC G1X 4S8, Canada
e-mail: afraser@laserax.com

J. Lee et al. (eds.), *Materials Processing Fundamentals 2020*, The Minerals, Metals & Materials Series, https://doi.org/10.1007/978-3-030-36556-1_17

Moreover, both dry ice blasting and chemical cleaning offer low levels of automation and often require a user to get involved. On the other hand, laser cleaning is a non-contact and non-abrasive process. It allows high repeatability and accuracy, does not produce hazardous waste, can be performed inline, is fast and efficient, and requires low maintenance. The main challenge of laser cleaning is to increase the contaminant removal rate since many applications require very fast cleaning.

To address this challenge, a more in-depth investigation of the various optical parameters, such as the line scan speed, the line spacing, the pulse energy, and the pulse diameter, was needed to achieve maximum laser cleaning efficiency. This paper presents the method used to optimize the laser cleaning process as well as the cleaning rate achieved for various contaminants. The contaminants studied are different types and thicknesses of paint, rust, and mill scale. The effect of the optical parameters on cleaning efficiency is also discussed.

Laser Ablation

Laser ablation is the removal of material from the substrate by direct absorption of laser energy. Each material is characterized by a physical parameter named the ablation threshold. The ablation threshold is the minimum energy intensity (fluence) which causes the sublimation of the material. It is this sublimation of material that is used to remove contaminants from a material surface. An interaction above the ablation threshold will affect the material permanently. When the intensity is below the ablation threshold, the absorbed energy will be converted to heat. The heat will then flow through the material through conduction. Rust and various contaminants each have an intrinsic ablation threshold. If the contaminant's ablation threshold is lower than the substrate's ablation threshold, it is then possible to remove the contaminant without damaging the surface of the substrate. Figure 1 schematizes the laser interaction during the contaminant removal process. The ablation threshold depends on the physical properties of the material to ablate and on laser parameters such as wavelength, pulse duration, and temperature. Typical fluence thresholds are between 1 and 10 J/cm^2 for metals, between 0.5 and 2 J/cm^2 for inorganic insulators, and between 0.1 and 1 J/cm^2 for organic materials [1].

However even if the density is below the threshold, substrate damage is still possible. As mentioned above, the energy will be converted into heat which will cause an increase in temperature. This increase of temperature generally leads to a decrease of the ablation threshold. Therefore, when multiple pulses are involved and when the repetition rate is high enough, the heat accumulation resulting from subsequent pulses may cause melting zones and even material ablation if the reduced threshold became lower than the pulse fluence. Typically, this can happen when the line scan speed is too slow, resulting in excessive pulse overlap. It is important to consider this kind of phenomenon since we want to limit the substrate's damage in cleaning applications.

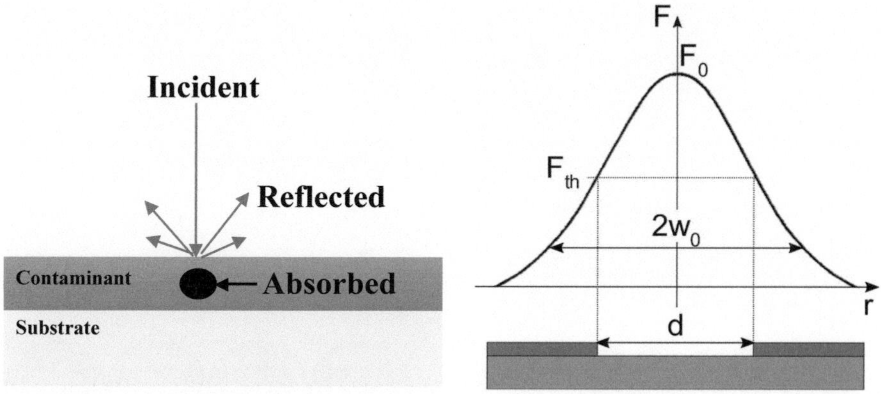

Fig. 1 Representation of the laser beam interaction with the material (left) and the Gaussian-like pulse energy distribution (right); F_{th}, F_0, $2w_0$, and d represent the fluence threshold, the peak fluence of the pulse, the pulse diameter, and the diameter of the cleaned ablated area, respectively

Surface Irradiation and Pulse Distribution

Pulsed lasers are generally used for laser cleaning. Subsequent pulses are directed at the surface at a given repetition rate, typically in the order of some hundred of kHz. The predominant methods to control the pulse distribution are beam steering by rotating polygon mirrors or galvanometer-mounted scanning mirror. Laser beams typically scan the surface by zigzags as illustrated in Fig. 2. The horizontal pulse overlap is characterized by the scanning speed and the pulse repetition rate. The repetition rate dictates the time interval between two subsequent pulses. The pulse overlap is then related to the line scanning speed v_x and to the laser repetition rate R by the relation $\Delta x = v_x/R$. The vertical overlap is characterized directly by the line spacing between two successive lines. Different combinations of scan speed and line spacing allow to obtain different operating regimes.

Experimental

Sample Preparation

Different types and thicknesses of paint were applied on Q-PANEL standardized steel substrate. Those plates were chosen since they offer a consistent and uniform surface finish. Table 1 summarizes the paint used. The paint thicknesses were measured using the coating thickness gauge PCE-CT 65.

The rust samples were prepared by leaving unpainted standardized steel substrates to external conditions. Those were left for different time intervals to obtain varying degrees of oxidation as illustrated in Fig. 3. The mill scale was obtained from various unprocessed steel parts.

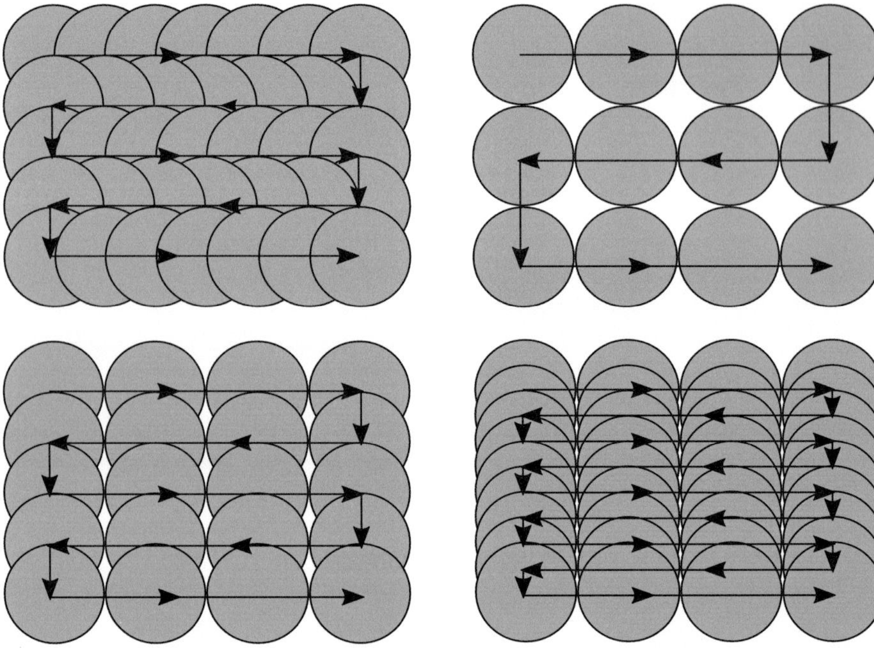

Fig. 2 Different possible scanning regimes

Table 1 Colors and thicknesses of the different types of paint tested

Paint type	Color	Thickness (μm)
Epoxy primer	Gray	30–180
Epoxy primer (powder)	Gray	80–145
Zinc-rich primer epoxy (powder)	Gray	60–70
Urethane primer	White	30–70
Polyester (powder)	Gray	30–120
Polyester (liquid)	Beige	15–35

Fig. 3 Various degrees of oxidation obtained for the rusted samples categorized as grade A (left), grade B (middle), and grade C (right)

Experimental Setup

The experiments were performed with a 100 and 500 W single mode laser (Laserax, LXQ series) as well as a 1 and 2 kW multimode laser (IPG Photonics, YLPN series). Those are all ytterbium-doped nanosecond Q-switched pulsed fiber lasers. The characteristics are presented in Table 2. The single mode laser beam has a round Gaussian-like energy distribution as illustrated in Fig. 1, while the multimode laser beam has a square top-hat uniform energy distribution. It is possible to modify the focused beam diameter by changing the optics and lens.

The experimental setups are schematized in Fig. 4. The scanning head contains optics like collimators and beam expanders as well as the galvanometric scanning system. The lenses used are F-theta lenses which are optimized for a flat field in the image plane and low distortion. The samples are placed on an elevation table allowing a precise positioning of the focal plane.

In order to find the optimum laser parameters for an efficient cleaning process, various zones of the samples were irradiated with different optical parameters. The different zones are arranged in an array as illustrated in Fig. 5 in order to obtain different operating regimes. For each set of parameters, the number of laser passes to completely clean the substrate is counted. The cleaning speed is determined by the scan speed, the line spacing, and the number of laser passes required to completely remove the contaminant as well as the scanner delays. Knowing the cleaning speed for each set of parameters, it is then possible to find the optimum one for each contaminant. The cleaning speed is determined by dividing the area cleaned by the total time for all laser cleaning passes. Similar arrays were used for different achievable beam spot sizes. Increasing the laser repetition rate above its nominal frequency has the effect of reducing the pulse energy. For thin contamination that is very easily removed in a single pass, it has been shown in previous experiments that this increase of the repetition rate allows to increase the scan speed and therefore the overall cleaning speed. For this study, we focus on contamination types for which nominal repetition rate and maximum pulse energy give the highest cleaning speed, so we kept these two parameters fixed to the nominal values shown in Table 2 unless the contaminant can be removed in a single pass like light rust.

The sample surfaces were visually inspected with an optical microscope to make sure that the contaminants were completely removed and that the substrate sustains no noteworthy damage.

Results

Paints

It was observed that the cleaning speed was inversely proportional to the paint thickness. This is not very surprising since a layer of paint twice as thick is expected to

Table 2 Pulsed fiber laser characteristics

	LXQ-100-2D	LXQ-500-2D	YLPN-100-70x100-1000	YLPN-150-100-2000
Laser type	Single mode	Single mode	Multimode	Multimode
Max. pulse energy	1 mJ	1 mJ	100 mJ	150 mJ
Max. average power	100 W	500 W	1000 W	2000 W
Pulse repetition rate (PRR)	5–500 kHz	50–1000 kHz	2–50 kHz	2–50 kHz
Nominal PRR	100 kHz	500 kHz	10 kHz	13.33 kHz
Pulse duration	100 ns	100 ns	70–100 ns	100 ns
Wavelength	1064 ± 2 nm	1064 ± 2 nm	1064 ± 2 nm	1064 ± 2 nm
Process fiber core	–	–	400 μm (square)	600 μm (square)
Collimator	–	–	85 mm, 100 mm, 120 mm	85 mm, 100 mm, 120 mm
Lens	255 mm, 420 mm	255 mm, 420 mm	160 mm, 255 mm	160 mm, 255 mm
Beam shape	Gaussian-like	Gaussian-like	Top-hat	Top-hat
Focused pulsed diameter	~50 μm/80 μm/120 μm/180 μm (@ $1/e^2$)	~70 μm/120 μm/180 μm (@ $1/e^2$)	533 μm/850 μm/1200 μm	800 μm/1275 μm/1800 μm

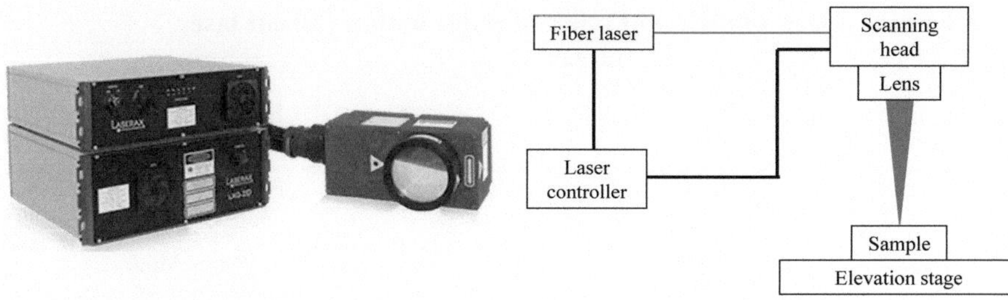

Fig. 4 Image of the 100 W single mode laser (left) and schematic of the experimental setup (right)

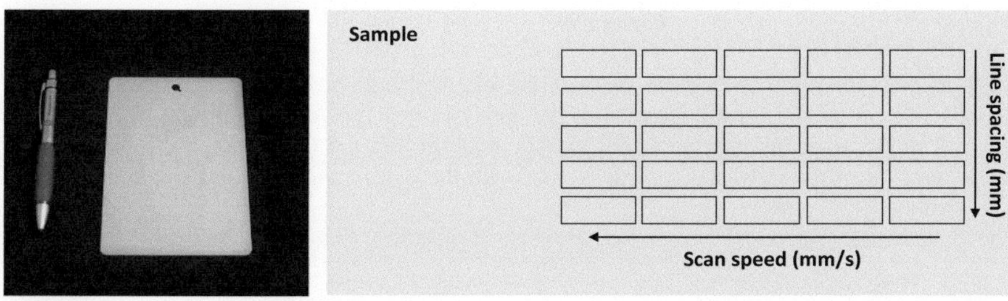

Fig. 5 Steel substrate used for the paint and rust samples (left) and matrix of tests with varying line spacing and scan speed (right)

take twice as long to remove. Figure 6 shows the typical shape of the relationship curve between the cleaning speed and the paint thickness. The cleaning speed is also linear with the laser power as shown in Fig. 7. It was observed that multimode laser has a slightly higher cleaning efficiency in general. Table 3 shows the maximal removal rate obtained for each type of tested paint. The cleaning speed is expressed in

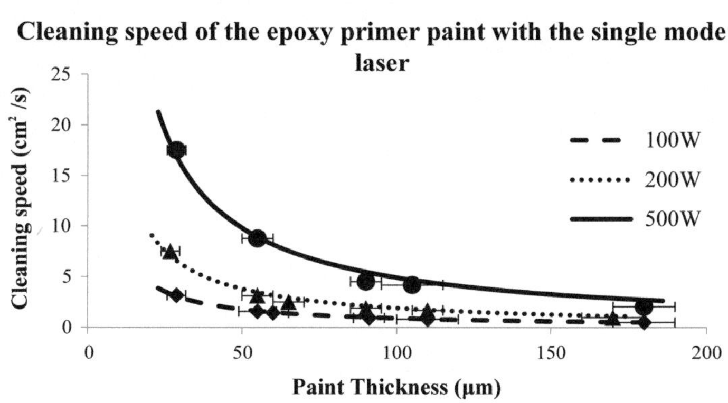

Fig. 6 Typical relation between the cleaning speed and the paint thickness; the removal rate is 89, 190, and 489 $cm^2/s\ \mu m$ with the 100, 200, and 500 W single mode lasers, respectively

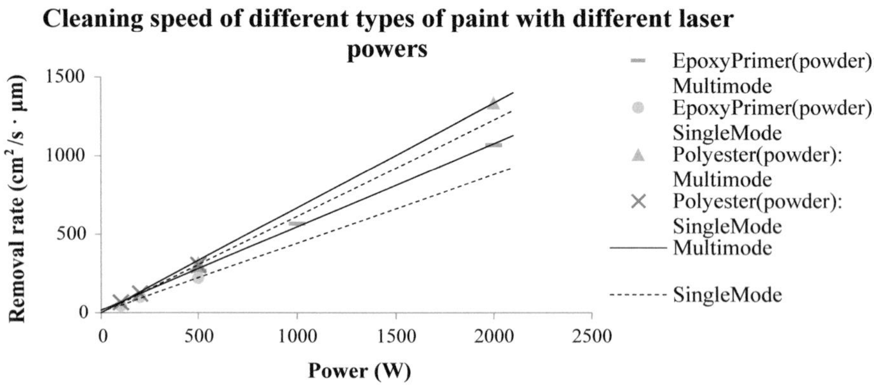

Fig. 7 Typical relation between the removal rate and the laser power; the relationships are 0.667 and 0.613 cm^2/s μm/W for the removal of the epoxy primer (powder) paint with the multimode laser and single mode laser, respectively, and the relationships are 0.540 and 0.452 cm^2/s μm/W for the removal of the polyester (powder) paint with the multimode laser and single mode laser, respectively

Table 3 Removal rate (cm^2/s μm) of each tested type of paint with different laser types and power

Paint type	Paint removal rate (cm^2/s μm)					
	100 W: single mode	200 W: single mode	500 W: single mode	500 W: multimode	1 kW: multimode	2 kW: multimode
Epoxy primer	89	190	489	–	739	1507
Epoxy primer (powder)	64	123	305	320	–	1337
Zinc-rich primer epoxy (powder)	54	121	253	–	–	898
Urethane primer	124	327	670	–	1878	4663
Polyester (powder)	43	101	222	267	567	1068
Polyester (liquid)	50	131	311	272	–	1512

cm^2/s μm which represents the volumetric removal rate of paint. So, a 100-μm-thick paint with a 1000 cm^2/s μm removal rate will be cleaning at a speed of 10 cm^2/s.

Table 4 Cleaning speed (cm^2/s) of the different rust grades with the single mode laser

Rust	Cleaning speed (cm^2/s)			
	100 W: single mode	200 W: single mode	500 W: single mode	2 kW: multimode
Grade A	12–21	25–42	50–75	324
Grade B	4–12	10–25	15–50	204
Grade C	1–4	2–10	5–15	102

Rust

The rusted surfaces of the samples were often non-uniform, and unlike the paint, there was no reliable and quick method to measure the rust layer thickness. Therefore, it was difficult to know if the removal rate of the rust was faster because the optical parameters were more efficient or because the rust layer was just thinner. For this reason, the rusted samples were categorized into different grades. Figure 3 shows the typical degree of oxidation for each rust grade. Table 4 shows the achievable cleaning speeds with the single mode and multimode laser. It was easy to remove the rust when it was compact and uniform. Bulk erosion, spikes, or flakes were harder to remove.

It was possible to measure the mill scale thickness with a surface profiler (model DEKTAK 150). Once a portion of the mill scale was removed, its thickness was measured by comparing the bare steel level with the mill scale level. Figure 8 shows the achievable cleaning speed for a 15-μm-thick layer of mill scale. We see that the multimode laser is almost twice as efficient as the single mode laser to remove the mill scale. Figure 9 shows typical microscope images of contaminants removed by laser from a steel substrate. Overall, it was possible to achieve faster cleaning speed than what is reported in the literature [2, 3].

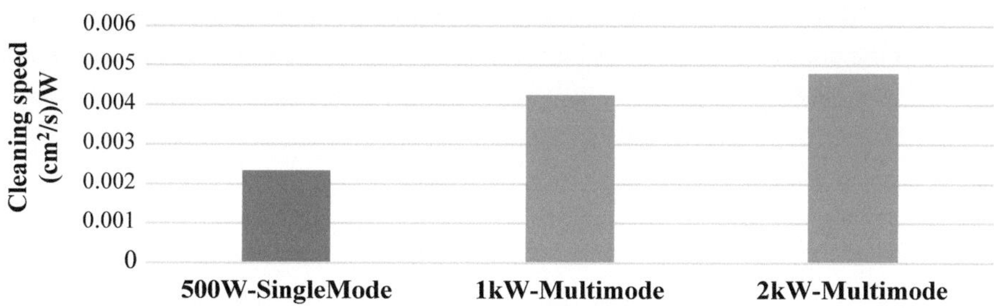

Fig. 8 Cleaning speed of a 15-μm-thick layer of mill scale. The corresponding cleaning speeds are 0.0023, 0.0043 (cm^2/s)/W, and 0.0048 for the 500 W single mode laser, the 1 kW multimode laser, and the 2 kW multimode laser, respectively

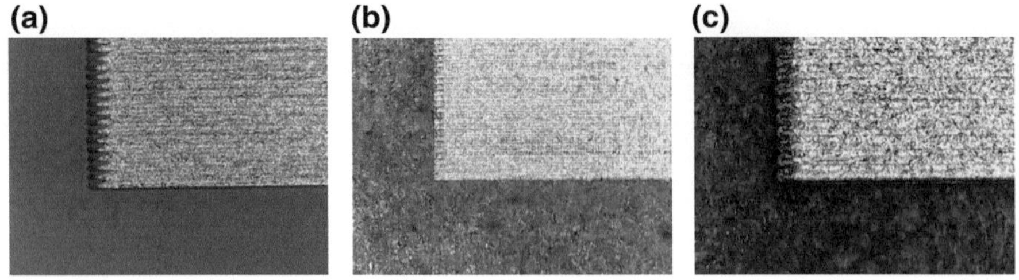

Fig. 9 Laser cleaning of epoxy gray paint (left) grade B rust (middle) and mill scale (right)

Analysis

We observed that the multimode laser is slightly more effective than the single mode laser for the paint removal and is almost twice as effective for the mill scale and rust removal. The energy distribution of the pulse probably explains this behavior. As listed in Table 2, the single mode fiber laser has a Gaussian-like pulse energy distribution while the multimode laser has a uniform top-hat pulse energy distribution. In Fig. 1, we see that to cause material ablation, the energy density must be above the ablation threshold. For a Gaussian-like distribution, the energy density is high at the pulse center and diminishes as we move away from the center. Therefore, there is a point where the energy density is not high enough to cause material removal. The absorbed energy is rather converted into heat. The conversion of some of the energy into heat reduces the material removal rate efficiency. On the other hand, with a top-hat energy distribution, almost all the pulse energy contributes to the material ablation and there is very little portion of the energy lost and converted into heat. This effect is expected to be more noticeable for material with a higher ablation threshold. This is exactly what has been observed. Since paint has a very low ablation threshold, there is very little energy lost in heat and the removal rate efficiency is similar for single mode and multimode lasers as shown in Fig. 10. However, in the case of mill scale, since it has a much higher ablation threshold, a bigger portion of the energy is lost for the single mode laser. Therefore, the top-hat pulsed multimode laser has a greater removal efficiency for the mill scale.

Fig. 10 Portion of the pulse energy lost as heat for a different energy distribution and ablation threshold

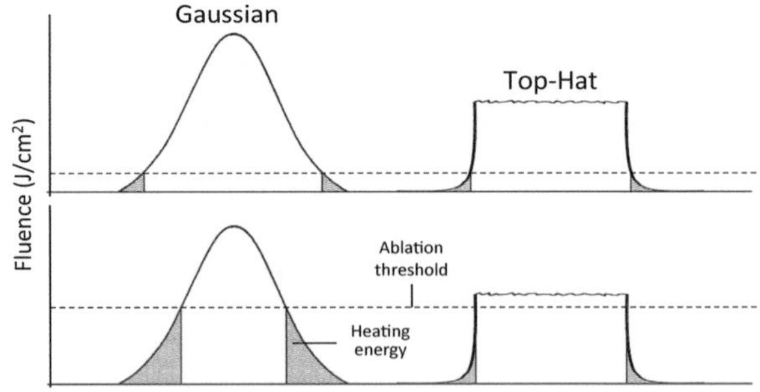

It is important to note that laser ablation is not the only mechanism that can contribute to the material removal. It was sometimes observed that parts of the material were ejected from the substrate without being vaporized by the laser. This probably results from the formation of shockwave in the material. Those shockwaves can be created by a process called spallation or by the rapid heating and cooling of the material [4]. With high enough amplitude, those shockwaves create shear stress large enough to crack and eject the material. This was observed for thinner paint like the liquid polyester paint and the urethane paint. Such removal mechanism possibly explains why those paint types do not follow the linear relationship between the removal rate and the laser power as well.

Another important aspect to discuss is the pulse fluence, which is the energy density in J/cm^2. For a fixed pulse energy, reducing the pulse diameter will increase the laser fluence. As illustrated in Fig. 11, the ablation depth per pulse (μm/pulse) increases for higher fluence until a saturation domain is reached [5]. In the saturation domain, increasing the pulse fluence does not lead to a much higher removal depth per pulse. The optimal fluence is therefore somewhere at the top of the linear ablation domain. If the pulse fluence is in the saturation domain, it is advantageous to increase the pulse diameter (reducing the fluence) since it will remove a similar depth of material but over a wider area. This was also observed for the paint removal with the multimode laser which was more efficient with a bigger spot size. On the other hand, if the pulse fluence is in the beginning of the linear ablation domain, it is advantageous to increase the fluence (reducing the pulse diameter) to enhance the material removal efficiency. This was also observed for the mill scale removal which was more efficient with a smaller spot size.

Another important aspect to note is the pulse dimension. As listed in Table 2, multimode laser has a much larger spot typically of about 1 mm while the single mode laser has a much smaller pulse of about 0.1 mm. Having a smaller pulse allows material removal in a much precise and selective manner. It is possible with the single mode laser, for example, to "mark" a logo or a Data Matrix code by removing paint or other material from its substrate. Since it also has a high fluence, it is possible with

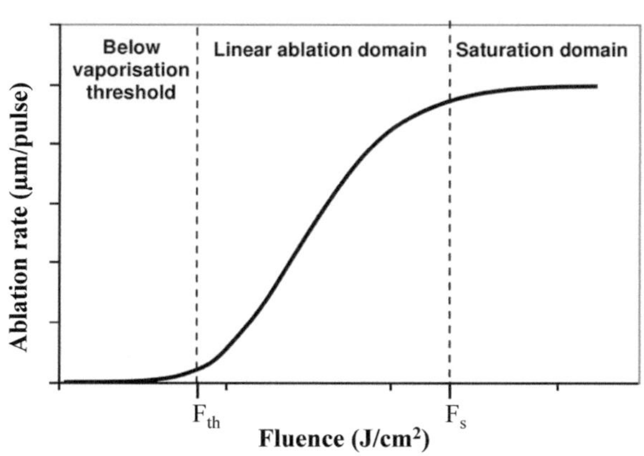

Fig. 11 Typical relation between the ablation rate and the pulse fluence [5]

Table 5 Summary of the advantages and limitations of both laser systems

	Single mode laser	Multimode laser
Advantages	• Small spot allows precise material removal • Can also do laser marking and texturing	• Can reach high average power • More efficient material removal rate due to its top-hat beam energy distribution
Limitations	• Reach lower average power • High-performance scanning system required because of the high pulse repetition rate • Slightly less efficient material removal rate due to its Gaussian-like beam energy distribution	• Bigger spot restricts the precision of the material removal • Cannot do laser marking

the single mode laser to do laser marking on metal. With its bigger spot size and lower fluence, the multimode laser cannot do laser marking and is more limited to laser cleaning. On the other hand, pulsed multimode laser can reach much higher laser power (typically up to 5 kW) than the single mode laser (typically up to 0.5–1 kW). Another important aspect to consider is the scanning system required for both laser systems. Since the single mode laser has low energy pulses (1 mJ), the high power is achieved by having a very high pulse repetition rate. The 500 W single mode laser has a 500 kHz nominal repetition rate, while the 2 kW laser has a 13.33 kHz nominal repetition rate. Therefore, the single mode laser scanning system must reach very high scanning speed in order to correctly space the pulse as illustrated in Fig. 2. Typically, the 500 W single mode laser must reach scanning speed of 50–75 m/s while the multimode laser scanning systems must reach scanning speed of 20–36 m/s. Table 5 summarizes the advantages and limitations of both laser systems.

Conclusion

This study has shown that it is possible to remove paint, rust, and mill scale from a steel surface with mid- to high-power nanosecond pulsed fiber laser without damaging the underlying substrate. The beam energy distribution has a great impact on the ablation efficiency. A top-hat distribution is preferred since it maximizes the amount of energy that leads to laser ablation and minimizes the energy loss converted into heat. Since single mode fiber laser has a Gaussian-like distribution, it is therefore recommended to use a beam shaper to convert the Gaussian energy curve to a flat-top distribution for cleaning application in order to maximize the ablation efficiency.

Laser contaminant removal can be considered a valid alternative to traditional technologies like sandblasting or chemical usage. Rust can be easily removed with lower power laser (100 W), while a higher power laser is preferred for paint removal to increase the process speed. Laser technologies can therefore be used for paint stripping of high-quality metal piece such as aeronautic or automotive components.

While multimode laser is more limited to cleaning applications, single mode laser can also be used for marking and surface treatment. Laser surface processing can also be a key element in several large-scale industrial manufacturing operations. As lasers continue to be developed, laser surface processing will continue to improve the performance of materials in existing applications and will open the door to new materials and novel applications that would not be possible without these unique processing capabilities.

References

1. Bäuerle D (2013) Laser processing and chemistry. Springer Science & Business Media
2. Brygo F, Dutouquet C, Le Guern F, Oltra R, Semerok A, Weulersse JM (2006) Laser fluence, repetition rate and pulse duration effects on paint ablation. Appl Surf Sci 252(6):2131–2138
3. Daurelio G, Chita G, Cinquepalmi M (1999) Laser surface cleaning, de-rusting, de-painting and de-oxidizing. Appl Phys A 69(1):S543–S546
4. Watkins KG (2000, Feb). Mechanisms of laser cleaning. In: High-power lasers in manufacturing, vol 3888. International Society for Optics and Photonics, pp 165–174
5. Siano S (2007) Principles of laser cleaning in conservation. In: Handbook on the use of lasers in conservation and conservation science, vol 26. http://www.science4heritage.org/COSTG7/booklet/chapters/prin_cle.htm. Accessed 9 Sept 2019

Part VI
Poster Session

Control Center Segregation in Continuously Cast GCr15 Bloom by Optimization of Solidification Structure

Hanghang An, Yanping Bao, Min Wang and Quan Yang

Abstract Complex electromagnetic stirring technique (M+F-EMS) and low superheat pouring can enlarge the equiaxed crystal zone and fine equiaxed grains, which are beneficial to improve center segregation of continuously cast high-carbon steel blooms. In this work, a cellular automaton-finite element (CAFE) coupling model has been established to predict the solidification process and analyze solidification structure evolution of continuously cast GCr15 bearing steel bloom with 220 mm × 260 mm, in which electromagnetic stirring was taken into consideration. The influences of M-EMS, superheat and casting speed on the solidification process and structure were numerically investigated, in addition, the compactness degree in the central equiaxed crystal zone was also evaluated. The results demonstrate that casting speed has the obvious effect on solidification end and central solid fraction in strand compared with superheat, which is closely related to F-EMS implement position. M-EMS and low superheat exhibit a significant increment on the central equiaxed crystal ratio and the compactness degree in the central equiaxed crystal zone in the bloom, which is closely related to the center solidification time. The industrial trials' results show that the central equiaxed crystal ratio and center segregation can be improved under optimized lower superheat and higher casting speed with M+F-EMS.

Keywords GCr15 bloom · CAFE coupled model · Solidification structure · Compactness degree · Center segregation

H. An (✉)
Collaborative Innovation Center of Steel Technology, University of Science and Technology Beijing, Beijing 100083, People's Republic of China
e-mail: hanghangan_ustb@outlook.com

Y. Bao · M. Wang
State Key Laboratory of Advanced Metallurgy, University of Science and Technology Beijing, Beijing 100083, People's Republic of China

Q. Yang
National Engineering Research Center of Flat Rolling Equipment, University of Science and Technology Beijing, Beijing 100083, People's Republic of China

J. Lee et al. (eds.), *Materials Processing Fundamentals 2020*, The Minerals, Metals & Materials Series, https://doi.org/10.1007/978-3-030-36556-1_18

Introduction

Macro segregation has been a major internal quality problem of high-carbon steel bloom due to wider mushy zone and longer solidification time during the solidification process [1, 2]. It is inherited in the following process and abnormal microstructure including network carbides and banded structure occur in roll bar, which seriously affect mechanical properties and fatigue performance of the final product [2]. Center segregation is chiefly caused by the interdendritic liquid rich in impurity element flowing towards the centerline owing to solidification shrinkage of the liquid pool in the region of the final solidification [3, 4], while center segregation is closely related to the solidification structure [5]. Center-equiaxed crystal can effectively prevent the formation of solidification bridges and redistribute the residual impure liquid, therefore, the center segregation decreases as the center-equiaxed crystal ratio and fine equiaxed grains increase [1, 6]. Center-equiaxed crystal is related to parameters in continuous casting process. Complex electromagnetic stirring technique (M+F-EMS) and low superheat pouring have proven to be effective ways to reduce macro segregation by a lot of studies [7, 8]. M-EMS and low superheat can provide wide center-equiaxed crystal zone, and F-EMS can refine center-equiaxed crystal. However, F-EMS implemented position depends on the center solid fraction in strand, which is related to the solidification parameters such as casting speed, secondary cooling, or superheat.

Desired solidification structure can be obtained by optimization of the solidification parameters. Over the decades, the effects of the solidification parameters with M+F-EMS on the solidification structure for high-carbon steel blooms have been investigated experimentally by a multitude of industrial trials [9, 10], which is costly. However, the regularity is not obvious owing to uncontrollable other factors in actual production, in addition, the solidification structure is difficult to evaluate and measure accurately by hot acid-etched method owing to existing serious center porosity and shrinkage cavity, and the complicated branches and interactions and the huge number of grains usually appear in the cross-section of a bloom. In comparison, it is more appropriate to investigate the evolution of solidification structure by numerical simulation. At present, the main models include deterministic model [11], stochastic model [12], and phase field model [13]. Deterministic models and phase field models are difficult to employ for simulating solidification structure during the continuous casting process owing to unique characteristics, while CAFE models, which is a stochastic model, has been firstly established by Rappaz and Gandin [14] and developed rapidly over the last couple of decades. In recent years, only the CAFE model has been successfully applied in investigating the solidification structure during the continuous casting process by several researchers [5, 15–22]. However, there are a few models for the solidification structure of continuously cast steel billets and blooms without considering the result of heat transfer models [5, 15, 17–19]. This research focuses on the evolution of the solidification structure during the continuous casting process for billet and bloom without considering the compactness degree of the equiaxed crystal zone, which greatly influenced macro segregation formation [17–20]. Complex electromagnetic stirring technique (M+F-EMS) has been widely applied in controlling center segregation of continuously cast high-carbon steel billets and blooms and parameters are optimized to control solidification structure and macro segregation by CAFE model, but the effect of M-EMS is only considered without regarding that of F-EMS [18, 22].

Although those researches provided a lot of useful information for practical production, in order to reduce macro segregation in the bloom, continuous casting of high-carbon steel with M+F-EMS, few attempts were made to optimize the solidification structure and the degree of compactness of the equiaxed crystal zone combined with F-EMS implemented position by theoretical model and experiments. The aim of the present work is to combine theoretical model with experiments to optimize major solidification parameters in order to reduce center segregation of GCr15 steel in the bloom continuous casting with combination of M+F-EMS. In the present study, a cellular automaton-finite element (CAFE) coupling model was established to predict the solidification process and analyze solidification structure evolution of continuous casting GCr15 bearing steel bloom with 220 mm × 260 mm, in which electromagnetic stirring was taken into consideration. The model was validated by surface temperature measurement, nail shooting test, and macro etch experiments, respectively. In addition, the effect of process parameters on solidification behavior, solidification structure, and the compactness degree of the equiaxed crystal zone has been investigated. In addition, combined with the effect of process parameters on F-EMS implemented position, process parameters were determined. Finally, industrial trials were carried out to validate the optimized parameters.

Model Description and Validation

In the present model, the cellular automaton method (CA) is combined with finite element method (FE) during the continuous casting process. The CAFE model is applied in the simulation of the solidification process and structure, which mainly includes a heat transfer model, a heterogeneous nucleation model, and a dendrite tip growth kinetics model. Temperature field during the solidification process was firstly solved by means of the heat transfer model, then the heterogeneous nucleation model and dendrite tip growth kinetics model were used to simulate the solidification structure based on it.

Heat Transfer Model Description

Governing Equations

Based on the slice moving method, a two-dimensional mathematical model of heat transfer is established by the finite element method to simulate the bloom solidification process in strand under the steady casting conditions. The governing equation is expressed as Eq. (1).

$$\rho C_{\mathrm{p}}\frac{\partial T}{\partial t} = \frac{\partial}{\partial x}\left(\lambda\frac{\partial T}{\partial x}\right) + \frac{\partial}{\partial y}\left(\lambda\frac{\partial T}{\partial y}\right) \tag{1}$$

where T is local temperature, °C; λ is thermal conductivity, W m^{-1} °C^{-1}; C_{p} is specific heat, J kg^{-1} °C^{-1}; ρ is the density, kg m^{-3}.

The effective specific heat method is used to deal with latent heat during the solidification process in this model, as shown in Eq. (2). The solid fraction is expressed by Eq. (3).

$$C_{\mathrm{eff}} = \frac{C_{\mathrm{s}} + C_{\mathrm{l}}}{2} + \frac{L_{\mathrm{f}}}{T_{\mathrm{l}} - T_{\mathrm{s}}} \tag{2}$$

$$f_{\mathrm{s}} = \begin{cases} 0, & T \geq T_{\mathrm{l}}; \\ \frac{T_{\mathrm{l}}-T}{T_{\mathrm{l}}-T_{\mathrm{s}}}, & T_{\mathrm{s}} < T < T_{\mathrm{l}}; \\ 1, & T \leq T_{\mathrm{s}}. \end{cases} \tag{3}$$

where C_{eff} is effective specific heat, J kg^{-1} °C^{-1}; C_{s} and C_{l} are solid and liquid specific heat capacity, J kg^{-1} °C^{-1}; L_{f} is latent heat, kJ kg^{-1}; T_{l} and T_{s} are liquidus temperature and solidus temperature, °C; f_{s} is solid fraction.

Initial and Boundary Conditions

In the present study, a five-strand bow type bloom caster with a section size of 220 mm × 260 mm is selected as the research object, and main technical parameters of the caster are shown in Table 1. A 220 mm × 260 mm × 13 mm slice model is developed and meshed with ProCAST software. By comprehensive consideration of calculation precision and calculation time, mesh and step are used together by heat transfer model and solidification structure model. The total nodes and cells in the meshed model are 251,002 and 250,000, respectively, as shown in Fig. 1.

The initial condition and boundary conditions have been described in detail in author's research [2]. The model is based on the hypothesis of heat transfer in the solid and liquid steel by conduction only and that convection at the solid–liquid interface is neglected. In addition, the heat transfer coefficient in foot roller zone and other zones can be expressed as Eqs. (4) and (5), respectively.

Table 1 Main technical parameters of caster

Parameters	Value
Caster type	Curved
Sectional dimension (mm^2)	220 × 260
Caster radius (m)	10
Metallurgical length (m)	30
Mold total length (m)	0.9
Mold effective length (m)	0.85
Length of secondary cooling zone 1–3 (m)	0.33, 1.75, 2.8
Installed location of F-EMS (m)	9.5

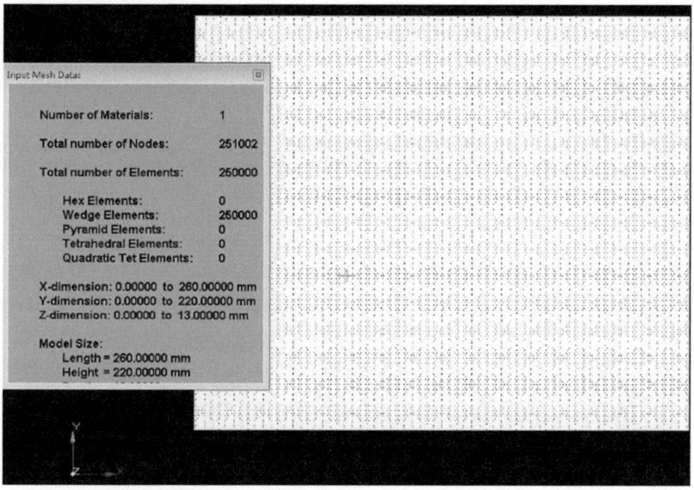

Fig. 1 Meshed slice of heat transfer model of the continuous casting bloom

$$h = 43W^{0.556} \tag{4}$$

$$h = 141 + 12.5W^{0.815} \tag{5}$$

where h is heat transfer coefficient in the secondary cooling zone, W m^{-2} °C^{-1}; W is water flow rate of the secondary cooling zone, L m^{-2} min^{-1}.

Material Parameters

In order to obtain more accurate material properties of GCr15 steel between solidus and liquidus temperatures range, ProCAST software is used in the research. Typical

Table 2 Values of typical chemical composition for bearing steel GCr15 (wt%)

C	Si	Mn	P	S	Al	Cr	Ni	Mo
0.97	0.2	0.35	0.01	0.005	0.02	1.37	0.1	0.05

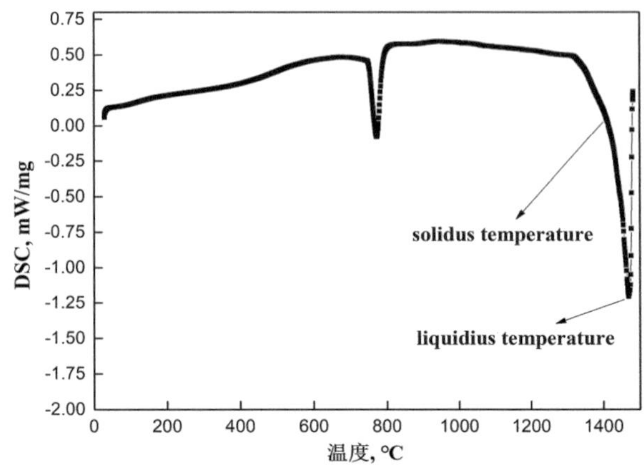

Fig. 2 The variation of heat with temperature during heat flux melting process of GCr15

composition of GCr15 steel is shown in Table 2, solidification latent heat and solidification temperature have a great impact on calculation result of the heat transfer model. Considering the diffusion of interstitial carbon in solid phase, Scheil model for the non-equilibrium solidification is applied in the calculation of solid fraction. Solidification latent heat and solidification temperature are determined by differential scanning calorimeter (DSC) method. The variation of heat with temperature during heat flux melting process of GCr15 steel is showed in Fig. 2. Figure 3 shows the variation in the thermal-physical parameters (solid fraction, density, specific heat, and thermal conductivity). Solidus and liquidus temperatures can be obtained in the DSC heating-up curve without regarding the nucleation undercooling.

Nucleation Model

Nucleation model established by Hou et al. [20] is adopted in the study. Moreover, nucleation density varies drastically between at the inner surface of mold and in molten steel. The surface and internal nucleation undercooling ($n_{\max,s}$, $n_{\max,b}$), standard deviation ($\Delta T_{n,s}$, $\Delta T_{n,b}$), and initial nucleation density ($\Delta T_{\sigma,s}$, $\Delta T_{\sigma,b}$) are listed. The nucleation parameter used in the CAFE model is shown in Table 3.

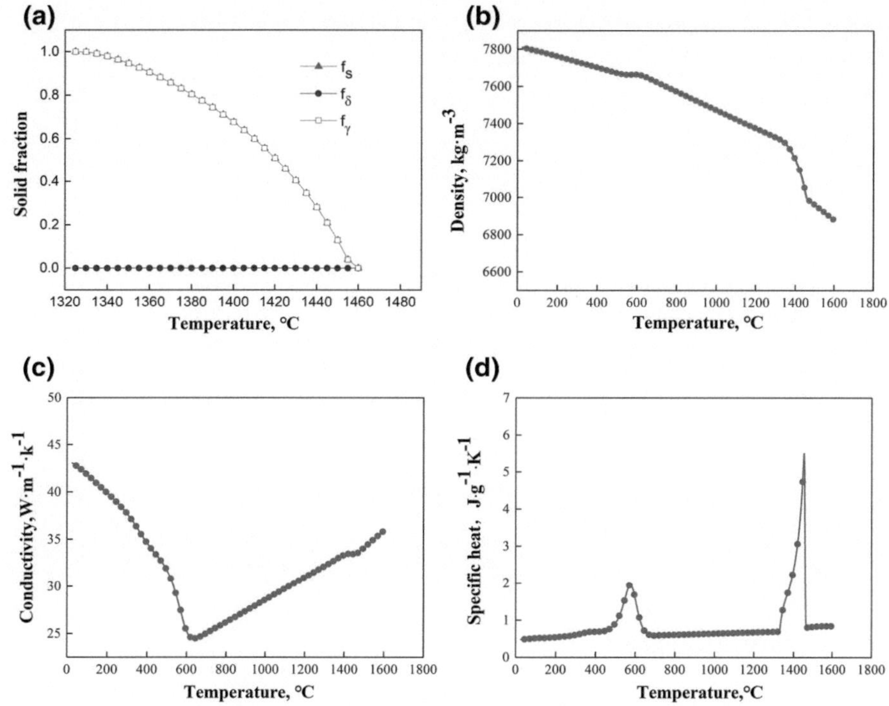

Fig. 3 Material properties of GCr15 steel: **a** solid fraction; **b** density; **c** conductivity; **d** specific heat

Table 3 Nucleation parameters of CAFE model

Parameters	$n_{max,s}$ (m^{-2})	$\Delta T_{n,s}$ (k)	$\Delta T_{\sigma,s}$ (k)	$n_{max,b}$ (m^{-3})	$\Delta T_{n,b}$ (k)	$\Delta T_{\sigma,b}$ (k)
Value	1.7×10^8	1.0	0.1	2.4×10^9	5	1.5

Dendrite Tip Growth Kinetics Model

In the present study, dendrite tip growth kinetics model is used in the literature [20]. In addition, the undercooling and the growth velocity of the dendrite tip were calculated first and then expressed in Eq. (6) by adopting the least-square method.

$$V(\Delta T) = a_2 \Delta T^2 + a_3 \Delta T^3 \tag{6}$$

where V is the growth velocity of the dendrite tip; a_2 and a_3 are the fitting coefficients.

In order to calculate the coefficients for the investigated steel, which is composed of various elements (as shown in Table 2), the Bobadilla, Lacaze, and Lesoult approach [22] is adopted. Therefore, a_2 and a_3 were determined as 0 and 4.59 ×

Table 4 Typical composition of GCr15 bearing steel and relevant parameters

Composition	C	Si	Mn	S	P	Cr
Mass fraction (%)	0.97	0.2	0.35	0.005	0.01	1.37
Partition coefficient (k)	0.34	0.52	0.78	0.035	0.13	0.86
Liquidus slope (m)	78	7.6	4.9	34.4	38	1.04
n	−1122	60	−12	140	160	13.4
Diffusivity in liquid (D)	2.0×10^{-8}	2.4×10^{-9}	2.0×10^{-8}	4.5×10^{-9}	4.7×10^{-9}	3.5×10^{-8}
Gibbs-Thomson (Γ)	1.9×10^{-7}	1.9×10^{-7}	1.9×10^{-7}	1.9×10^{-7}	1.9×10^{-7}	1.9×10^{-7}

10^{-6} using ProCAST software, respectively, and the relevant parameters are listed in Table 4.

Calculation Procedure

The flowchart of the simulation process is shown in Fig. 4. The three-dimensional grid model is first established and meshed by GeoMesh. Then initial and boundary conditions are set. Finally, the thermal field is calculated by ProCAST, and the solidification structure is simulated by the CAFE model calculation. The CAFE model is one module of ProCAST software. The FE mesh is used to calculate the heat transfer process, and the CA cell is used to calculate the grain growth process. Firstly, interpolation coefficients are defined between FE mesh and CA cells in order to combine the FE and CA calculation in a single model, then nucleation model and dendrite tip growth kinetics are used for simulating the solidification structure based on the cellular automaton platform. Crystallographic orientation of crystal growth is randomly chosen from predefined orientation classes, and the crystallographic orientation <100> is selected preferentially.

Model Verification

Validation of Heat Transfer Model

Accuracy of the heat transfer model is verified by a comparison of shell thickness and surface temperature where the distance from the meniscus is appropriate between calculated and measured data. Nail shooting test is one of the most accurate methods for measuring the shell thickness. Macrostructure method is used to obtain accurate

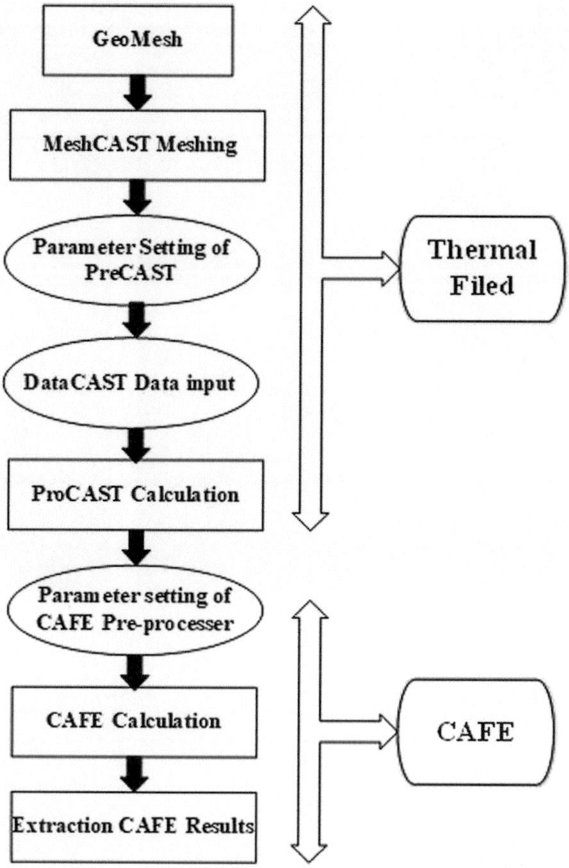

Fig. 4 Flowchart of simulation process

shell thickness of the sample from nail shooting test. Figure 5 reveals nail shooting device and morphology of nail into a bloom. Simulated results for fraction solid distributions and morphology of bloom samples with steel nails are shown in Fig. 6. Table 5 shows the detailed information for the experiments. The comparison between predicted and the measured results with the casting speed of 1.1 m min^{-1} is displayed in Fig. 7. As shown in Fig. 7, the relative error between the predicted and the measured temperature is less than 0.3%, while the relative error between the predicated shell thickness and nail shooting results is less than 5%. Therefore, it is concluded from the above validation analysis that the initial and boundary conditions and materials properties have proven to be sound and the present model is suitable for predicting the solidification process of the bloom continuous casting.

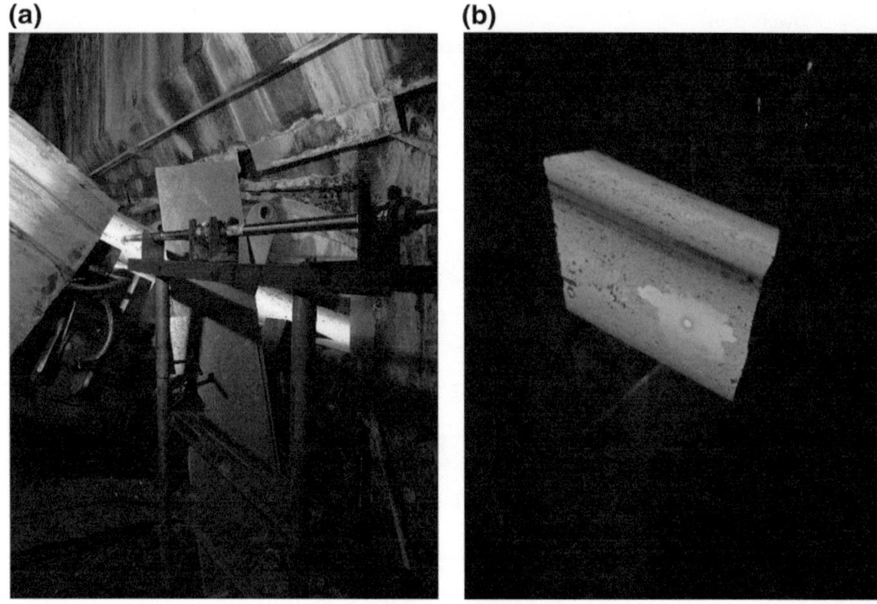

Fig. 5 **a** Nail shooting gun and **b** morphology of nail into the bloom

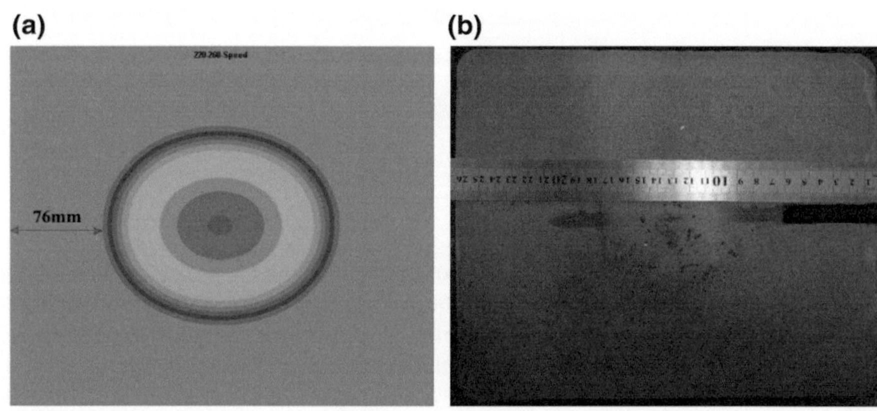

Fig. 6 Simulated results of **a** fraction solid distributions and **b** morphologies of bloom samples with steel nails

Table 5 Operating parameters and results of nail shooting tests

Casting speed (m min^{-1})	Nail shooting gun location (m)	Measured shell thickness (mm)	Simulated shell thickness (mm)	Error (%)
0.75	13	128	125	2.3
1.1	13	79	76	3.8

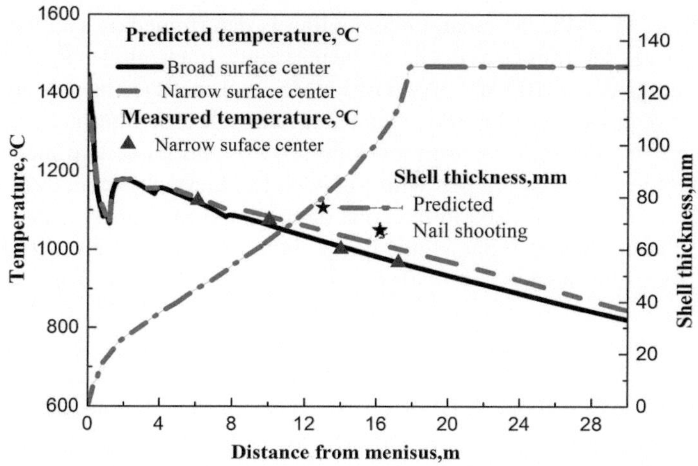

Fig. 7 Comparison between the predicted and measured surface temperature and shell thickness

Validation of Solidification Structure

Figure 8 shows the simulated results for solidification structure without M-EMS and with M-EMS using the CAFE method, main process parameters are shown in Table 6.

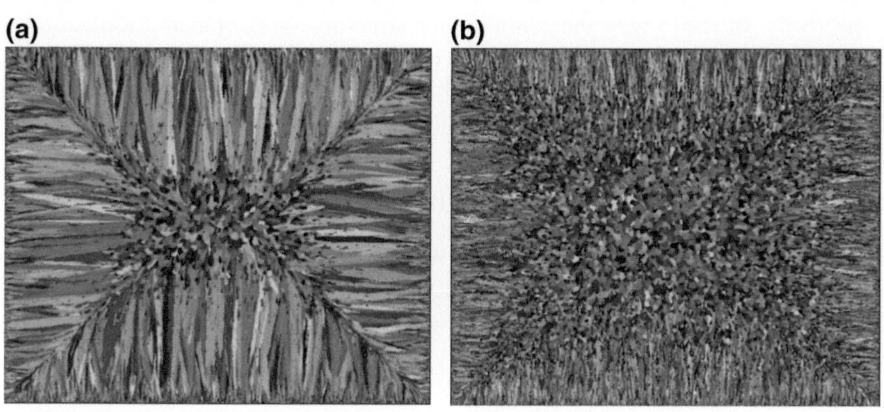

Fig. 8 Simulated solidification structure: **a** without M-EMS; **b** with M-EMS

Table 6 Operating parameters and results of nail-shooting tests

Casting speed (m min^{-1})	Nail gun location (m)	Measured shell thickness (mm)	Simulated shell thickness (mm)	Error
0.75	13	128	125	2.3%
1.1	13	79	76	3.8%

When without M-EMS, the mean undercooling is determined as 1 k, the standard deviation ΔT_σ is determined as 0.1 k and the maximum nucleation density of volume is determined as 2.1×10^9 m^{-3}. As for M-EMS, two methods have been reported for incorporating its effect into the calculation: changing the thermal conductivity of liquid and the nucleation density of volume [15], and all of two methods have been adopted in the model. When with M-EMS, the thermal conductivity of solid, i.e., 32.7 W m^{-1} °C^{-1}, is used in the zone of solid and the mushy zone with solid fraction (f_s) larger than 0.7, but a three times higher thermal conductivity is used in the liquid zone and mushy zone with $f_s < 0.3$, and in the mushy zone with $0.3 \leq f_s \leq 0.7$, thermal conductivity is assumed to change with fs linearly. At the same time, the maximum nucleation density of volume is changed from 2.1×10^9 m^{-3} to 2.6×10^9 m^{-3}. While F-EMS has little effect on the center-equiaxed crystal ratio. As seen from Fig. 8, without M-EMS, coarse columnar and obvious CET (columnar to equiaxed transition) zone exists, and the center-equiaxed crystal ratio is merely 7.8%. After stirring, CET zone becomes faint, columnar grains become fine and compact, and the center-equiaxed crystal ratio increases to 31.1%. This is mainly because EMS enhances the effect of flow of molten steel on breaking off columnar grains and increases the thermal conductivity in the liquid region.

The industrial blooms are polished and hot acid-etched, then macrostructure can be obtained. Figure 9 shows the comparison of the numerical and experimental solidification structure with M-EMS+F-EMS, respectively. Main process parameters are shown in Table 7. Center-equiaxed grains ratio from the industrial bloom and the calculated is about 24.2 and 25.3% by using the image analysis software Image-Pro, respectively. In above way, the solidification structures under the above-mentioned condition is simulated, and the simulated results agree well with the actual results, as shown in Fig. 9. Consequently, the established model is reasonable for simulating the solidification process of the actual continuous casting bloom.

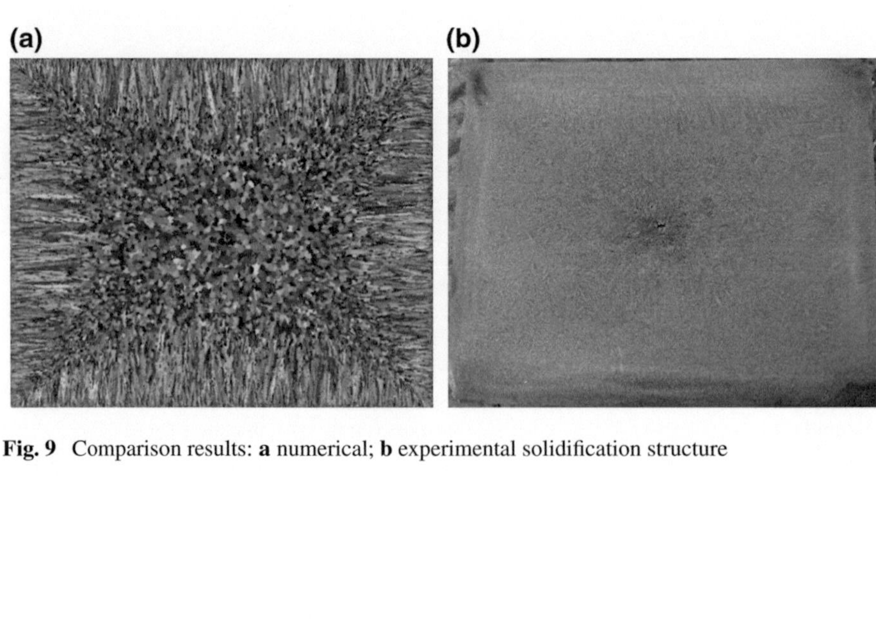

Fig. 9 Comparison results: **a** numerical; **b** experimental solidification structure

Table 7 Main process parameters

Parameters	Value
Casting speed (m min^{-1})	0.7
water amount in mold (m^3 h^{-1})	100
Temperature difference in mold (°C)	7.5~8.5
Secondary cooling intensity (L Kg^{-1})	0.2
Distribution ratio of cooling water in Zone 1~3 (%)	25,38,37
Current and frequency of M-EMS (A/Hz)	150/2
Current and frequency of F-EMS (A/Hz)	200/6

Results and Discussion

In the present study, grain density of whole cross-section in the continuous casting bloom and the center-equiaxed crystal ratio are determined as quantitative indexes to evaluate the solidification structure, a bloom is chosen to evaluate the compactness degree of the equiaxed crystal zone. Grain number is indirectly indicative of grain size. Grain density can be obtained by CAFE module of ProCAST software, and center-equiaxed crystal ratio is calculated by using the image analysis software Image-Pro.

The verified coupling model is used to investigate the influence of process parameters on evolution law of the solidification structure and the compactness degree in the equiaxed crystal zone in combination with the solidification process, and consequently optimum process is proposed in order to improve the solidification structure and center segregation.

Effect of Superheat and Casting Speed on Solidification Behavior

The effects of superheat and casting speed on solidification behavior are investigated. Bearing steel GCr15 has high crack sensitivity owing to high-carbon-chromium contents, and the specific water ratio of 0.2 L kg^{-1} has be proven to be reasonable for the bloom with 220 mm × 260 mm [9]. Different superheats are corresponding to the casting speed of 0.75 m min^{-1} and the specific water ratio of 0.2 L kg^{-1}, and different casting speeds are corresponding to 25 °C and the specific water ratio of 0.2 L kg^{-1}, respectively. Figure 10 presents wide surface and center temperature distribution along the strand; eliminating superheat position, solidification position, and length of mush zone is shown in Fig. 11.

Figure 12 reveals centerline solid fraction along the strand. As seen from Figs. 10a, 11a, and 12a, superheat has little effect on surface and centerline temperature distribution, centerline solid fraction along the strand. As superheat increases from 10 to 40 °C, eliminating superheat position and solidification position slightly enlarge,

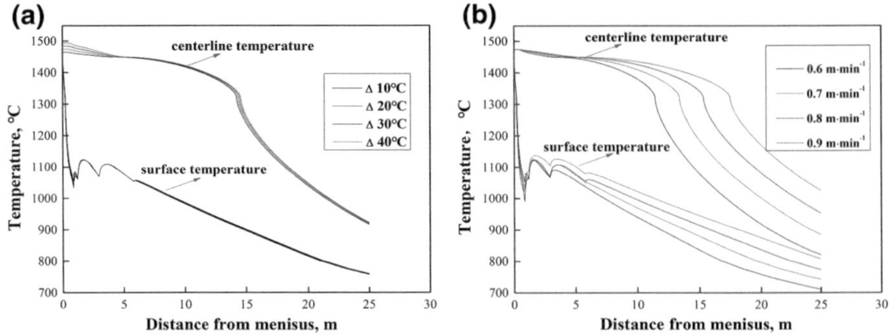

Fig. 10 Surface and centerline temperature distribution along the strand: **a** different superheats; **b** different casting speeds

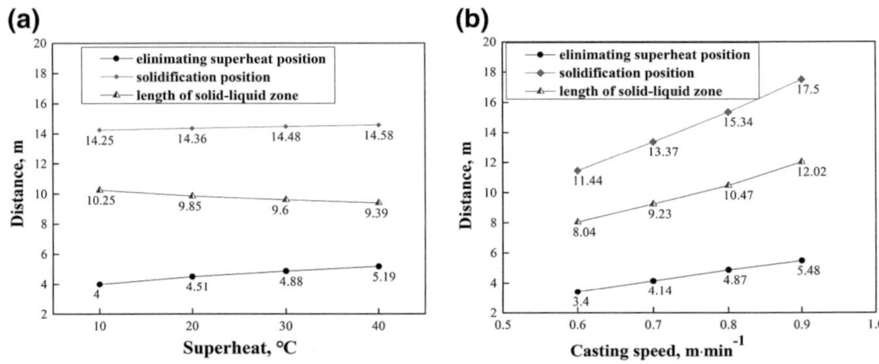

Fig. 11 Eliminating superheat position, solidification position and length of solid-liquid zone: **a** different superheats; **b** different casting speeds

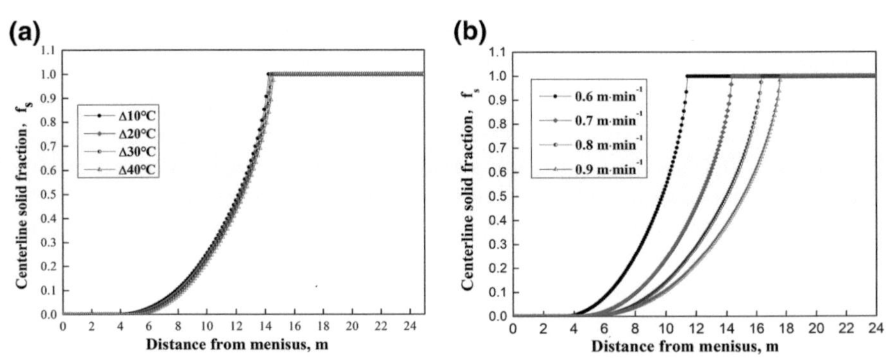

Fig. 12 Centerline solid fraction along the strand: **a** different superheats; **b** different casting speeds

while length of mushy zone slightly decreases. The increase of 10 °C results in the change of 5 °C in bloom surface temperature and 0.01 in centerline solid fraction. With the increase of 10 °C, eliminating superheat position and solidification position is prolonged by about 0.51 and 0.11 m, and length of mushy zone decreases by 0.25 m, respectively.

As can be seen in Figs. 10b, 11b, and 12b, casting speed has obvious effect on solidification behavior compared with superheat. As the casting speed increases from 0.6 to 0.9 m min^{-1}, wide surface, centerline temperature distribution, and solid fraction along the strand increase. Besides, eliminating superheat position, length of mushy zone, and solidification position enlarge. The increase of 0.1 m min^{-1} result in the change of 23 °C in bloom surface temperature; with the increase of 0.1 m min^{-1}, eliminating superheat position, length of mushy zone, and solidification position is prolonged by about 0.7 m, 1.2 m, and 2.0 m, respectively. With the increase of 0.1 m min^{-1} from 0.6 to 0.9 m min^{-1}, centerline solid fraction increases by 0.04, 0.11, and 0.23, respectively.

Effect of Superheat and Casting Speed on Solidification Structure

Figure 13a–d shows the simulated results for solidification structure under different superheats with the casting speed of 0.75 m min^{-1} and the specific water ratio of 0.2 L kg^{-1}. Figure 13e–h presents the simulated results for the solidification structure under different casting speeds with the superheat of 25 °C and the specific water ratio of 0.2 L kg^{-1}. Central equiaxed crystal ratio of the bloom is presented in Fig. 14. Figure 16 shows total grain number of whole cross-section of the bloom under different superheats and casting speeds. Meanwhile, M+F-EMS are used under different superheats and casting speeds.

As seen from Figs. 13a–d to 14a, the center-equiaxed crystal ratio markedly increases with the decrease of the superheat. When superheat increases from 10 to 40 °C, center-equiaxed crystal ratio decreases from 38.6 to 24.4%. In addition, columnar grains grow coarsening and columnar crystal zone enlarges, which easily causes intergranular micro segregation of solute elements. The total grain number of the whole cross-section increases with the decrease of the superheat, as shown in Fig. 15a. When superheat increases from 10 to 40 °C, the total number of grains decrease from 920,572 to 723,400. Whereas, both center-equiaxed crystal ratio and the total grain number vary slightly between 30 and 40 °C.

As revealed by Figs. 13e–h and 14b, the center-equiaxed crystal ratio increases with the increase of casting speed. When casting speed increases from 0.6 to 0.9 m min^{-1}, the center-equiaxed crystal ratio increases from 35.8 to 41.5%. Besides, columnar grain hardly becomes bigger and columnar crystal zone basically remains unchanged. The total grain number of the whole cross-section increases with the increase of casting speed, as shown in Fig. 15b. When casting speed increases from

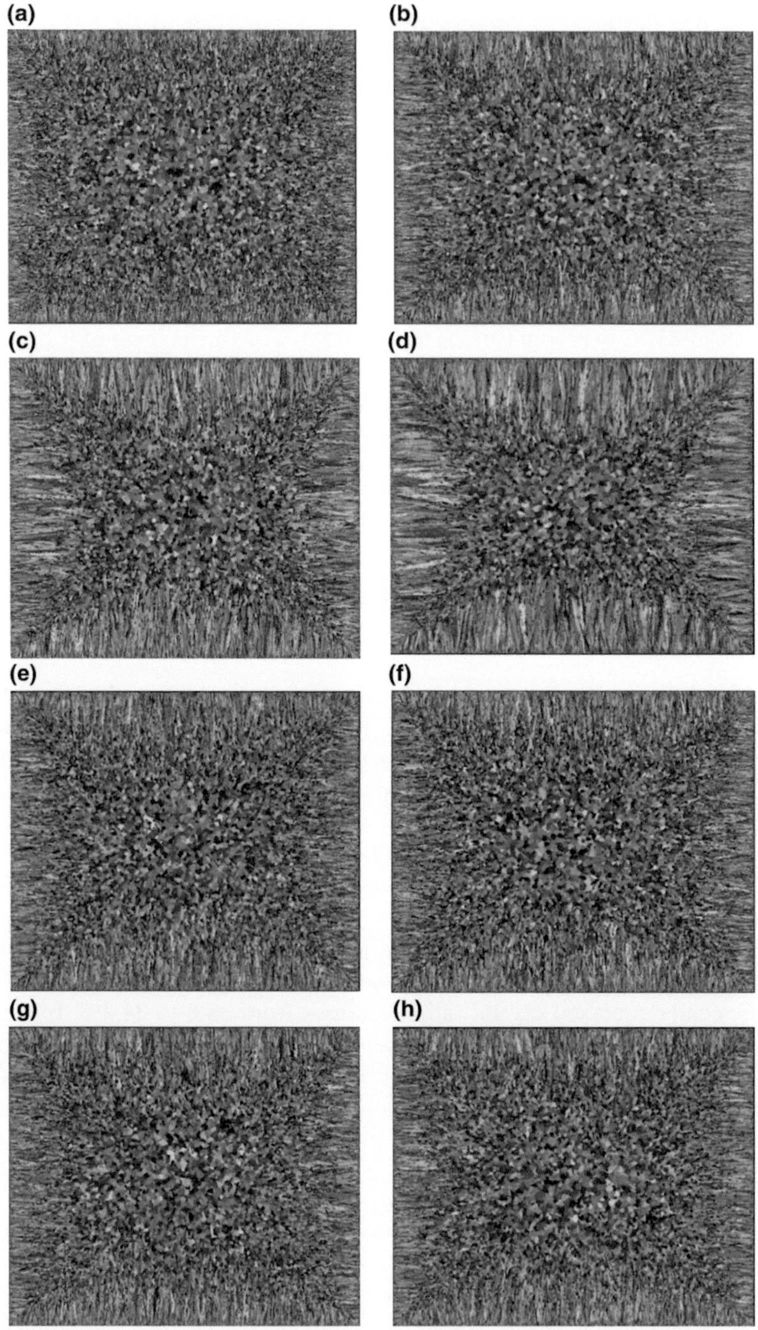

Fig. 13 Simulated solidification structure under different superheats and different casting speed:**a** 10 °C, 0.75 m min^{-1}; **b** 20 °C, 0.75 m min^{-1}; **c** 30 °C, 0.75 m min^{-1}; **d** 40 °C, 0.75 m min^{-1}; **e** 0.6 m min^{-1}, 25 °C; **f** 0.7 m min^{-1}, 25 °C; **g** 0.8 m min^{-1}, 25 °C; **h** 0.9 m min^{-1}, 25 °C

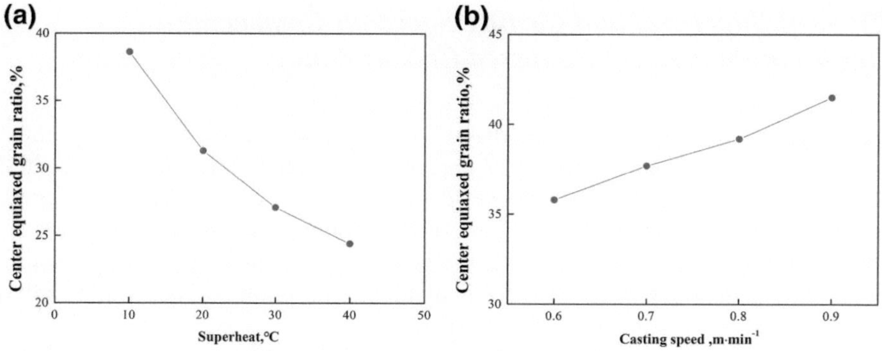

Fig. 14 Center equiaxed grain ratio of whole cross-section of the bloom: **a** different superheats; **b** different casting speeds

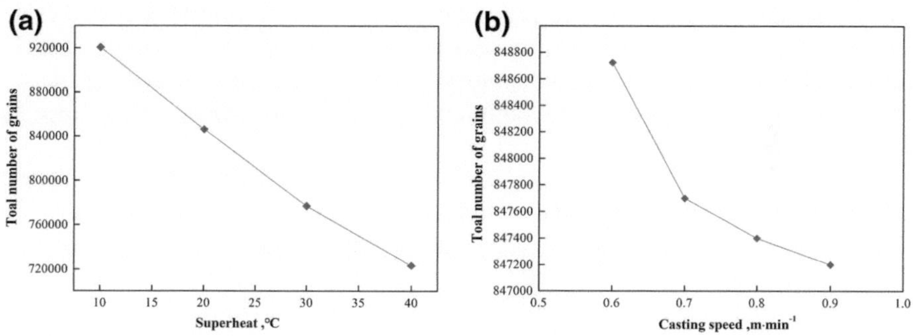

Fig. 15 Total grain number of whole cross-section of the bloom: **a** different superheats; **b** different casting speeds

0.6 to 0.9 m min^{-1}, the total number of grains decreases from 848,723 to 847,400. Whereas, the slight change in the center-equiaxed crystal ratio and the total grain number occur between 0.8 and 0.9 m min^{-1}.

The temperature gradient at liquid–solid interface during the solidification process decreases with the increase of superheat and the decrease of casting speed, as shown in Fig. 10. Therefore, the effects of superheat and casting speed on solidification structure are mainly attributed to the increase in the number of nucleation cores of the whole cross-section and a larger temperature gradient at liquid–solid interface occur during the solidification.

As can be seen in Figs. 8 and 13, M-EMS and superheat have more obvious effects on the center-equiaxed crystal ratio and total grain number compared with casting speed. In view of comprehensive consideration of castability and solidification structure in the actual production of GCr15 steel, superheat should be controlled below 30 °C, and casting speed should be below 0.8 m min^{-1}.

Effect of Superheat and Casting Speed on Compactness Degree of the Central Equiaxed Crystal Zone

Figure 16 shows the simulation results of the central equiaxed crystal zone without M-EMS and with M-EMS. Figure 17 presents the simulation results in the central equiaxed crystal zone under different superheats and casting speeds. Total grain number of central equiaxed grain zone of the bloom is presented in Fig. 18. As seen from Fig. 16, M-MES has evident effect on total grain number in the central equiaxed crystal zone. After M-MES is applied, total grain number in the central equiaxed crystal zone increase from 1498 to 2064.

As can be seen in Figs. 17a–d and 18a, total grain number in the central equiaxed crystal zone increases with the increase of superheat. The difference of total grain number between 15 and 20 °C is very smaller. When superheat increases from 20 to 40 °C, total grain number of central equiaxed grain zone increases from 2018 to 2087; as seen from Figs. 17e–h to 18b, total grain number of central equiaxed grain zone increases with the decrease of the casting speed. When casting speed increases from 0.6 to 0.9 m min^{-1}, total grain number of central equiaxed grain zone decreases from 2157 to 1995. Whereas, the slight change in the total grain number appears between 0.8 and 0.9 m min^{-1}.

As revealed by Fig. 11, eliminating superheat position, length of mushy zone, and solidification position correspond to eliminating superheat time, whole solidification time and center solidification time [20]. For the central equiaxed crystal zone of a bloom, with the increase of superheat and the decrease of casting speed, the center solidification time increases. When increasing the center solidification time during solidification, the nucleation cores form in the relatively high temperature region will have longer time to grow, which may lead to smaller space for the formation of new cores in the relatively low temperature region. Consequently, the grain number of

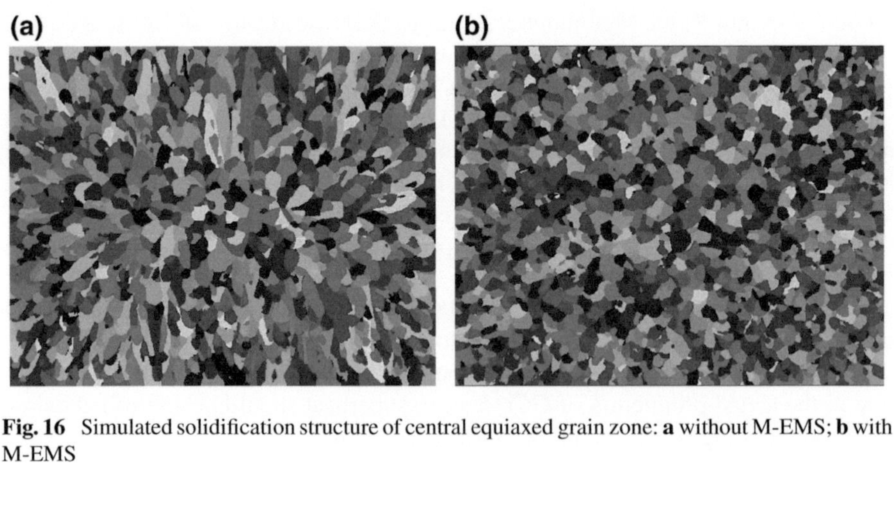

Fig. 16 Simulated solidification structure of central equiaxed grain zone: **a** without M-EMS; **b** with M-EMS

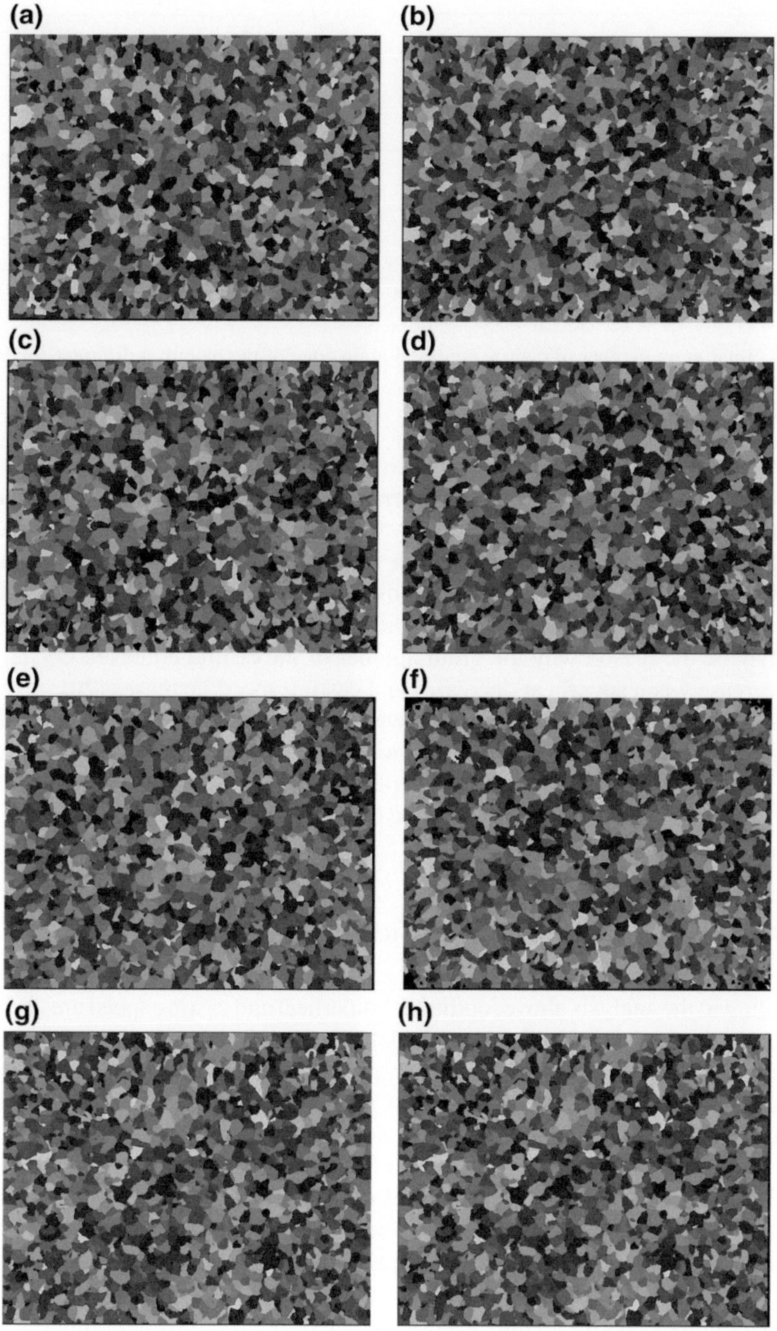

Fig. 17 Simulated solidification structure of central equiaxed grain zone under different superheats and different casting speeds: **a** 10 °C, 0.75 m min^{-1}; **b** 20 °C, 0.75 m min^{-1}; **c** 30 °C, 0.75 m min^{-1}; **d** 40 °C, 0.75 m min^{-1}; **e** 0.6 m min^{-1}, 25 °C; **f** 0.7 m min^{-1}, 25 °C; **g** 0.8 m min^{-1}, 25 °C; **h** 0.9 m min^{-1}, 25 °C

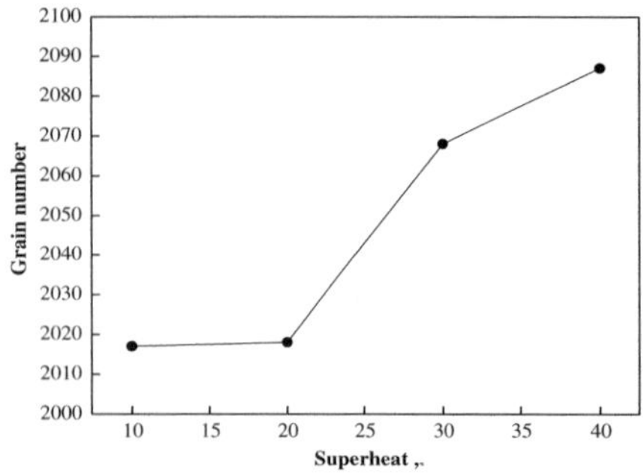

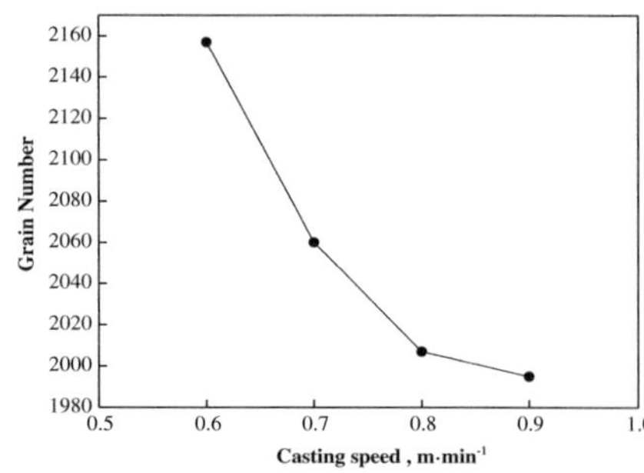

Fig. 18 Total grain number of central equiaxed grain zone of the bloom: **a** different superheats; **b** different casting speeds

the central equiaxed crystal zone will decrease with the increase of the center solidification time. When increasing the center solidification time during solidification, the nucleation cores form in the relatively high temperature region will have longer time to grow, which may lead to smaller space for the formation of new cores in the relatively low temperature region. Consequently, the grain number in the central equiaxed crystal zone will decrease with the increase of the center solidification time. This may be the reason why the grain number of the central equiaxed crystal zone will decrease when superheat decreases and casting speed increases [20].

M-EMS and superheat have more obvious effects on total grain number in the central equiaxed crystal zone compared with casting speed. Considering the castability in actual production, reasonable superheat should be controlled below 20 and 30 °C, and casting speed should be below 0.8 m min^{-1}.

Optimization of Superheat and Casting Speed

According to the analysis above, optimized superheat and casting speed are 25 °C and 0.8 m min^{-1} with M+F-EMS, respectively. Solidification structure in cross-section of the bloom before and after optimization is shown in Fig. 19. Center-equiaxed crystal ratio increases from 27 to 38%. As shown in Fig. 20, center solid fraction at F-EMS position varies from 0.25 to 0.15 at a distance of 9.5 m from the meniscus, where F-EMS is implemented. By heat transfer model and plant trials [2, 9, 10], the effect of F-EMS for high-carbon blooms on reducing macro segregation has been investigated. The criterion of center solid fraction in strand around 0.1–0.2 [2, 9] was mainly applied in determining optimum F-EMS implemented position. In consequence, it can be deduced that reduction of center segregation in continuously cast GCr15 bloom with 220 mm × 260 mm can be achieved by adopting optimized superheat and casting speed.

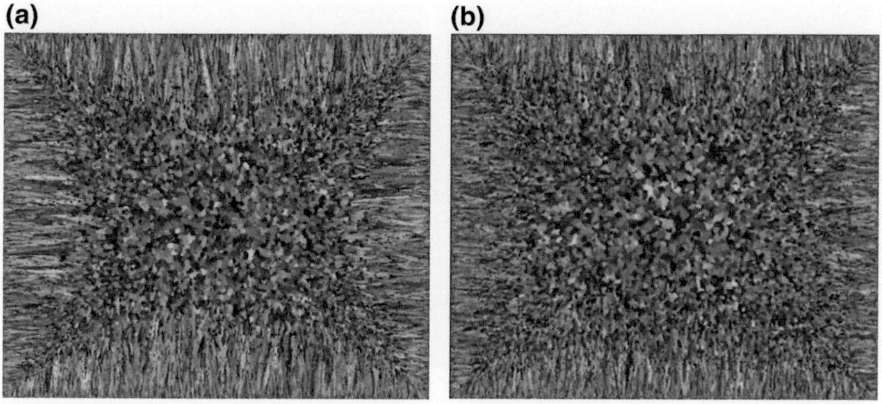

Fig. 19 Simulated solidification structure before and after optimization: **a** 0.7 m min^{-1}, 35°C; **b** 0.8 m min^{-1}, 25°C

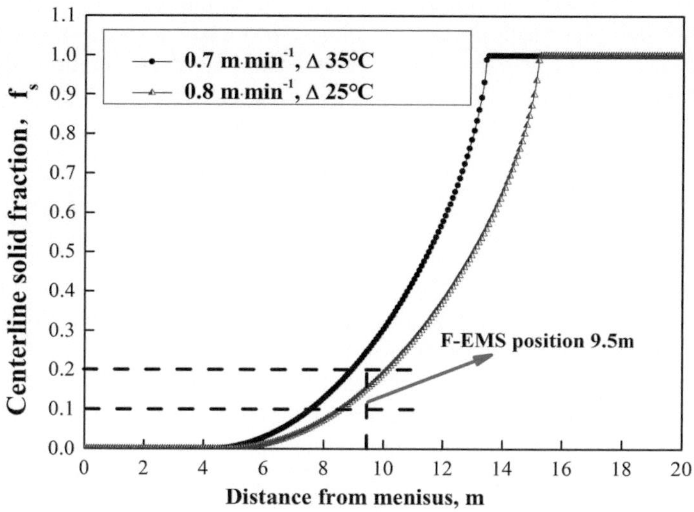

Fig. 20 Centerline solid fraction along the strand before and after optimization

According to the above calculation and experiment results, plant trials were carried out to validate the theoretical optimized results, which were evaluated comprehensively by macrostructure and center segregation ratio. Center segregation ratio was analyzed by the method in author' previous research [9]. Figure 21 gave the comparison between the cross-section morphologies of typical defects for GCr15 bloom before and after optimization. As shown in Fig. 21, compared with the original parameters with casting speed of 0.7 m min^{-1} and superheat of 35 °C, the bloom central soundness has been improved remarkably under optimized parameters. As an example, defects such as oversized center porosity disappeared in as-cast bloom. The

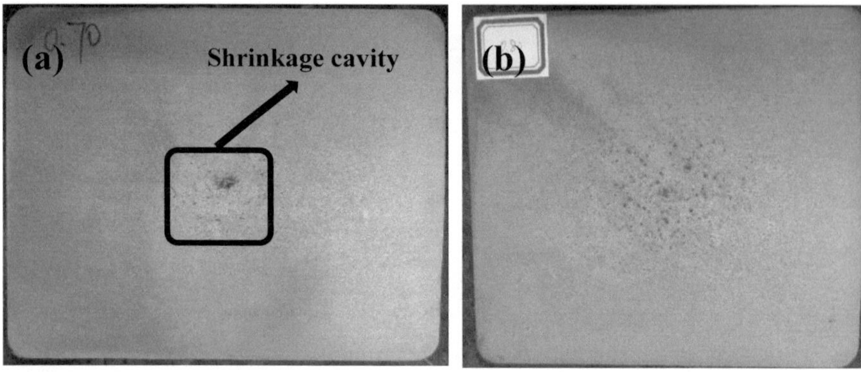

Fig. 21 Morphologies of macrostructure from experimental results: **a** before optimization; **b** after optimization

grades of center porosity decreased from 1.5 to 0, in addition, the center-equiaxed crystal ratio increased from 26 to 37% under with casting speed of 0.8 m min^{-1} and superheat of 25 °C. The maximum center carbon segregation ratio in the as-cast bloom decreased from 1.32 to 1.12.

Conclusion

In the present work, a CAFE coupling model has been established to simulate the solidification process and solidification structure in the continuous casting bloom of bearing steel GCr15 with 220 mm × 260 mm, in which M+F-EMS was taken into consideration. Based on the validated model by experimental data, the effects of superheat and casting speed on the solidification structure and the compactness degree in the central equiaxed crystal zone of the bloom were investigated. Thereafter, industrial trials were carried out to validate the optimized parameters. The results were summarized as follows.

(1) A coupled CAFE model is developed and validated by surface temperature measurement, nail shooting test, and macro etch experiments. The simulated results are well consistent with experimental results, and the solidification process and evolution of solidification structure in the continuous casting bloom under different conditions can be exactly simulated.
(2) Compared with casting speed, M-EMS, and superheat exhibit a significant increment on the central equiaxed crystal ratio and compactness degree. The central equiaxed crystal ratio increases with the decrease of superheat, and decreases with the decrease of casting speed.
(3) The compactness degree of central equiaxed grain zone in the billet increases obviously when M-EMS is applied. In addition, it also increases with the

increase of the superheat, and the decrease of the casting speed, which is closely related to the center solidification time.

(4) The industrial trials' results show that optimized parameters with higher casting speed of 0.8 m min^{-1} and lower superheat of 25 °C in the continuous casting bloom of GCr15 bearing steel with 220 mm × 260 mm, which is taken into M+F-EMS, the center-equiaxed crystal ratio increased by 37% and the maximum center carbon segregation ratio at the as-cast bloom decrease from 1.32 to 1.12.

Acknowledgements The work was financially supported by the National Natural Science Foundation of China (No. 51874021) and Foundation of State Key Laboratory of Advanced Metallurgy in University of Science and Technology Beijing (41602014).

References

1. Choudhary KS, Ganguly S (2007) Morphology and segregation in continuously cast high carbon steel billets. ISIJ Int 47(12):1759–1766
2. An HH, Bao YP, Wang M, Zhao LH (2017) Reducing macro segregation of high carbon steel in continuous casting bloom by final electromagnetic stirring and mechanical soft reduction integrated process. Metall Res Technol 114:1–12
3. Flemings MC (2000) Our understanding of macro segregation: past and present. ISIJ Int 40(9):833–841
4. Lesoult G (2005) Macro segregation in steel strands and ingots: characterization, formation and consequences. Mater Sci Eng A 413–414:19–29
5. Straffelini G, Lutterotti L, Tonolli M, Lestani M (2011) Modeling solidification microstructures of steel round billets obtained by continuous casting. ISIJ Int 51(9):1448–1453
6. Ludlow V, Normanton A, Anderson A, Thiele M, Ciriza J, Laraudogoitia, van der Knoop W (2005) Strategy to minimize central segregation in high carbon steel grades during billet casting. Ironmaking Steelmaking 32(1):68–74
7. Li JC, Wang BF, Ma YL, Cui JZ (2006) Effect of complex electromagnetic stirring on inner quality of high carbon steel bloom. Mater Sci Eng A 25(1–2):201–204
8. An HH, Bao YP, Wang M, Zhao LH (2018) Effects of electromagnetic stirring on fluid flow and temperature distribution in billet continuous casting mould and solidification structure of 55SiCr. Metall Res Technol 115:1–12
9. An HH, Bao YP, Wang M, Yang Q, Huang YS (2019) Improvement of center segregation in continuous casting bloom and the resulting carbide homogeneity in bearing steel GCr15. Ironmaking Steelmaking. https://doi.org/10.1080/03019233.2019.1604614
10. Du WD, Wang K, Song CJ, Li HG, Zhai QJ, Zhao P (2008) Effect of special combined electromagnetic stirring mode on macro segregation of high strength spring steel blooms. Ironmaking Steelmaking 35(2):153–156
11. Nastac L (2014) A 3D stochastic mesoscopic model for prediction of microstructure evolution during solidification of dendritic alloys. Metall Res Technol 111:311–319
12. Anand G, Datta S, Chattopadhyay PP (2013) Deterministic approach for microstructurally engineered formable steels. Int J Metall Eng 2(1):69–78
13. Rezende J, Senk D, Hüttenmeister D (2015) Phase-field modeling of the dendrite growth morphology with influence of solid-liquid interface effects. Steel Res Int 86(1):65–72
14. Rappaz M, Gandin CA (1993) Probabilistic modelling of microstructure formation in solidification processes. Acta Mater 41(2):345–360
15. Yamazaki M, Natsume Y, Harada H, Ohsasa K (2006) Numerical simulation of solidification structure formation during continuous casting in Fe–0.7 mass% C alloy using cellular automaton method. ISIJ Int 46(6):903–908

16. Gandin CA, Desbiolles JL, Rappaz M, Thevoz P (1999) A three-dimensional cellular automation-finite element model for the prediction of solidification grain structures. Metall Mater Trans A 30(2):3153–3165
17. Luo YZ, Zhang JM, Wei XD, Xiao C, Hu ZF, Yuan YY, Chen SD (2012) Numerical simulation of solidification structure of high carbon SWRH77B billet based on the CAFE method. Ironmaking Steelmaking 39(1):26–30
18. Dou K, Yang ZG, Liu Q, Huang YH, Dong HB (2016) Influence of secondary cooling mode on solidification structure and macro-segregation behavior for high-carbon continuous casting bloom. High Temp Mater Proc 36(7):741–753
19. Gao XZ, Yang SF, Li JS, Liao H (2016) Numerical simulation on optimization of center segregation for 50CrMo structural alloy steel. High Temp Mater Proc 35(6):583–589
20. Hou ZB, Jiang F, Cheng GG (2012) Solidification structure and compactness degree of central equiaxed grain zone in continuous casting billet using cellular automaton-finite element method. ISIJ Int 52(7):1301–1309
21. Fang Q, Wang B, Zhang H, Ye F (2017) Effects of EMS induced flow on solidification and solute transport in bloom mold. Metals 72(7):1–13
22. Bai L, Wang B, Zhong HG, Zhai QJ, Zhang JY (2016) Experimental and numerical simulations of the solidification process in continuous casting of slab. Metals 53(6):1–12

Effects of Welding Conditions and Post-Weld Heat Treatment on Precipitation of Widmanstätten-Austenite of Duplex Stainless Steels

Yunxing Xia, Xiaofu Zhang, Fumikazu Miyasaka and Hiroaki Mori

Abstract In the case of welding for duplex stainless steel, the welds change in the morphology and ferrite/austenite phase ratio due to the welding conditions and the alloy composition of the material itself. Especially when using the laser beam welding what has the ultra-high heat input and rapid cooling, the change in the welds is much pronounced. Therefore, in this study, after etching the welds of the samples with 10% KOH solution, we have observed the microstructure changes of three different duplex stainless steels under different welding conditions. The results of the experiment of duplex stainless steels, super duplex, and lean-alloy duplex showed that the structure of the welds was significantly different between the steel grades; it was found that the increase in welding velocity significantly reduced the formation of the austenite phase. To counterbalance the ratio of ferrite/austenite phase by using the laser surface treatment after welding, it has become possible to promote a part of the content of the austenite phase even in the welds. In addition, these results could become a basis and verification for the simulation of the duplex stainless steels' welds microstructure in the future.

Keywords Laser beam welding · Post-weld heat treatment · Precipitation behavior · Phase ratio · Duplex stainless steels

Introduction

In recent years, with the development of industrial technology, materials with high strength and high corrosion resistance have been increasingly demanded. Therefore, duplex stainless steels are superior to austenitic stainless steels in terms of material properties and cost performance which have received wide attention.

Y. Xia (✉) · X. Zhang · H. Mori
Management of Industry and Technology, Graduate School of Engineering, Osaka University, 2-1 Yamadaoka, Suita, Osaka 565-0871, Japan
e-mail: xsia-y18@mit.eng.osaka-u.ac.jp

F. Miyasaka
Department of Adaptive Machine Systems, Graduate School of Engineering, Osaka University, 2-1 Yamadaoka, Suita, Osaka 565-0871, Japan

J. Lee et al. (eds.), *Materials Processing Fundamentals 2020*, The Minerals, Metals & Materials Series, https://doi.org/10.1007/978-3-030-36556-1_19

Duplex stainless steel is in a structure with an austenite/ferrite phase ratio of approximately 1:1 and has mechanical properties and corrosion resistance superior to those of conventional products. It is widely applied in facilities and machines that are exposed to various corrosive environments. However, the properties of duplex stainless steels may deteriorate when the phase ratio at the welds collapses during the welding process, and the amount of ferrite phase increases significantly when the heat input is large. Since the late twentieth century, a lot of studies have been conducted on the selection of different welding methods and welding materials in order to investigate the corrosivity of the welds [1, 2]. However, due to the processing performance of the duplex stainless steels there were not being used very widely in most applications at that time. After that, lean-alloy duplex stainless steel with lower cost and good characteristics was developed. There are not many studies about welds on this new type of stainless steel, and also, there are fewer discussions on the three steel grades under the same welding conditions.

Therefore, the purpose of this study was that we have summarized three types of duplex stainless steels in order to investigate the effect of welding velocity of laser beam welding with rapid cooling effect on the precipitation morphology of austenite precipitated after ferrite single phase solidification. Moreover, we also used the same laser beam welding heat source to perform post-weld heat treatment on the welds. Based on the above results, we analyze the promotion of precipitation in austenite phase. In addition, we have also investigated the elemental distribution within the microstructure of the welds and have examined the change and precipitation relationship. And besides, the second purpose of this study was to apply these relationships as a basis for the establishment of a simulation model.

Experimental Conditions

Material

In this experiment, the following three types of duplex stainless steels were subjected to a bead on plate test by laser beam welding: the duplex stainless steels (S31803), super duplex stainless steels (S32750), and lean-alloy duplex stainless steels (S32101). The specification of the sheet is 100 mm (length) × 100 mm (width) × 3 mm (thickness), and the chemical compositions thereof are shown in Table 1.

Laser Beam Welding and Post-Weld Heat Treatment Condition

Process parameters for laser beam welding and post-treatment are shown in Tables 2 and 3. In order to make a better comparison, the output of each weld is a fixed value.

Table 1 Chemical compositions of materials used (mass%)

Material	C	Si	Mn	P	S	Ni	Cr	Mo	Cu	N
S31803	0.014	0.32	0.9	0.026	–	5.3	22.6	3.1	–	0.18
S32750	0.01	0.16	0.78	0.024	–	6.4	25.7	3.4	–	0.31
S32101	0.021	0.72	5.06	0.20	0.01	1.57	21.35	0.31	0.25	0.215

Table 2 Processing conditions of the laser beam welding

		Laser output (w)	Velocity (mm/min)	Focus distance	Shielding gas
Laser beam weld	No. 1	2000	1000	31.4 mm (J.F.)	Ar.
	No. 2	2000	1500	31.4 mm (J.F.)	
	No. 3	2000	2000	31.4 mm (J.F.)	

Table 3 Processing conditions of the post-weld heat treatment

		Laser output (w)	Velocity (mm/min)	Focus distance (mm)	Shielding gas
Heat treatment	No. 4	1000	1000	131.4	Ar.

We have adjusted the welding velocity differently to change the heat received by the welds per unit time.

For post-weld heat treatment, when we used the same laser heat source as the welding process, because the energy of laser beam welding is too concentrated, while we changing the laser output, welding velocity, and in the same time we also adjusted the focus distance.

Microstructure Observation

To observe and study the microstructure of the welds, after polished these samples we etched them with a 10% KOH solution. In order to investigate the influence of the alloy composition of the welds, we have performed EPMA analysis on the JAX-8700.

Results and Discussion

The Microstructure in Welds by OM

Figure 1 shows the microstructure of the welds in three kinds of duplex stainless steels under different welding conditions. When the laser output is constant, the welding velocity is faster and the welds become narrower. Further, the amount of precipitation of the austenite in the welds (including precipitated in the ferrite grains) decreases as the welding velocity increases. In the comparison of different steel grades, we have found that the precipitation of austenite both at the grain boundary and at the ferrite grains in S82750 and S32101 is more than those in S31803. According to the calculation results of Cr/Ni equivalent, S31803 and S32101 are F solidification

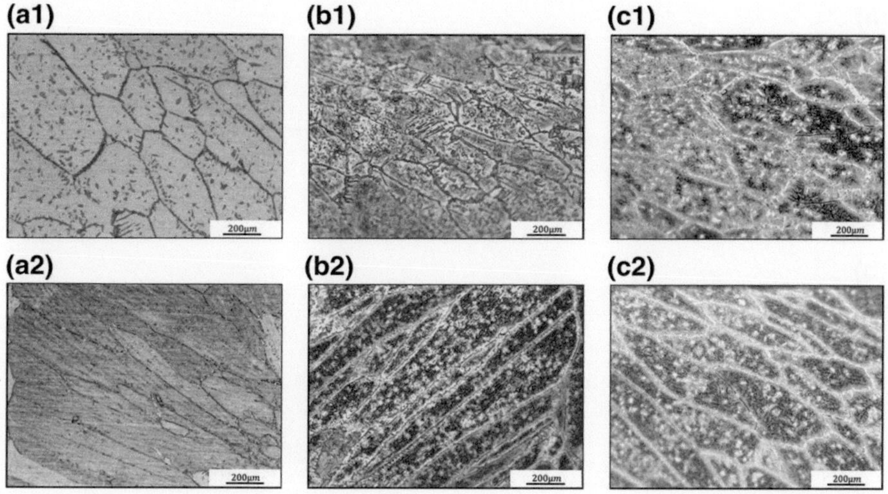

Fig. 1 Effect of laser beam welding velocities on microstructure of S31803 (**a**), S82750 (**b**), and S32101 in welds (1: 1000 mm/min, 2: 2000 mm/min)

model, while S82750 should be with FA solidification mode, but there is almost no difference with S32101 under these laser beam welding conditions.

EPMA Analysis in Welds

In normal conditions, we know the concentration distribution of nickel, nitrogen can promote the precipitation of austenite, and chromium has a decisive influence on ferrite precipitation. However, observation by EPMA analysis (Fig. 2), we have found that although nickel and chromium have partial concentration differences, they are basically presented throughout the welded portion. Nitrogen is relatively concentrated in the austenite precipitation, whether in the grain boundary or in the ferrite grain. It can be inferred that under the laser beam welding, due to the very fast cooling rate, only nitrogen with a faster diffusion rate could be diffused in a certain amount during phase transformation. Nickel and chromium have hardly spread in this short period of time, and they just solidified along the direction of welding and temperature distribution.

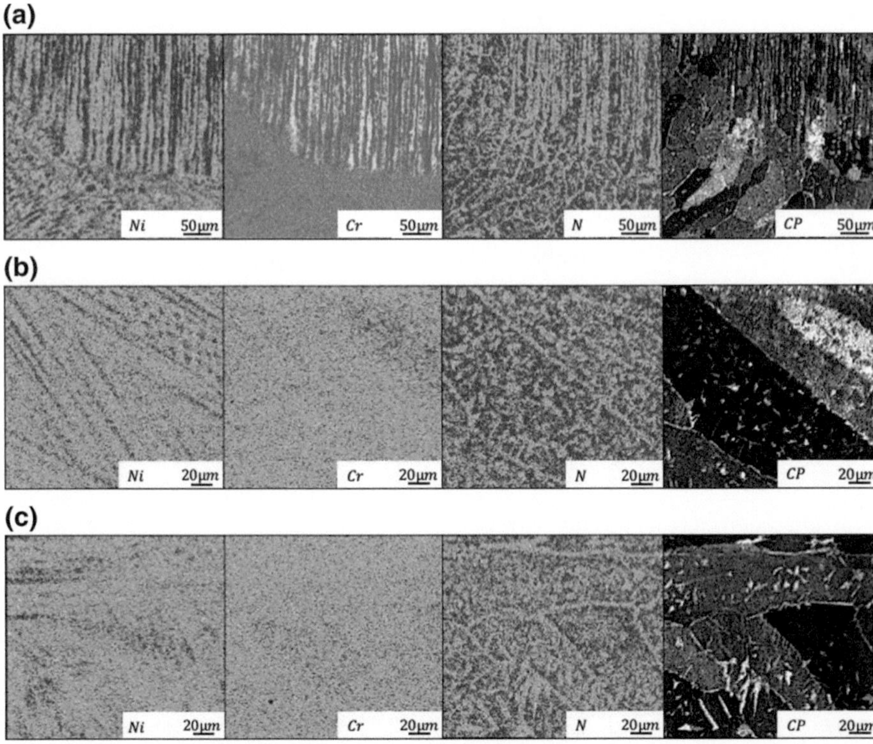

Fig. 2 EPMA analysis in welds of S32101 (**a** interface with parent metal, **b** upper part of the welds, and **c** middle part of the welds)

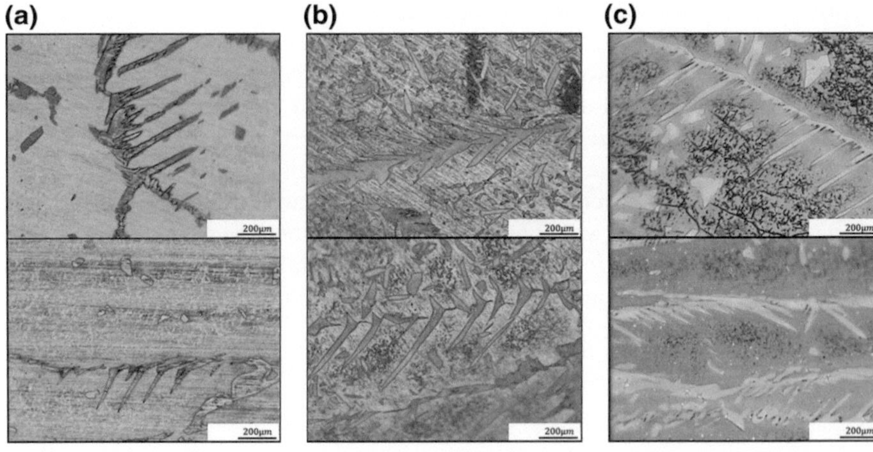

Fig. 3 Widmanstätten-austenite in welds of S32101 (**a**), S32750 (**b**), and S31803 (**c**)

Precipitation and Structural Observation of Widmanstätten-Austenite

Under high power microscope, Fig. 3, we have found that the precipitation of Widmanstätten-austenite was found in the grain boundary part of the welds. According to the comparison of three different steel grades, we have observed that the S32101 and S32750 have precipitated much Widmanstätten structure than S31803, and from this result, it is presumed to be that the precipitation of Widmanstätten structure is proportional to the amount of precipitation of austenite and definitely has a certain correlation with nitrogen content.

Most of the research on Widmanstätten structure so far has been about carbon steel [3]. Referring to the precipitation and growth conditions in carbon steel, we can speculate that the austenite of Widmanstätten structure precipitated in the duplex stainless steel is also related to the heat input and the certain cooling rate. The only difference is that the effect of the alloy composition on Widmanstätten-ferrite is the carbon content of the component and its rate of diffusion; however, Widmanstätten-austenite is dominated by nitrogen. And we suspect that the nitrogen content in austenite at different temperatures will also have a certain effect on it. Through the observation of the shape of Widmanstätten-austenite, we can find that it has the same shape as Widmanstätten-ferrite [4]. Some are grown along the grain boundary into the one adjacent side ferrite grain, and some are grown on both of the two adjacent sides; furthermore, some of them are continuous and some are disconnected. Most of the Widmanstätten structure as triangular or needle shaped, while some of them also have a certain angle, similar to the shape of shark teeth. Although the welding velocity has a certain influence on the precipitation of Widmanstätten-austenite in the welds, it is difficult to find a regular change in the shape of Widmanstätten-austenite. And in the super duplex stainless steel with the highest nitrogen content, we can find that Widmanstätten-austenite is generally slightly larger than the ordinary duplex stainless steel with the lowest nitrogen content.

Due to the heterogeneity and preferential growth direction of Widmanstätten-austenite, when we take the cross section for observation, there definitely has a Widmanstätten-austenite that is not parallel to the cut surface. If the cut portion is at its elongated needle tip, it will be look like many small austenite which were precipitated in the ferrite grains. Therefore, in order to make a distinction, we will use EBSD to analyze the grain orientation of intragranular austenite and Widmanstätten-austenite after this paper. Moreover, we believe that if there has a way which can conduct three-dimensional observation of Widmanstätten-austenite, it will help us better explain these problems and understand their true shape.

Furthermore, in Fig. 3b, c we can observe that there are many black areas which were being precipitated in the ferrite grain that what can be presumed to be CrN and Cr2N from [5]. Their precipitation is affected by the rapid cooling effect after laser beam welding, which leads to a decrease in the solubility of nitrogen in the ferrite. Indicated from the results of this experiment, once CrN or Cr2N is precipitated, it will reduce the possibility of precipitation of austenite in the ferrite grains and also

Fig. 4 Effect of post-weld heat treatment on microstructure of S31803 in welds (**a** after electric furnace heating, **b** no affected area after laser heating, and **c** affected area after laser heating)

reduce the growth ratio with the Widmanstätten-austenite. Therefore, how to prevent the precipitation of nitride in the ferrite grain is one of the important methods which adjust the welds of the duplex stainless steel when we use the laser beam welding.

Post-Weld Heat Treatment

In order to adjust the phase ratio of the welds after laser beam welding, we tried to use the electric furnace at the beginning. According to the austenitic precipitation temperature range of the duplex stainless steel, we set the heating temperature to 1100 °C [6] and the heating time was set to 30 h. Although the results of this experiment show that the phase ratio at the welds is restored to the parent metal level, due to the heating time is too longer and also the electric furnace was not only heated the welds. Therefore both of the parent metal and the welds are become coarse grains (Fig. 4a). Therefore, for the convenience of the overall heat treatment process and the material properties of the parts other than the welds were not affected by the heat treatment, we have decided to try to perform post-weld heat treatment with the same heat source as the welding.

For further observation, the post-weld heat treatment condition (No. 4) is based on the least amount of austenitic precipitation laser beam welding condition. The metal structure of the welds after the post-weld heat treatment is shown in Fig. 4c. Regarding the result, although the phase ratio was not as effective as by electric furnace, for the processing velocity of the laser post-weld heat treatment and the result that the base material was not affected, we confirm it will be feasible and effective in practical applications. However, after the post-weld heat treatment, due to the high density of the laser heat source, a ferrite single phase structure part appeared on the surface of the welds. The surface was re-melted by the overheat, after that it has become to ferrite structure by rapidly cooled.

On the other hand, if reducing too much output of the laser beam welding, the heat treatment temperature may become too lower to change the phase ratio of welds. Moreover, how to solve the problem of the depth of influence of post-weld treatment is also a point that must be considered when selecting conditions. Unfortunately, at this time, there are still nearly 2/3 parts in welds that are not affected as shown in Fig. 4b.

Conclusions

The summary of the microstructure that the three types of the duplex stainless steels under the laser beam welding and the post-weld heat treatment is as follows:

(1) Under the same laser output, the faster the welding velocity, the less obvious the precipitation of the austenite in the welds.
(2) The amount of nitrogen and the concentration distribution determine the precipitation of austenite. The concentration distribution of nickel and chromium in the welds is not obvious.
(3) The shape of Widmanstätten-austenite is similar to that of Widmanstätten-ferrite. Its growth and shape are less affected by temperature, and it is more obvious in steels with higher nitrogen content. The growth and shape are affected by the influence of chromium nitride.
(4) It is feasible to apply a laser as a heat source for post-weld heat treatment. Under appropriate conditions, the area affected by the heat treatment promotes the precipitation and growth of austenite. However, the depth of the affected part is shallow. Also, if the heat input is too large, the surface will re-melted by the overheat and converted into ferrite single phase solidification.

References

1. Ogawa T, Koseki T (1988) Welding technology trend of duplex stainless in oil and gas industry applications. J Jpn Weld Soc 57(2):92–105
2. Omura T, Kushida T, Komizo Y (2000) Microstructural features and corrosion properties in laser welded duplex stainless steels. Weld Int 14(4):257–260
3. Todorov RP, Khristov KhG (2004) Widmanstatten structure of carbon steels. Met Sci Heat Treat 46(1):49–53. https://doi.org/10.1023/B:MSAT.0000029601.58461.bd
4. Yin J, Hillert M, Borgenstam A (2017) Morphology of proeutectoid ferrite. Metall Mater Trans A 48(3):1425–1443. https://doi.org/10.1007/s11661-016-3903-y
5. Kokawa H, Okada J, Kuwana T (1993) Nitrogen absorption and microstructure of duplex stainless steel weld metal. Weld Int 7(5):384–389
6. Fukui F (1981) Weldability duplux stainless steel. J Jpn Weld Soc 50(3):235–240. https://doi.org/10.2207/qjjws1943.50.235

Experimental and Numerical Investigation on Surface Damage of Cold Rolled Sheet Caused by Inclusion Movement

Xin Li, Min Wang, Lidong Xing, Jianhua Chu and YanPing Bao

Abstract In present work, samples were cut from ultra-low carbon steel slab to investigate the formation of surface damage induced by inclusions during cold rolling. Surface damage such as dent, peeling, and pockmark was reproduced by rolling experiment. The inclusions such as Al_2O_3, SiO_2, and $MgO{\cdot}x CaO$ were found in surface damages. FE method was employed to reveal the relationship between surface damage and Al_2O_3 inclusion movement. The initial position and size of inclusions are significant to the formation process of surface damage. 50 μm Al_2O_3 inclusion located at the 1/8 of sheet thickness is easily rolled to the surface and causes the surface damage. Morphology of Al_2O_3 inclusion is difficult to change due to the higher hardness and larger elastic modulus. However, the Al_2O_3 inclusion can be crushed into pieces, which increase the damage of steel matrices. Surface damage was formed through the coalescence behavior of microscopic damages.

Keywords Surface damage · Inclusion movement · Cold rolling process · Ultra-low carbon steel · Damage coalescence

Introduction

With the development of Chinese automobile industry, surface quality requirements of cold rolled ultra-low carbon steel sheets have been gradually improved. Surface damage is a long-standing problem occurring in the cold rolled ultra-low carbon steel sheets, although the process flows are carefully controlled. Surface damage in cold rolled steel sheets is not a special case, but a common phenomenon in steel industry. Through recent ten years literature investigation, surface damage in cold rolling process was induced by inclusions [1–5], mould slag particles [6–9] and bubble damage [10–12]. For example, Al_2O_3 and FeO_x particles were found under the pockmarked and scratch damage using scanning electron microscope (SEM) and

X. Li · M. Wang · L. Xing · J. Chu · Y. Bao (✉)
State Key Lab of Advanced Metallurgy, University of Science and Technology Beijing, Beijing 100083, China
e-mail: 15030561550@163.com

J. Lee et al. (eds.), *Materials Processing Fundamentals 2020*, The Minerals, Metals & Materials Series, https://doi.org/10.1007/978-3-030-36556-1_20

energy dispersive spectrometer (EDS) [2]. Peeling damage in cold rolled sheets was mainly induced by mould slag particles, Al_2O_3 clusters, and bubbles [6, 11].

Many works indicated the inclusions were harmful to the rolled sheets, and the related technologies schemes were applied in continuous casting and rolling. A series of studies on surface damage indicated the amount and size of inclusions have a negative effect on the control of surface damage [13]. A novel mould slag was employed to reduce the large inclusions sources, which was contribution to control surface damage [14]. Pickling, surface cleaning process, rolling speed control process, and rolling temperature control process were carried out to reduce the appeared probability of surface damage during hot rolling [15–17]. In addition, surface damages were reproduced by rolling experiments to understand the formation and evolution process of surface damage [18–20]. Although many measures have been taken in casting and rolling, surface damage cannot be solved perfectly. Most researchers suppose that subcutaneous inclusions were the main inducing factors during the formation of surface damage, especially in the sheet rolling process [21–25]. Inclusions were randomly distributed within the sheet matrices before cold rolling. With the reduction of sheet thickness, the distance between the sheet surface and inclusions was continuously reduced. It is like those inclusions moving toward the sheet surface, which has a negative effect on surface quality. However, the relationship between inclusion movement and surface damage was less reported.

In present work, the size and position of inclusion in hot rolled plates were detected using ultrasonic device and electrolytic extraction method. Then, surface damages induced by inclusions were reproduced through cold rolling experiments and characterized by SEM and EDS. To further reveal the relationship between inclusion movement and surface damage, a 2D FE model was established. At last, the formation of surface damages caused by inclusion movement was discussed.

Experiment

Ultra-low carbon steel main composition (wt%) was C: 0.0018, Si: 0.01, Al: 0.03, Mn: 0.0095, O: 0.0015. Physical rolling experiments were carried on rolling machine as shown in Fig. 1. Hot rolled plates were originated from head slab containing greater amount and larger size of inclusions to ensure the formation of surface damage in experiments. Before cold rolling experiments, hot rolled plates were detected through ultrasonic device to determine the size and position of inclusions. Ultrasonic device testing thickness was about 2 mm, and the minimum detection size of inclusion was about 100 μm. Electrolytic extraction method was used to extract inclusions from steel to verify the ultrasonic device testing results. Cold rolling samples were cut from hot rolled plates and put in acid solution more than 10 h. And then samples were cleaned by mechanical brush to ensure the sheet surface cleanness and smoothness. The sheet thickness is 6, 3.0, 1.5, and 1.0 mm, respectively, during cold rolling processes. Finally, the cold rolled sheets with surface damage were observed using SEM and EDS.

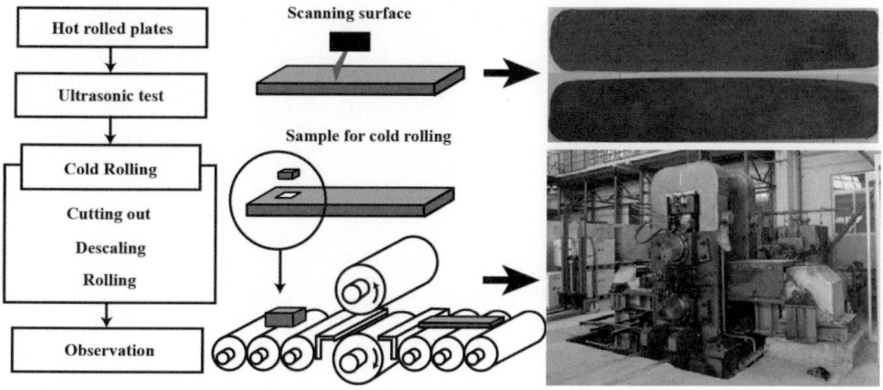

Fig. 1 Schematic of experimental process

Semi-geometric model was established in ANSYS-LS platform to track the inclusion movement, as shown in Fig. 2a. The rolls diameter (Dr) is 400 mm, and the rolling direction and sheet thickness direction are—*X* and *Y*, respectively. The model element type is selected 2D solid-162. Rigid material model and bilinear isotropic material model are employed for the rolls and inclusions, respectively. Friction coefficient between inclusions, matrix, and rolls is assumed 0.3. Mapped method is used for meshing the geometrical model as shown in Fig. 2b. Material parameters are listed in Table 1.

Figure 2a shows L_i is the distance between inclusion top and sheet surface. With the reduction of sheet thickness, the inclusion move towards the sheet surface. Top node of inclusion was set as tracking node (A) in Fig. 2b. Figure 2c shows a case of acquiring the inclusion position by tracking the top node position. L_i can be calculated by the difference of inclusion position and sheet thickness.

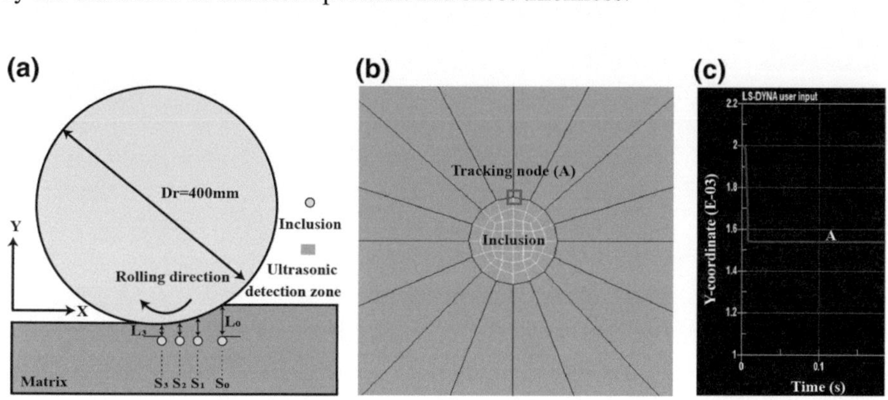

Fig. 2 Schematic diagram of rolling: **a** FE model, **b** meshing of matrix with inclusion, **c** node tracking process

Table 1 Material parameters in rolling

Parameters	Roll	Sheet	Inclusion
Density (kg m^{-3})	7850	7800	3800
Young' modulus (GPa)	210	193	350
Poisson' ratio	0.3	0.32	0.24
Initial deformation resistance (MPa)	–	210	264
Harding coefficient	–	0.112 [22]	–

Results and Discussion

Source of Inclusions

Figure 3 illustrates inclusion distribution in hot rolled plates detected through ultrasonic device and electrolytic extraction. Inclusions were enriched in the subcutaneous zone with 2 mm thickness on both sides of the plate, as shown in Fig. 3b. Metallographic samples were cut from the normal zone and inclusion enriched zone for electrolytic extraction experiment. Figure 3a indicates most Al_2O_3 and SiO_2 particles less than 20 μm are found in the normal zone. But the SiO_2, Al_2O_3, and $MgO{\cdot}xCaO$ particles are more than 100 μm in the inclusion enriched zone, as shown in Fig. 3c. These large inclusions with irregular morphology were originated from mold power particles, while the spherical Al_2O_3 and block SiO_2 less than 20 μm were deoxidation products [26].

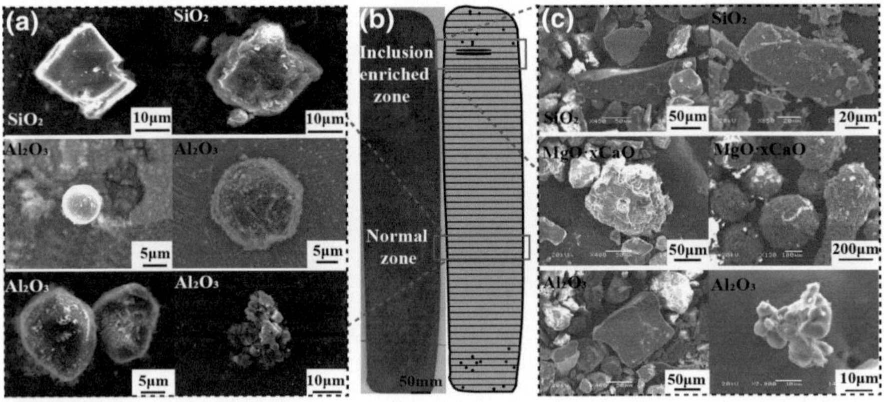

Fig. 3 Inclusions in hot rolled plate: **a** Inclusions in normal zone, **b** ultrasonic testing results, **c** Inclusions in enriched zone

Surface Damage Reconstruction

Cold rolling sample 1 (S1) was cut from the normal zone without macro inclusions, but sample 2 (S2) was originated from the inclusions enriched zone, as shown in Fig. 4a. After cold rolling, serious surface damages were distributed on both sides of the S2, while slight surface damage on S1, as shown in Fig. 4a. Although the inclusion size was smaller in S1 than that in S2, surface damage in cold rolled sheets was still induced by the inclusions, as shown in Fig. 4a, b. These results indicate surface damage is strongly influenced by the size and position of inclusion. Dent damage was induced by Al_2O_3 and SiO_2 as shown in Fig. 4c. Figure 4d illustrates a large number of Al_2O_3, SiO_2, and $MgO{\cdot}xCaO$ inclusions less than 10 μm were existed in the peeling damage. Pockmark damage was induced by the Al_2O_3 and $MgO{\cdot}xCaO$ inclusions, as shown in Fig. 4e. In summary, a lot of inclusions such as Al_2O_3, SiO_2, and $MgO{\cdot}xCaO$ less than 10 μm is the main factor of surface damage formation, but these inclusions are far smaller than the inclusions in inclusion enriched zone. Thus, crushing behavior of inclusions was appeared in the sheet deformation processes.

Inclusion Movement and Surface Damage

Surface damage induced by inclusions can be proved by physical rolling experiments, but the behavior of inclusions was hardly observed in sheets. FE method was employed to track the inclusion movement in sheet during rolling. According to the morphology analysis of surface damage [20] and stress gradient distribution around single inclusion [26], we assume that surface damage can be formed when L_i less than the inclusion diameter. Inclusion sizes in normal zone are usually less than 20 μm, thus the inclusion size was set to 20 μm. Figure 5a depicts the relationship between sheet thickness reduction and inclusion movement under various initial positions (L_0). L_0 was set to 0.37 mm, 0.75 mm, 1.12 mm, and 1.50 mm, respectively, before rolling, and the L_3 is about 0.04, 0.19, 0.26, and 0.34 mm after three-pass rolling. Non-center Al_2O_3 inclusions are gradually close to the sheet surface during rolling processes, but both of them cannot induce surface damage. Thus, inclusions in the 0.37 mm thick zone below the sheet surface may be the source of surface damage. Figure 5b shows the Al_2O_3 movement behavior of 5 μm, 10 μm, 20 μm, and 50 μm, respectively, in the 0.37 mm thick zone. Al_2O_3 with a size of 50 μm is almost moved to the sheet surface, which can act as the induction factor of surface damage. From the above analysis, the size of inclusions controlled within 20 μm can reduce the probability of surface damage formation. During deformation process, the deformation of surface steel matrices is larger than the center matrices [27, 28], and the deformation around hard inclusion is not homogeneous for the resistance of inclusion [21]. Thus, large inclusions near the surface can be easily transferred to the sheet surface. According to the simulation results, surface damage can be induced by Al_2O_3 inclusions, but the initial position of smaller inclusion is further close to the surface.

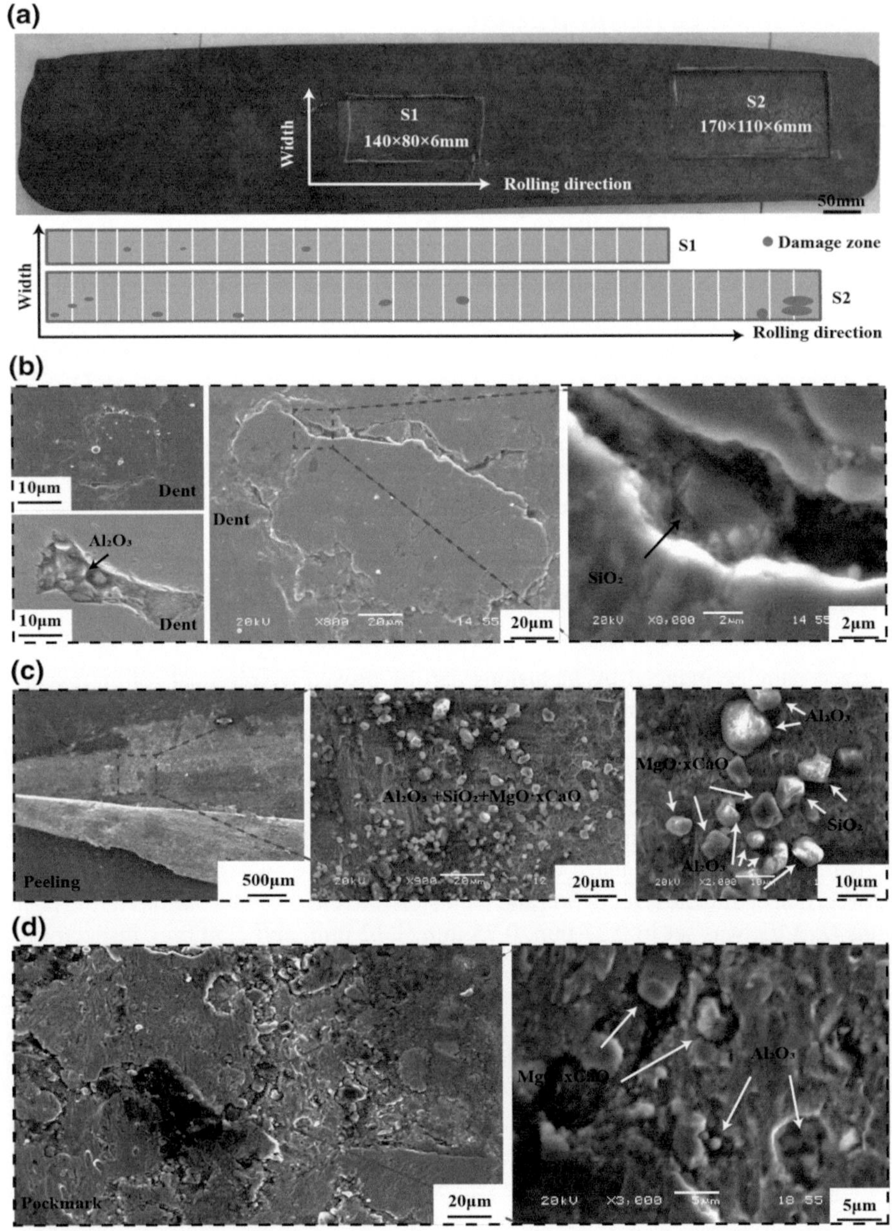

Fig. 4 Surface damage distribution: **a** Samples cutting and cold sheet testing, **b** dent damage, **c** peeling damage, **d** pockmark damage

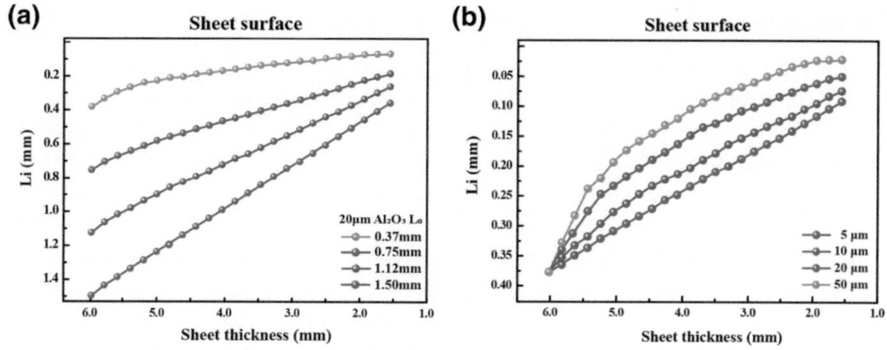

Fig. 5 Sheet thickness and inclusion position: **a** various relative positions, **b** various sizes

Microscopic damages around inclusions were existed in the vertical section of surface damage, as shown in Fig. 6a. Due to the higher hardness and larger elastic modulus of Al_2O_3 [29], the deformation of steel matrices around Al_2O_3 inclusion is hindered. Microscopic damage such as voids and cracks appears around inclusions. Combined with the inclusion movement behavior, the formation mechanism of surface damage can be described in Fig. 6b: Hard and brittle inclusions are existed near the sheet surface. As the sheet thickness reduction, microscopic damages were generated around inclusions in rolling direction. In addition, large brittle inclusions can be rolled into small pieces increasing the probability of damages formation [30]. With the inclusion moving towards the surface gradually, the damages also propagate along the rolling direction. Microscopic damages were coalesced with around damages, and the larger damages were formed. When the damage spread to the surface, the surface damage is formed.

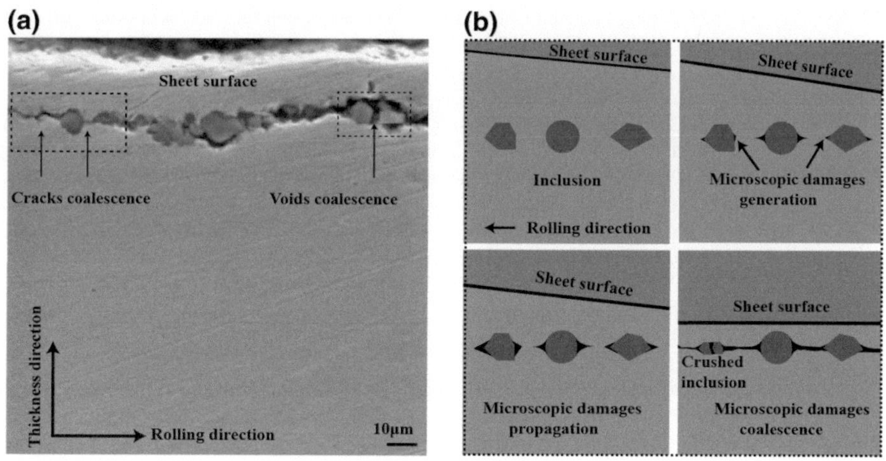

Fig. 6 Damages and inclusions: **a** vertical section of surface damage, **b** formation mechanism of surface damage

Conclusions

Source, morphology, size, and position of inclusion were investigated to provide a foundation for surface damage research. Surface damage such as dent, peeling, and pockmark is reproduced by rolling experiments. Al_2O_3, SiO_2, and $MgO{\cdot}xCaO$ originated from hot rolled plate are found in surface damages. FE method is employed to reveal the relationship between inclusion movement and surface damage. The results obtained in this study are as follows: At the same initial position, the inclusion is more easily move to the surface and induce the surface damage with the increase of inclusion size during rolling. Surface damage can be induced by various sizes of inclusion, but the initial position of smaller inclusion is further close to the surface. Propagation and coalescence of microscopic damages around inclusion are the vital way for microscopic damages transition to macroscopic surface damages.

Acknowledgements This work was supported by the National Natural Science Foundation of China (No. 51774031), the Open Project of State Key Laboratory of Advanced Special Steel, Shanghai University (No. SKLASS 2017-13).

References

1. Zhu GS, Yu HX, Wang WJ, Wang XH (2004) Study of surface defects of cold-rolled IF steel sheet. Iron Steel 4:53–56
2. Guo WF (2006) Research on surface defects of cold rolled sheet. Steelmaking 22:22–25
3. Zhang WW, Li XW, Lü CF (2009) Surface sliver defect of cold-rolled if steel sheet. J Iron Steel Res 21:59–62
4. Peng QC, Tian J, Zhang XH, Tang ST, Zhang W (2009) Progress of research on inclusion induced surface defect of cold rolled sheet. Steelmaking 25:73–77
5. Wang Y, Qin ShY, Zhang QJ, Zhu LG, Ma JH (2017) Research on alice skin defect on cold rolled sheet surface. Foundry Technol 38:1643–1645
6. Li DK, Yuan XM (2007) Analysis of surface bubble defects in cold rolled strip. J Iron Steel Res 35:23–25
7. Song JY, Zhao Y, Chen LSh, Wei YL, Tian YQ, Yang D (2013) Analysis on occurrence and cause of linear defects on cold-rolled sheet. Iron Steel Vanadium Tittanium 34:107–112
8. Ji YQ, Wang XH, Deng XX, Yang D (2014) Study on the cause and controlling of defects for cold-rolled sheet. Steelmaking 30:22–25
9. Dong ShP, Song JY, Chen YX, Zhang BQ, Chen LSh, Tian YQ (2012) Types and forming reasons of cold rolled steel plate surface defects. Mater Mech Eng 66:100–104
10. Shan QL, Peng GP, Zhang BL (2015) Study on surface sliver defect of cold-rolled sheet. Steelmaking 31:67–68
11. Cui JZ, Gao TZ (2016) Analysis and improvement measures of surface upwarping defects of cold-rolled plate. Hot Work Technol 45:172–174
12. Zhang ZQ (2018) Analysis of surface slivers on cold rolled sheet for the second slab in first heat. Contin Cast 43:77–79
13. Wang HF, Li YH (2011) Control technology of subsurface inclusion defects of continuous casting slab. Contin Cast S1:367–369
14. Wu XY, Zhu LG, Mei GH, Yan ChL, Gao ShL, Zhang QJ (2017) Analysis and research for surface sliver defect of steel plate. Iron Steel Vanadium Titanium 38:157–162

15. Wang XH (2013) Non-metallic inclusion control technology for high quality cold rolled steel sheets. Iron Steel 48:1–7
16. Chu JW, Sun MJ, Qiu ZS (2015) Analysis on quality problems about hot rolled IF steel sheet and research on the key technologies. Steel Roll 32:18–21
17. Gao RF, An DY, Guan JD, Wang Ch, Yu Y, Wu QL (2018) Discussion and control on surface sacle on hot rolled auto plates. Metal Mater Metall Eng 1:19–24
18. Peng K, Liu YZ, Xie B, Zhou LY, Cui J (2007) Analysis of sacr defect of hot rolled strip. Iron Steel 42:44–46
19. Jing YA, Han Y, Hu XD, Zhu KY (2007) Evolution of rolling scratch on surface of steel sheet. J Univ Sci Technol Liaoning 35:412–417
20. Cui H, Wu HJ, Yue F, Wu WS, Wang M, Bao YP (2011) Surface defects of cold-rolled Ti-IF steel sheets due to non-metallic inclusions. J Iron Steel Res Int 18(S2):335–340
21. Luo C, Stånhlberg U (2001) Deformation of inclusion during hot rolling of steels. J Mater Process Technol 114:87–97
22. Yu HL, Liu XH, Bi HY, Chen LQ (2009) Deformation behavior of inclusion in stainless steel strips during multi-pass cold rolling. J Mater Process Technol 209:455–461
23. Luo C (2001) Evolution of voids close to an inclusion in hot deformation of metals. Comput Mater Sci 21:360–374
24. Ervasti E, Stahlberg U (2005) Void initiation close to a macro-inclusion during single pass reductions in the hot rolling of steel slabs: a numerical study. J Mater Process Technol 170(1–2):142–150
25. Yu HL, Liu XH, Li XW (2008) FE analysis of inclusion deformation and crack generation during cold rolling with a transition layer. Mater Lett 62(10–11):1595–1598
26. Yu HL, Bi HY, Liu XH, Tu YF (2008) Strain distribution of strips with spherical inclusion during cold rolling. T Nonferr Metal Soc 18:919–924
27. Liu C, Hartley P, Sturgess CEN, Rowe GW (1987) Finite-element modelling of deformation and spread in slab rolling. Int J Mech Sci 29:271–283
28. Serajzadeh S (2006) A model for prediction of flow behavior and temperature distribution during warm rolling of a low carbon steel. Mater Des 27:529–534
29. Li X, Bao YP, Wang M (2018) Genetic evolution of inclusions in interstitial-free steel during the cold rolling processes. Trans Indian Inst Met 71:1067–1072
30. Stiénon A, Fazekas A, Buffière JY, Vincent A, Daguier P (2009) A new methodology based on x-ray micro-tomography to estimate stress concentrations around inclusions in high strength steels. Mater Sci Eng, A 513:376–383

Heterogeneous Grain Microstructure Reducing/Eliminating Edge Breaks in Low Carbon Steels

Tihe Zhou, Hatem Zurob, Peng Zhang and Sang Hyun Cho

Abstract Edge breaks initiating at both edges of steel strip are strain lines/Lüders lines which are unacceptable for surface critical and semi-critical applications. Some steel manufacturers are using de-gassing technology by producing interstitial-free steel to reduce or eliminate edge breaks and associated yield point elongation (YPE). Nonetheless, in low carbon steel with free interstitials (C, N), it is necessary to develop other approaches to reduce edge breaks and YPE. This paper proposed that heterogeneous grain microstructures can reduce or eliminate YPE in steels containing free interstitials. Heterogeneous grain microstructures can be obtained in two ways during cold mill processing low carbon steels: (1) through partially recrystallized microstructure using a low annealing temperature, and (2) creating a bi-modal grain size distribution in a full-recrystallized material. The mechanisms of heterogeneous grain microstructure on the formation of edge breaks/Lüders lines, as well as the effects of chemical composition, reduction ratio, annealing temperatures, and mechanical properties are discussed.

Keywords Heterogeneous grain structure · Yield point elongation · Edge breaks · Lüders lines · Low carbon steel

Introduction

The surface quality of steel sheet is becoming more critical to the modern steel industry. Edge breaks are surface defects which are recognized unacceptable for surface critical and semi-critical applications such as exposed automotive sheets, appliance panels, grain bins. Edge breaks (EB) are defined as transverse Lüders lines, which are normally initiated at two edges of the steel strip. In some cases, these Lüders lines can go through the full width of the strip. Zhou et al. [1] summarized the different

T. Zhou (✉) · P. Zhang · S. H. Cho
Algoma Inc., 105 West Street, Sault Ste. Marie, ON P6A 7B4, Canada
e-mail: tom.zhou@stelco.com

H. Zurob
Department of Materials Science and Engineering, McMaster University, 1280 Main Street West, Hamilton, ON L8S 4L8, Canada

J. Lee et al. (eds.), *Materials Processing Fundamentals 2020*, The Minerals, Metals & Materials Series, https://doi.org/10.1007/978-3-030-36556-1_21

appearances of edge breaks, which include angular, radial, dispersion, wrinkled, and straight Lüders lines. Typically, edge breaks appear on steel strips when they are exiting the cold mill processing unit. The challenge for most steel manufacturers is that the initiating point for edge breaks cannot be identified during processing at cold mill and is not caught until finishing the last processing step of temper rolling. Even though edge breaks have different names and different appearances, the mechanism of this defect is Lüder band formation when steel strips are processed at different processing units. According to [2, 3], Lüders bands have been studied since 1840s and are often noted in mild steel and other body-centered cubic (BCC) poly crystalline metal. Lüders bands are a result of non-uniform yielding during the localized heterogeneous transition from the elastic to plastic deformation process. This non-uniform yielding process is also described as the yield point phenomena. Hall [4] and Srinivasan et al. [5] explored the non-uniform yielding mechanism using the interaction of solute atmosphere and dislocation movement. As shown in Fig. 1, the up yield point (UYP) is corresponding to the unpinning of local dislocation from their solute atmosphere which leads to the Lüders band formation. The low yield point (LYP) and the coming load plateau are corresponding to the yielding of the region ahead of the Lüders band which leads to increasing mobile dislocations density. This collective and self-organized movement of these mobile dislocations manifest as Lüder band front propagation on the surface of the steel strip. Non-uniform yielding is normally measured by the yield point elongation (YPE). YPE is the difference in strains between the strain at UYP and the strain at the end of LYP region.

Most research on this topic is concerned with dislocation dynamics of Lüders bands initiation/propagation and Lüders bands characterization [6–9]; however, less attention was given to minimize Lüders bands by tailoring material chemistry, process design, and operating practices. This paper will target at adjusting microstructures by revising the reduction ratio and annealing temperature to reduce edge breaks/Lüder lines, i.e., minimize YPE during the cold mill processing different low carbon steel

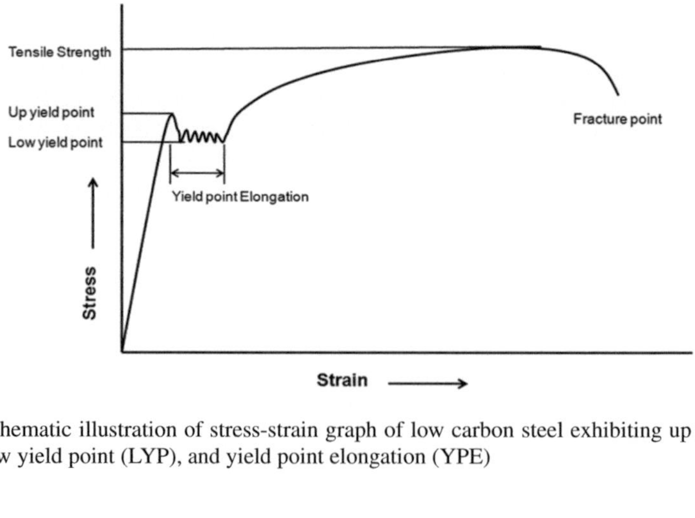

Fig. 1 Schematic illustration of stress-strain graph of low carbon steel exhibiting up yield point (UYP), low yield point (LYP), and yield point elongation (YPE)

grades. This endeavor not only will support steel industry to cut down internal rejection, customer claim, un-planned inspection, and steel retreat due to the suspected edge break defect but also can reduce energy consumption and environmental impact.

Materials and Experimental Procedure

Experimental specimens were taken from the hot-rolled steel strip at the hot strip mill before being transferred to the cold mill processing unit. Table 1 showed the chemical compositions used in this study. The specimens of different chemistry were cold rolled to 60, 70, and 80% reduction ratios. Table 2 summarized the annealing cycles used in this investigation. Annealing cycles 1, 2, and 3 (AC1, AC2 and AC3) were using industrial bath annealing furnace, while annealing cycle 4 (AC4) was using a programmable tube furnace with controlled 7% hydrogen and 93% nitrogen mixed atmosphere which is similar to the industrial batch annealing (HNX) atmosphere. All specimens were furnace cooled to 100 °C after annealing treatment, then air cooled to room temperature. A Lloyd Instruments T5100 Tensile Testing machine was used for tensile testing according to ASTM E8. Cold reduced and annealed specimens were prepared by standard metallographic techniques and etched by using 2% nital etching solution. The microstructure was studied using optical microscopy, and grain size was performed using the NIS-Elements software. In order to obtain reliable statistics, five micrographs were analysed for each specimen.

Results

An example of specimen 4032 microstructure evolution with 60, 70, and 80% reduction ratio is summarized in Fig. 2. The cold reduced optical microstructures consist of ferrite (bright phase) with a small percentage of cementite carbide (dark phase). Figure 2 displayed highly deformed ferrite grains and a preferential grain boundary distribution of the directional carbides.

Example of optical microstructures and their corresponding tensile stress-strain behavior with different chemical compositions and reduction ratios annealed by AC1, 2, 3, and 4 are summarized in Figs. 3, 4, 5, and 6, accordingly. Figures 3, 4, 5, and 6 show the bright phase is ferrite and dark phase is cementite.

AC1 with high annealing temperature (cold spot 690 °C) is used for drawing quality (DQ) applications. Figure 3a shows specimen 4039 (with boron addition) cold reduced 60% has uniform recrystallized ferrite microstructure, with the average grain size of 14.1 μm. The UYP and LYP are well defined with 6.1% YPE, which is shown in Fig. 3b. Specimen 4032 (with 60% reduction ratio) without boron addition shows coarser ferrite grains (Fig. 3c), whose average grain size is about 27.4 μm. However, the stress-strain curve shown in Fig. 3d has about 4.5% YPE.

Table 1 Chemical composition of experimental materials (wt%)

Sample ID	C	MN	S	P	Si	Al-T	B	Ca	N
4032	0.044	0.25	0.008	0.010	0.015	0.025	0.0002	0.0022	0.0050
4039	0.041	0.17	0.004	0.007	0.023	0.031	0.0053	0.0034	0.0040

Table 2 Summary of annealing cycles (AC)

Set-points		Cycle 1 (°C)	Cycle 2 (°C)	Cycle 3 (°C)	Cycle 4[a] (°C)
Final	Hot spot	720	685	650	700
	Cold spot	690	620	560	700
Step 3, 2 h	Gas stream	600	600	550	0
Step 2, 4 h	Gas stream	490	480	450	0
Step 1, 8 h	Gas stream	375	375	345	0

[a]Annealing Cycle 4 holding time is 20 h, the temperatures of hot spot and cold spot are the same for a programmable tube furnace

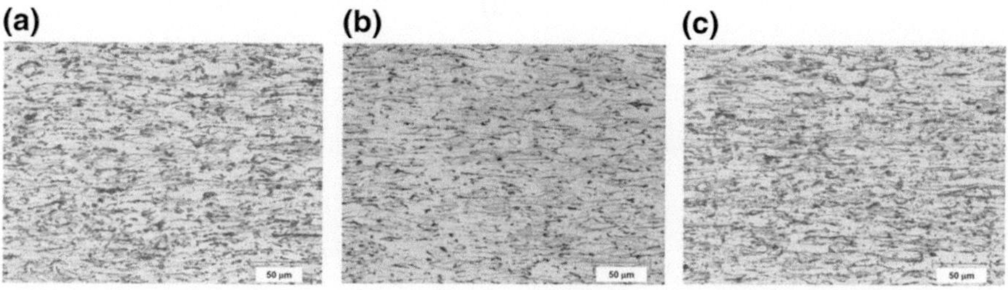

Fig. 2 Specimen 4032 microstructure evolution with different reduction ratios. **a** 60%, **b** 70%, and **c** 80%

AC2 with cold spot 627 °C is designed for commercial quality (CQ) applications. Specimen 4032 (60% reduction) has very small uniform distributed ferrite grains with average diameter of 16.8 μm (Fig. 4a). Figure 4b shows that the yield strength is around 214 MPa, with YPE 3.8%. Figure 4c shows the microstructure of specimen 4039 (reduction ratio 60%), the average grain size is 14.9 μm, with approximately YPE 5.5% (Fig. 4d). The UYP and LYP of 4032 and 4039 are both well defined as well.

AC3 is designed for non-formable commercial quality application with low cold spot temperature of 590 °C. Figure 5a, c shows a heterogeneous, partially recrystallized microstructure with reduction of 60% and 70%, respectively, and stress-strain curves (Fig. 5b, d) show 0.1 and 0.3% YPE accordingly.

AC4 has a holding temperature of 700 °C for 20 h. Both Fig. 6a, c show specimens 4039 with reduction of 60 and 70% have heterogeneous grain microstructure. Abnormal grain growth developed after the recrystallization during the grain growth process. The UYP and LYP are well defined in Fig. 6b, c, the YPE of both graphs, however, is less than 1% (around 0.7%) [1].

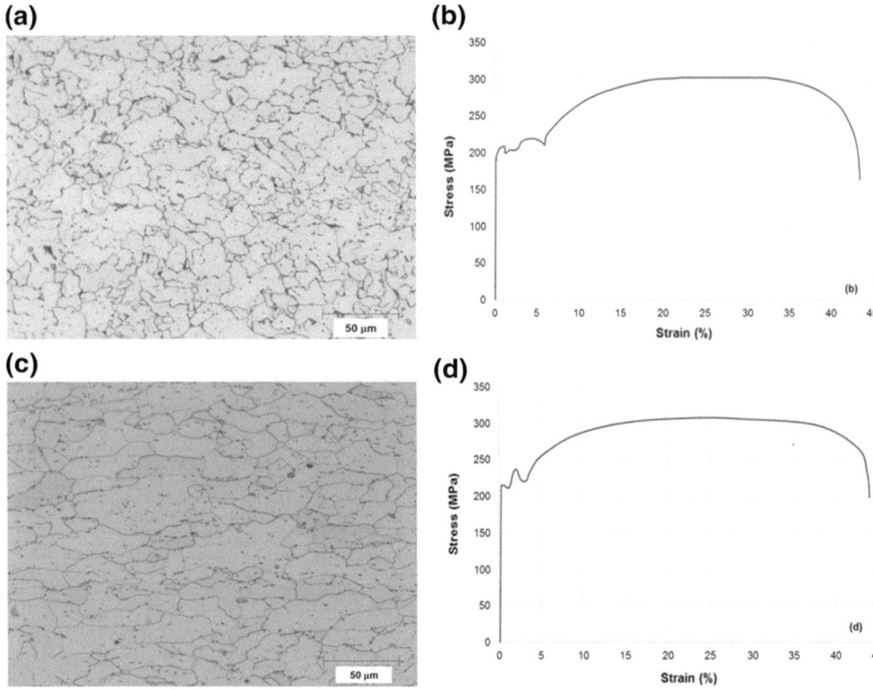

Fig. 3 Optical microstructure and tensile behavior of specimens are annealed by using AC1. **a** 4039 microstructure with 60% reduction ratio, **b** correlated stress-strain curve, **c** 4032 microstructure with 60% reduction ratio, and **d** correlated stress-strain curve

Discussion

In present study, YPE is varied from 0.1 to 6.1%. The most favorable outcomes are using specimen 4032 which has heterogeneous partial recrystallized microstructure annealed by AC3 (YPE, 0.1%) and using specimen 4039 which has bi-model (heterogeneous) grain size distribution annealed by AC4 (YPE, 0.7%). Both heterogeneous grain microstructures are the results of the chemical composition, reduction ratio, and annealing cycle. The heterogeneous grain microstructure and reduction ratio are analyzed first; then followed by discussion of the effects chemical composition, annealing cycles, and the effects on the mechanical properties.

Heterogeneous Grain Structure

Figure 5 shows those specimens 4032 with 60 and 70% reduction and annealed by using AC3 have YPE as low as 0.1%. The cold deformed microstructure would

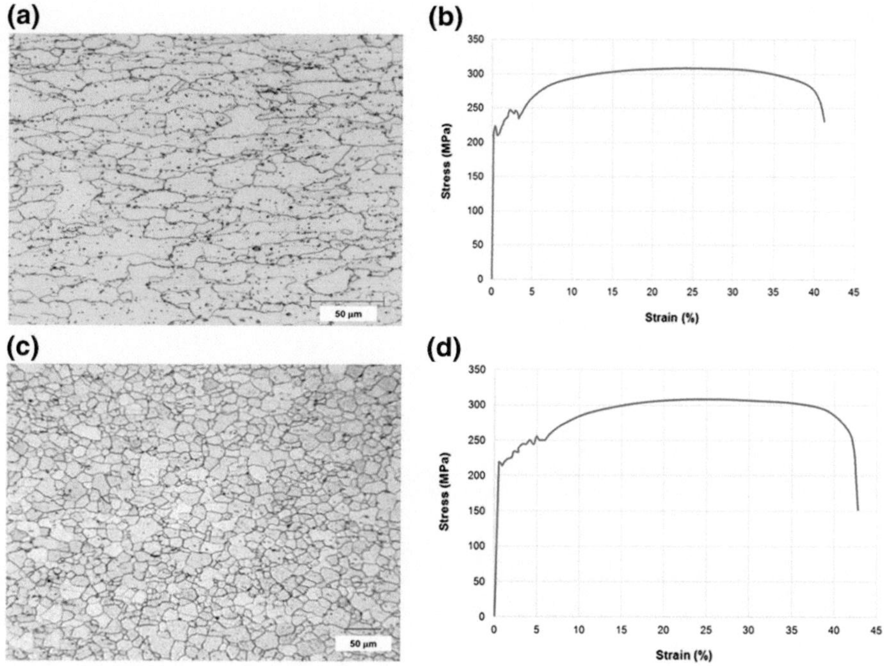

Fig. 4 Optical microstructures and tensile behavior of specimens are annealed by using AC2. **a** 4032 microstructure with 60% reduction ratio, **b** correlated stress-strain curve, **c** 4039 microstructure with 60% reduction ratio, and **d** correlated stress-strain curve

experience recovery, recrystallization, and grain growth during the annealing process. However, specimen 4032 has heterogeneous partial recrystallized microstructure after annealing owning to the low annealing temperature (cold spot temperature 590 °C). Elongated ferrite grains inherited from cold reduction still can be seen in both Fig. 5a, c. These elongated ferrite grains indicate the dislocation density is still very high in partial recrystallized microstructure after annealing. It is well established that the yield stress directly depends on the dislocation density and the number of dislocations increased after initiation. According to Hahn's theory [8], an increase of dislocation density would increase the chances of homogeneous deformation during tensile testing. Itoh et al. [9] also confirmed that further increasing dislocation density would increase work hardening and the plastic zone around the yield stress. Therefore, the stress-strain curves of specimen 4032 using AC4 showed almost continuous-yielding.

Figure 6 shows those specimens 4039 with 60 and 70% reduction and annealed by using AC4 also have a significantly reduced YPE. The YPE of these specimens is approximately 0.7%. According to Zhou et al. [1], specimen 4039 has 0.0053 (wt%) boron addition, the AlN and M_2B precipitates will pin ferrite grain growth after recrystallization during annealing process. However, the AlN and M_2B precipitates

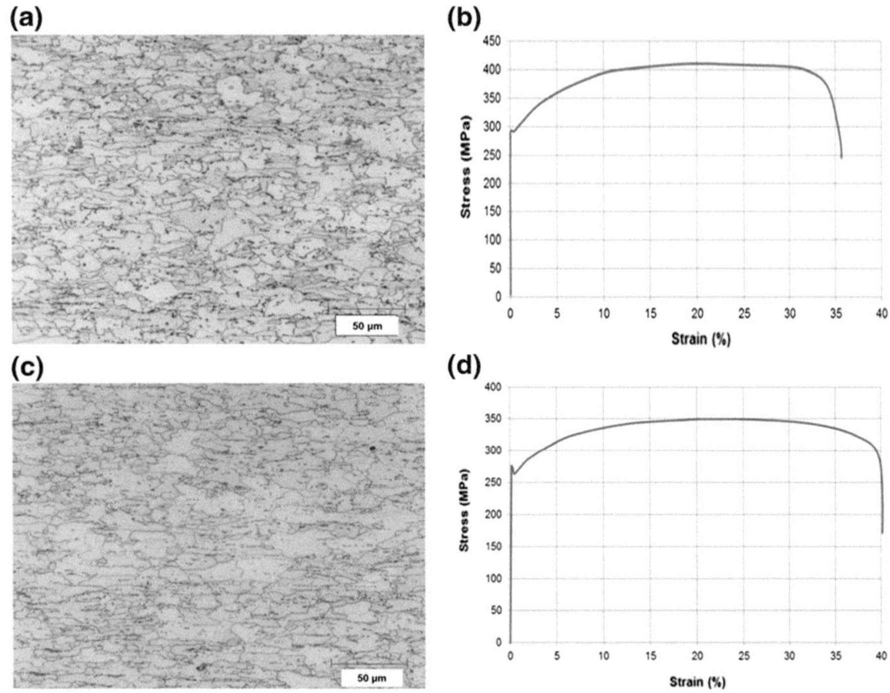

Fig. 5 Optical microstructures and tensile behavior of specimens are annealed by using AC3. **a** 4032 microstructure with 60% reduction ratio, **b** correlated stress-strain curve, **c** 4032 microstructure with 70% reduction ratio, and **d** correlating stress-strain curve

tend to dissolve with increasing time and temperature which will lead to the unpinning of some ferrite grains; these unpinning grains will then devour the original small grains by the abnormal grain growth process [10, 11]. Thus, to achieve a heterogeneous grain size distribution, it may be necessary to add boron in low carbon steel to trigger abnormal grain growth. The present study confirmed that a heterogeneous grain structure can greatly reduce YPE; however, the mechanism of heterogeneous grain microstructure decreasing Lüders band formation is unknown. Lüders bands are nucleated at stress concentration and will propagate when a nucleated Lüders band deforms the nearby materials [12]. The moving deformation region is called the band front, and slip lines develop prior to macroscopic yielding and form at the band front. The slip lines are the result of the generation of mobile dislocations forming at grain boundaries, preceding the Lüders bands. An increase in the number of mobile dislocations will reduce the grain's resistance to flow, which is known as the Lüders strain. The larger dislocation density with small grain size introduced larger Lüder strain [13, 14]. However, in heterogeneous grain microstructure (bi-modal grain size distribution), the larger grains have lower of surface area to volume, which means a lower ratio of grain boundary to dislocations. The less grain boundaries

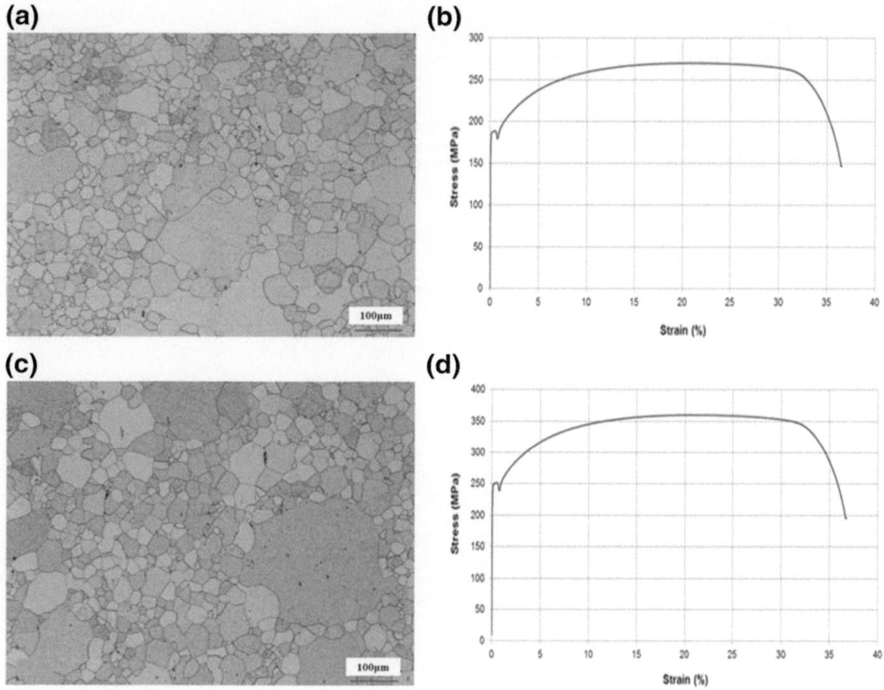

Fig. 6 Optical microstructures and tensile behavior of specimen 4039 are annealed by using AC4. **a** Microstructure with 60% reduction ratio, **b** correlated stress-strain curve, **c** microstructure with 70% reduction ratio and **d** correlating stress-strain curve [1]

would introduce the fewer obstacles for dislocation to move. Thus, the YPE of specimen 4039 annealed by AC4 is less than 1% comparing with specimens annealed by AC1 and 2 with uniform finer grain microstructure.

Reduction Ratio

Figure 2 shows specimen 4032 microstructure evolution with 60, 70, and 80% reduction ratios before annealing. Ferrite grains were heavy deformed and carbide were distributed along the grain boundaries. However, it is difficult to discern the microstructural differences with increasing reduction ratio before annealing. Table 3 summarized the effect of reduction ratios on average grain size and YPE by using different annealing cycles. During cold rolling process, most of the energy expended in cold work dissipates in the form of heat and only a small fraction (<10%) is stored in the reduced strip as strain energy. The releasing of stored energy provides the driving force for recovery and recrystallization during the subsequent annealing process [15]. From Table 3, specimen 4032 annealed by using AC1 and AC2 show an increase

Table 3 Summary of the effect of reduction ratio on average grain size and YPE using different annealing cycles

	Specimen ID	Average grain diameter (μm)	YPE (%)
AC1	4032, reduction ratio 60%	27.4	4.5
	4032, reduction ratio 70%	21.5	5.6
	4039, reduction ratio 60%	14.1	6.1
	4039, reduction ratio 70%	13.3	6.0
AC2	4032, reduction ratio 60%	16.8	3.8
	4032, reduction ratio 70%	14.7	4.5
	4039, reduction ratio 60%	14.9	5.5
	4039, reduction ratio 70%	13.7	5.9
AC3	4032, reduction ratio 60%	N/A	0.1
	4032, reduction ratio 70%	N/A	0.3
	4039, reduction ratio 60%	N/A	0.5
AC4	4039, reduction ratio 60%	26.7	0.7
	4039, reduction ratio 70%	31.8	0.7
	4032, reduction ratio 80%	18.9	6.0

in YPE and decrease of average grain size once reduction ratio is increased from 60 to 70%. More reduction ratio at reduction mill can provide more nucleation sites for recrystallization during annealing process to achieve a more uniform finer grain microstructure. A more uniform microstructure, specifically at the edges, may reduce edge breaks occurrence. However, a study on the effect of cold rolling on the yielding phenomenon of AA5182 aluminum alloy showed that increasing the reduction ratio increases the strain which yield point elongation occurs over, even after annealing. The reason for the increase in YPE with increase in reduction is due to dislocation glide [16]. Different sized grains do not have any difference with respect to how dislocations move to the grain boundary, since specimens with smaller grain sizes have a larger number of grains along its whole length, the number of steps will be larger, and the total YPE for smaller grain size specimens is larger. Therefore, a larger grain size microstructure will be required to decrease or eliminate Lüders bands/edge breaks. Table 3 has good agreement with Hu [13] and Johnson [14] observations: specimen 4032 using AC1 with a grain size of 27.4 μm (reduction ratio 60%) has a 4.5% YPE; however, with a smaller grain size of 21.5 μm (reduction ratio 70%) has a 5.6% YPE. Even though specimens 4039 with different reduction rations have heterogeneous grain microstructure by using AC4, it is affirmed that the YPE is reducing with increasing grain size. AC3 shows deformed grain size which cannot comparable with other annealing cycles. It is important to couple the influences of chemical composition, annealing temperature, and time, when studying the effect of grain size and grain microstructure on YPE, i.e., edge breaks.

Chemical Composition

It is well established that the Lüders band formation mechanism in low carbon steel is related to the unpinning of dislocations from interstitial solute atmospheres such as C and N [17, 18]. TCFE7 database of Thermo-calc was used to calculate the free C and N in specimens 4039 and 4032 [19]. The free C and N mass fraction in ferrite as function of temperature are shown in Fig. 7, respectively.

According to the calculation, there is a noticeable amount of free C and N atoms in the ferrite matrix when specimens were annealed at different cycles. Specimen 4039 with 0.0053 (wt%) boron has a noticeable decrease in free N in the ferrite matrix compared to specimen 4032 which has no boron addition. The mass fraction of free N in specimen 4039 (with B addition) reaches 2.35×10^{-7} at annealing temperature 700 °C, however, at the same temperature, 4032 (the boron-free steel) had a free N mass fraction of 1.32×10^{-6}. B will tie up N to form BN and M2B precipitates in the 4039 specimens. However, specimens of 4039 and 4032 still contain free C as shown in Fig. 7a. In order to eliminate or reduce YPE in low carbon steel, the essential point is to minimize the free interstitial elements C and N, for industrial without de-gasser facility, titanium could be added to tie up N and C at high temperature and produce steel strip which is similar to IF steel (called quasi-IF steel).

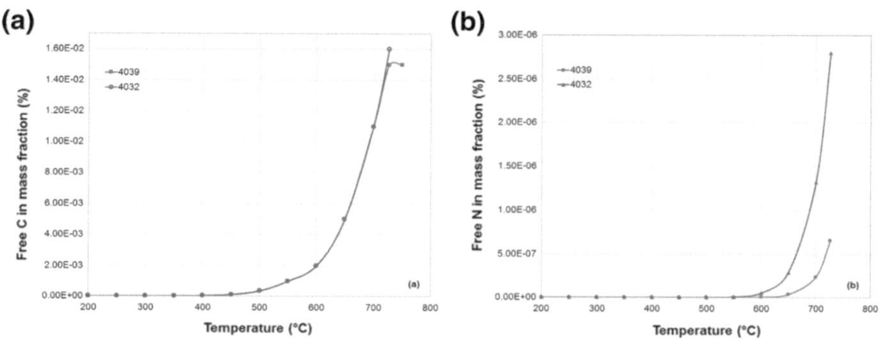

Fig. 7 Specimens 4039 and 4032 **a** free interstitial C and **b** free interstitial N in ferrite matrix versus temperature calculated by Thermo-calc of using TCFE7 database [19]

Annealing Temperature

Due to the significant impact of holding time and annealing temperature on the recovery, recrystallization, grain growth, and precipitates formation during annealing process, the design of annealing temperature and holding time is essential to control the final annealed microstructure. AC1 is designed for drawing quality applications. From Fig. 7, it can be seen that the concentration of free C and N interstitials increases with the increase of reheating temperature. Once the temperature reaches the cold spot set point, the reheating process is completed, and then continues with the subsequent cooling process. AlN and BN particles in specimen 4039 with B addition precipitate with decreasing of temperature. Owing to the Zenner pinning effect [10], these fine particles retarding grain growth results in an uniform finer ferrite grain microstructure as shown in Fig. 3a, with an average grain size 14.1 μm and a 6.1% YPE. It is concluded that steels with B addition cannot be used for drawing quality applications by using AC1. Specimen 4032 without boron addition subjected to AC1 has larger grain size with less YPE. AC2 with cold spot temperature 681 °C was designed for commercial quality non-surface critical application. Specimen 4032 showed YPE in the range of 3.8–5.5% with different reduction ratios. The AC3 shows partial recrystallization microstructure due to the low annealing temperature 590 °C, which introduce heterogeneous grain microstructure result in the YPE is less than 0.3%. AC4 was selected to maximize the formation of nitrides, carbonitrides and carbides, which will assist to remove free C and N interstitials in the ferrite matrix [20]. Specimen 4039 annealed by AC4 has an YPE less than 1% (about 0.7%).

Mechanical Properties

Table 4 summarized mechanical properties as a result of different reduction ratios and annealing cycles. Based on ASTM Al008 specification, steel strip for CQ applications recommend minimum yield strength of 140 MPa and maximum of 275 MPa, with minimum elongation of 30%. While DQ applications suggest meeting yield strength between 150 and 240 MPa, elongation of 36% minimum, as well as, n value in the range of 0.17–0.22, and r value between 1.3 and 1.7. From Table 4, it can be seen that specimens with reduction ratio of 60, 70, and 80% using AC1 and AC2 can meet CQ mechanical specification, among which, only specimen 4032 with AC1 can meet DQ n value and r value requirements. The yield strength annealed by AC3 are close to the high end of ASTM A1008 yield strength specification, therefore, this design can only be applied to commercial quality without high formability applications. Specimens with reduction ratio of 60 and 70% annealed by AC4 met both CQ and DQ tensile, elongation, and n value specification; unfortunately, r value calculation is not available.

Table 4 Summary of mechanical properties with different reduction ratios annealed by different annealing cycles

	Sample ID	Cold reduction (%)	Yield strength (MPa)	Tensile strength (MPa)	Elongation (%)	*n* value	*r* value
AC1	4032	60	214	311	43	0.244	1.95
	4032	70	226	315	42	0.234	1.84
	4032	80	230	319	43	0.203	1.72
	4039	60	192	304	43	0.232	1.14
	4039	70	189	315	41	0.228	1.13
	4039	80	201	310	40	0.226	1.38
AC2	4032	60	214	310	42	0.199	2.16
	4032	70	220	324	41	0.194	1.68
	4039	60	214	304	41	0.250	1.73
	4039	70	232	354	42	0.210	2.41
AC3	4032	60	287	415	35	0.210	1.73
	4032	70	249	354	40	0.225	2.18
	4039	60	296	437	32	0.199	1.71
AC4	4032	60	206	324	39	0.197	N/A
	4039	60	175	270	36	0.194	N/A
	4039	70	235	360	36	0.199	N/A

Conclusions

Heterogeneous grain microstructures can greatly reduce YPE. This can be achieved by two ways: one is to produce partial recrystallization microstructure; the other is to create abnormal grain growth after recrystallization. Both approaches can reduce YPE to less than 0.7%, while partial recrystallized heterogeneous grain microstructure almost eliminate YPE. The mechanisms of the heterogeneous grain microstructure greatly reducing YPE were discussed using dislocation density theory; however, these mechanisms will need further investigation. AC3 introduced partial recrystallized heterogeneous microstructure due to the low annealing temperature. While AC4 triggered abnormal grain growth creating bi-model grain size distribution owing to the dissolution of M_2B and AlN particles during annealing process. YPE increases with reduction ratio increasing from 60 to 70%. More reduction ratio provides more nucleation sites for recrystallization during annealing to achieve a more uniform and finer grain microstructure. Uniform and finer grain microstructure have lager YPE.

References

1. Zhou T, Zurob H, Zhang P, Kuuskman K, Cho SH, Burella D (2019) Control of edge breaks during cold mill processing of commercial and drawing quality low-carbon steels. Ironmak Steelmak 46(7):656–662
2. Moon DW (1971) Consideration on the present state of Lüders band studies. Mater Sci Eng 8:235–243
3. Hutanu R, Clapham L, Rogge RB (2005) Intergranular strain and texture in steel Lüders bands. Acta Mater 53:3517–3524
4. Hall EO (1970) Yield point phenomena in metals and alloys, 1st edn. Plenum Press, New York, pp 1–71
5. Srinivasan N, Raghu N, Venkatraman B (2014) Study on the deformation band characteristics in mild steel using digital image correlation. JMEST 1(5):400–403
6. Iricibar R, Mazza J, Cabo A (1975) The microscopic strain profile of a propagating Lüders band front in mild steel. Scr Metall 9:1051–1058
7. Verel DJ, Sleeswyk AW (1973) Lüders band propagation at low velocities. Acta Metall 21:1087–1098
8. Hahn GT (1962) A model for yielding with special reference to the yield-point phenomena of iron and related BCC Metals. Acta Metall 10:727–738
9. Itoh M, Yoshida F, Yamashita Y, Ohmori M (1992) Finite element analysis for non-uniform yielding processes in mild steel plates under stretching. Jpn Soc Mech Eng-Int J 35:70–77
10. Hillert M (1965) On the theory of normal and abnormal grain growth. Acta Metall 13:227–238
11. Zhou T, Zurob H (2011) Abnormal and post-abnormal austenite grain growth kinetics in Nb–Ti microalloyed steels. Can Metall Q 50(4):389–395
12. Ananthan VS, Hall EO (1987) Surface features of microscopic shear bands at Lüders fronts in mild steel. Scr Metall 21:1699–1704
13. Hu H (1983) Effect of solutes on lüders strain in low-carbon sheet steels. Metall Trans A 14:85–91
14. Johnson DH (2012) Lüders bands in RPV steel, Ph.D. thesis, Cranfield University, URI: http://dspace.lib.cranfield.ac.uk/handle/1826/8039
15. Humphreys FJ, Hatherly M (2004) Recrystallization and related annealing phenomena, 2nd edn. Elsevier Ltd., Oxford
16. Wen W, Morris JG (2204) The effect of cold rolling and annealing on the serrated yielding phenomenon of AA5182 aluminum alloy. Mater Sci Eng A 373(1–2):204–216
17. Cottrell AH, Bilby BA (1949) Dislocation theory of yielding and strain ageing of iron. Proc Phys Soc 62:49–62
18. Ballarin V, Soler M, Perlade A, Lemoine X, Forest S (2009) Mechanisms and modeling of bake-hardening steels: part I. Uniaxial tension. Metall Mater Trans A 40:1367–1374
19. Available from THERMO-CALC software. www.thermocalc.com
20. Iron and Steel Institute (1963) Recent developments in annealing. The Institute, London

Investigation on the Flow Field of Molten Steel in Ultrahigh-Speed Billet Continuous Casting Mold

Pei Xu, Dengfu Chen, Peng Liu, Qinzheng Wang, MuJun Long, Huamei Duan, Jie Yang and Qimin Wang

Abstract In this paper, a multiphase mathematical model coupling with the air and a single-phase model was, respectively, established to investigate the flow field of molten steel in ultrahigh-speed billet continuous casting mold with 160 mm × 160 mm cross section. The results show that the casting speed has a little effect on the flowing pattern of molten steel in the mold, and the flow field between the multiphase model and the single-phase model is not much different. Moreover, with the increase of casting speed, the impacting depth of molten steel stream and free surface velocity rise, and the area of high-speed region augments, the level fluctuation is more intense, and the wave height gets large. And the most level fluctuations at four casting speeds are basically within ±3 mm, and the average wave height is less than 3 mm. Although there exists an exception at the 30 s of casting speed 4.5 m/min and the wave height is 5.654 mm, no slag entrapment happens. Thus, the flow field of ultrahigh-speed billet continuous casting mold is appropriate at the current operation parameters.

Keywords Ultrahigh-speed continuous casting · Fluid flow · Numerical simulation · Level fluctuation

Introduction

The behavior of molten steel flow and level fluctuation in the continuous casting mold strongly affects the quality of finial steel products, such as air entrainment and slag entrapment caused by them are the important source of resulting in the re-oxidation of molten steel and the large inclusion particles inside the slab so much as the slab surface cracks and other defects, which are significantly unfavorable for

P. Xu · D. Chen (✉) · P. Liu · Q. Wang · M. Long · H. Duan · J. Yang · Q. Wang
College of Materials Science and Engineering, Chongqing University, Chongqing 400044, China
e-mail: chendfu@cqu.edu.cn

P. Xu
e-mail: xupeibest@126.com

Chongqing Key Laboratory of Vanadium-Titanium Metallurgy and New Materials, Chongqing University, Chongqing 400044, People's Republic of China

J. Lee et al. (eds.), *Materials Processing Fundamentals 2020*, The Minerals, Metals & Materials Series, https://doi.org/10.1007/978-3-030-36556-1_22

guaranteeing the high casting speed and producing high-quality products. Therefore, how to reasonably control the molten steel flow and level fluctuation in the mold has been paid high concern and attention by many domestic and foreign researchers.

For example, Thomas et al. [1] and Lait et al. [2] conducted the relevant numerical simulation about the effect of submerged entry nozzle (SEN), solidified shell and blowing on the behavior of molten steel flow in the mold by establishing a three-dimensional mathematical model, and obtained good calculation results, and these make people furtherly understand the flow characteristic of molten steel. Arcos-Gutierrez et al. [3] employed RSM turbulence model and VOF multiphase model to study the effect of the internal SEN design on the flow pattern in a funnel thin slab caster and concluded that the jets oscillation had its origin from the inside of the SEN. Miranda et al. [4] chosen $k - e$ turbulence model and VOF model to study the flow pattern prevailing at the steel-slag interface in a water model of a slab continuous casting mold and obtained a good agreement between simulated and experimented data. Miaoyong Zhu, Dengfu Chen, et al. [5–7] researched the influence of different casting speeds, SEN structure, and parameter on the behaviors of molten steel flow in the mold by numerical simulation and also conducted the corresponding water modeling experiment to verify numerical simulation. Wang and Zhang [8] employed VOF algorithm to simulate the interface evolution between the molten steel and air during cast start of steel continuous casting process. Tan et al. [9] used VOF method to simulate the shape of free surface in a continuous casting mold and analyzed the mechanism of the surface oscillation. Although these studies are very effective, they are basically concentrated on the conventional casting speed, and so the research on the flow field of ultrahigh-speed continuous casting mold is still lacking. In addition, the level fluctuation of free surface of billet continuous casting mold also was rarely investigated.

Therefore, aiming at the behavior of molten steel flow and level fluctuation of free surface in the billet mold with 160 mm $\times$ 160 mm cross section from casting speed 3.5–6.5 m/min, this paper conducted the corresponding numerical simulation. And by doing them, the understanding of the flow field in the mold of ultrahigh-speed billet continuous casting has been strengthened, which is of certain significance to the development of ultrahigh-speed billet continuous casting that its range is above 5.0 m/min generally.

Mathematical Formulation

In the current study, a three-dimensional (3D) single-phase model was established to simulate the fluid flow inside the mold, and a two-dimensional (2D) multiphase model coupling with the air was used to investigate level fluctuation on the top surface of molten steel.

The 3D Single-Phase Model

(a) Assumptions

In the mathematical model, the molten steel in mold was assumed to be a three-dimensional, steady-state, incompressible Newtonian fluid, and regard as homogenous phase. The strand bulging, oscillation, mold taper, air gap, meniscus fluctuation, and flux slag were neglected.

(b) Governing equations

The liquid steel flow in the liquid pool was achieved through solving the continuity and momentum equations. And the popular standard $k - \varepsilon$ turbulence model with the standard wall function of near wall treatment was chosen. The equations are as follows.

$$\text{The continuity equation:} \frac{\partial u_i}{\partial x_i} = 0 \tag{1}$$

$$\text{the momentum equation:} \frac{\partial(\rho u_i u_j)}{\partial x_j} = -\frac{\partial P}{\partial x_i} + \frac{\partial}{\partial x_i}\left[\mu_{\text{eff}}\left(\frac{\partial u_i}{\partial x_j} + \frac{\partial u_j}{\partial x_i}\right)\right] + \rho g_i + S_P \tag{2}$$

$$\text{the effective viscosity:} \mu_{\text{eff}} = \mu_0 + \mu_t = \mu_0 + \frac{\rho C_\mu k^2}{\varepsilon} \tag{3}$$

$$\text{turbulent kinetic energy:} \frac{\partial}{\partial x_i}\left(\rho\mu_i k - \frac{\mu_{\text{eff}}}{\sigma_k} \cdot \frac{\partial k}{\partial x_i}\right) = G - \rho\varepsilon \tag{4}$$

$$\text{turbulent energy dissipation:} \frac{\partial}{\partial x_i}\left(\rho\mu_i \varepsilon - \frac{\mu_{\text{eff}}}{\sigma_\varepsilon} \cdot \frac{\partial \varepsilon}{\partial x_i}\right) = \frac{C_1 \varepsilon G - C_2 \rho \varepsilon^2}{k} \tag{5}$$

where u_i and u_j are the velocity components, x_i and x_j are the spatial coordinates, ρ is the density of liquid steel, P is the pressure, μ_{eff} is the effective viscosity, and S_P is the sink term of velocity. C_u is constant. G is the generation of turbulence kinetic energy due to the mean velocity gradient. σ_k and σ_ε are the turbulent Prandtl numbers for k and ε, respectively. C_1 and C_2 are constants. The values of C_u σ_k, σ_ε, C_1, and C_2 are empirical constants, respectively, and are recommended by Launder and Spalding [10] in the paper.

(c) Computational Domain

Due to the symmetry of fluid flow in the mold, only a 1/2 of the strand was modelled to minimize computation. The casting parameters are shown in Table 1. Figure 1 shows the geometrical model and the corresponding mesh of the strand. The maximum size of mesh is 5 mm, the mesh at the outlet of SEN was refined to 2 mm, and the number of mesh is about 450,000. For the selection of mesh size, there exist several considerable factors, for example, computation cost, computation accuracy

Table 1 Dimensions and parameters

Parameter	Value
Mold section	160 mm × 160 mm
Mold length	1000 mm
Mold radius	8000 mm
Clearance height of mold	70 mm
Computational domain length	1930 mm
SEN type	Straight through nozzle
Casting speed	3.5–6.5 m/min
Density of molten steel	7200 kg/m^3
Viscosity of molten steel	0.0062 kg/(m s)
Density of air	1.225 kg/m^3
Viscosity of air	1.7894×10^{-5} kg/(m s)
The height of air	50 mm
Surface tension of molten steel	1.80 N/m

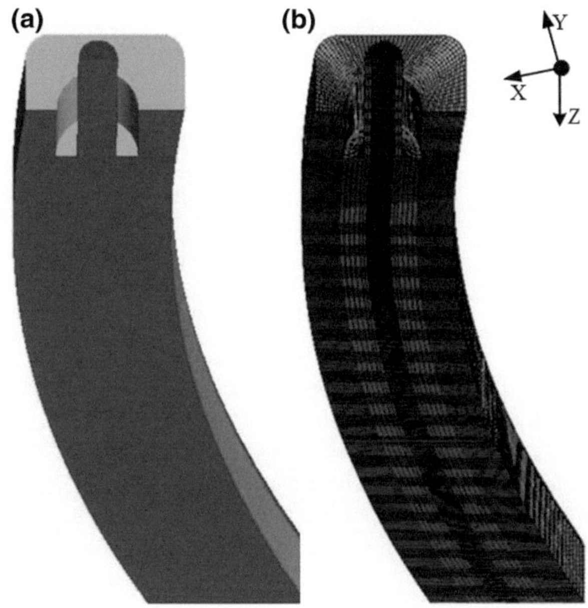

Fig. 1 Three-dimensional mathematical model. **a** Geometrical model, **b** mesh model

and research objective, and so on, and if the mesh size is too large, the mathematical model wall cannot catch accurately the information in the mold, while if it gets very small, the accuracy will increase, but the computation cost will augment dramatically, because the computational accuracy does not increase proportionally with the decrease of mesh size. Based on them, in the section, the 3D model was mainly used to simulate the flow field in the mold, the mesh size 5 mm can absolutely satisfy the accuracy, and the cost is also appropriate. To the validation of the 3D model, this paper compared the flow field in the mold of 3D model with it of water modeling experiment.

(d) Boundary conditions

At the inlet of SEN, the velocity inlet boundary condition was used. The inlet velocity was calculated through the mass conservation based on the casting speed. The pressure outlet boundary condition was applied at strand exit. The top free surface was set as zero shear stationary wall and heat insulation. At the symmetry plane, the normal velocity components and normal gradients of all other variables were assumed to be zero.

The 2D Multiphase Model

(a) Assumptions

To the 2D multiphase model, except for the abovementioned, the additional assumptions are that only two phases are considered, air and molten steel, respectively, as well as the density and viscosity of the both are assumed to keep constant.

(b) Governing equations

The 2D multiphase model was solved through the volume of fluid (VOF) model with the continuum surface force (CSF) model. For the VOF model, the tracking of the interface is achieved by the solution of a conservation equation for the volume fraction of both phases:

$$\frac{\partial \alpha_q}{\partial t} + v\frac{\partial \alpha_q}{\partial x_i} = 0 \tag{6}$$

where α_q is the volume fraction of phase "q" (percent), and $\boldsymbol{v}$ is the vector of velocity components.

A constraint of the system implies that the sum of the volume fractions of both phases in the shared cells should be equal to 1, namely

$$\alpha_l + \alpha_g = 1 \tag{7}$$

Thus, material properties in each point in the domain are represented using mixture equation of α_l and α_g, such as the averaged properties of density and viscosity of the mixture are given by

$$\rho = \alpha_1\rho_1 + \alpha_g\rho_g \tag{8}$$

$$\mu = \alpha_1\mu_1 + \alpha_g\mu_g \tag{9}$$

The continuity and momentum equations are the same as Sect. 2.1, the difference is that the source term S_P in Eq. (2) becomes F_σ, and F_σ is the force source due to surface tension given by the following equation, which is modeled using the CSF model of Brackbill et al. [11].

$$F_\sigma = \sigma_{g-1} \frac{\rho \kappa \nabla \alpha_g}{0.5(\rho_g + \rho_l)} \tag{10}$$

Here, σ_{g-l} is the surface tension coefficient. κ is the local curvature of this interface.

(c) Computational Domain

Due to the complexity of multiphase numerical simulation, the three-dimensional model during the computation occurs easily divergence and its relevant problems are solving. Therefore, in order to investigate the level fluctuation on the top surface of ultrahigh-speed continuous casting mold, the simplified 2D multiphase mathematical model was adopted. The casting parameters are shown in Table 1, and the geometrical model and the corresponding mesh of the strand are shown in Fig. 2. The maximum size of mesh is 2 mm, the mesh near the interface between air and molten steel was refined, and the number of mesh is about 70,000. For the validation of the 2D model, this paper also compared the flow field in the mold of 2D model with it of water modeling experiment. However, it is worth noting that the comparison figure did not display repeatedly, because the flow field between the 2D and 3D is mostly similar.

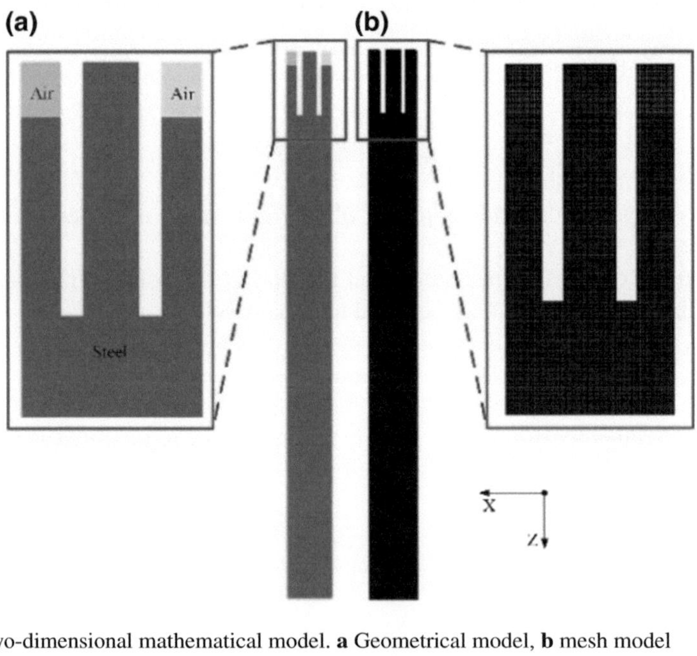

Fig. 2 Two-dimensional mathematical model. **a** Geometrical model, **b** mesh model

(d) Boundary Conditions

A constant velocity was imposed as the nozzle inlet condition, and the velocity was calculated by casting speed according to mass balance. A pressure outlet condition as zero pascal gauge pressure was used at the top surface of air region. The fixed velocity was employed at the domain exit corresponding to the casting speed. And the wall was set as non-slip boundary condition with zero normal velocity.

Solution Strategy

The governing equations were discretized in ANSYS Fluent by using an implicit, first-order up-winding scheme and SIMPLE algorithm for pressure-coupling. The convergence criterion was obtained when the residuals of the output variables reached values equal to or smaller than 1×10^{-6}. The time step was 0.01 s. To the 2D multiphase simulation, the total observation time is 30 s.

Results

(1) To the 3D model, the flow field of mold at four casting speeds was investigated by numerical simulation. And Fig. 3 implies the velocity contour and streamline of the central symmetry plane. In Fig. 3a, this paper thinks that the distance from the bottom of velocity isoline where its velocity value equal to the casting speed to meniscus defined the impacting depth. Except for the flow field of mold at the central symmetry plane, the velocity fields on the free liquid surface at four casting speeds are shown in Fig. 4.
(2) For verifying the validation of 3D and 2D mathematical models, the corresponding water modeling experiments were carried out and then compared two methods. Figure 5 shows the flow field comparison between the 3D model and

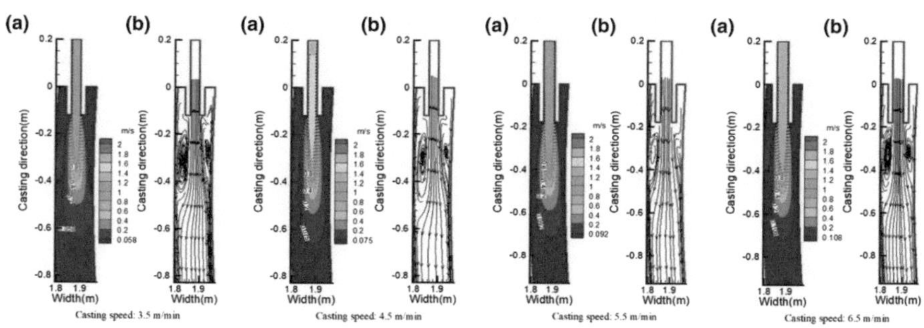

Fig. 3 Velocity contours and streamlines of central symmetry plane at four casting speeds. **a** Velocity contour, **b** streamline

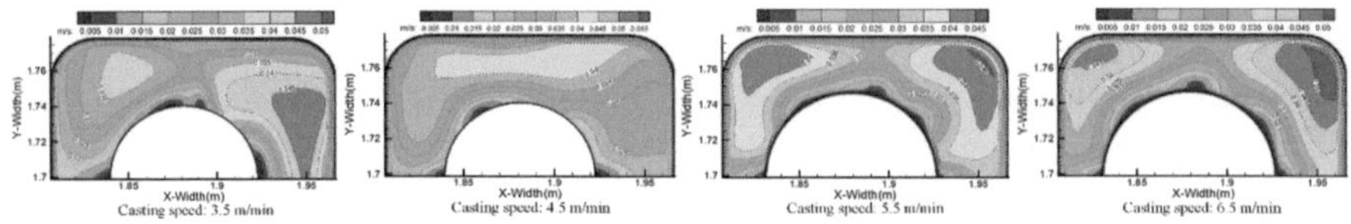

Fig. 4 Velocity contours of free surface in the mold at four casting speeds

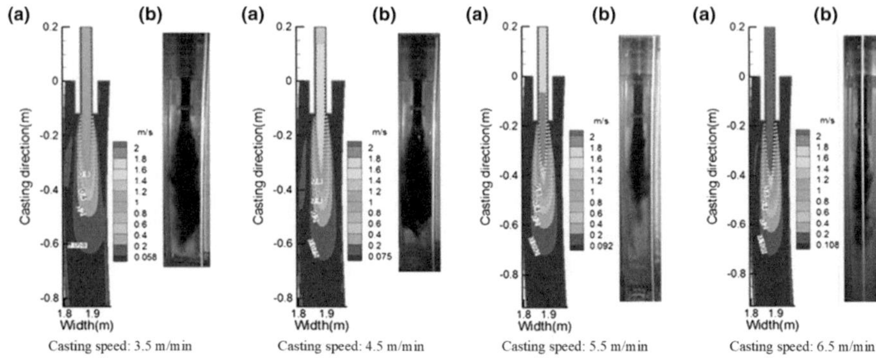

Fig. 5 Comparison of flow field on the central symmetry plane. **a** Numerical simulation, **b** physical modeling

water modeling at four casting speeds. And due to the similarity of flow field between 3D and 2D model, this paper only presented one comparison figure.

(3) To the 2D multiphase model, the level fluctuation of free surface in the mold was simulated. Figure 6 shows the typical level fluctuation contours of different times at casting speed 6.5 m/min. Figure 7 presents the fluctuation level value of different simulation times at four casting speeds.

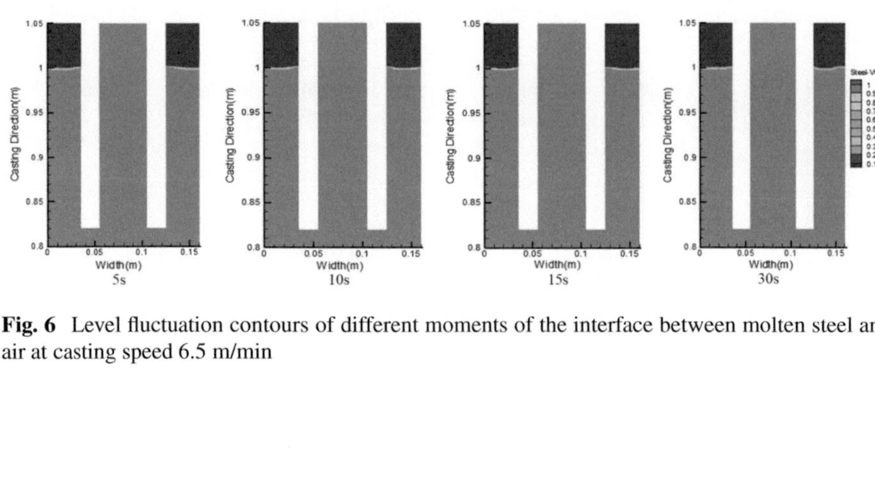

Fig. 6 Level fluctuation contours of different moments of the interface between molten steel and air at casting speed 6.5 m/min

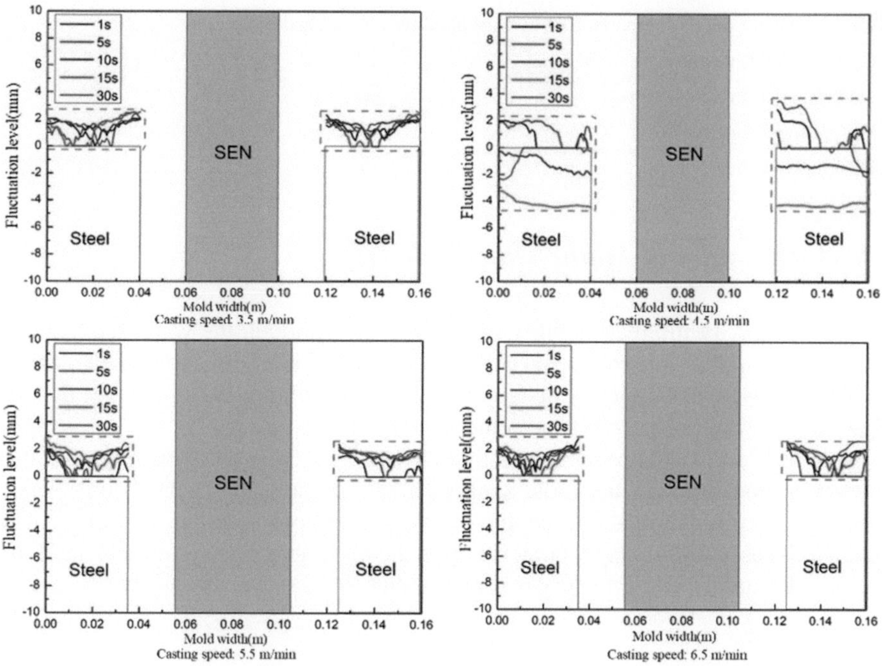

Fig. 7 Fluctuation level of different moments along the centerline of air/molten steel interface

Discussion

The Study of Flow Field in the Mold

The flow field in the mold will influence the inclusions floatation and mold meniscus fluctuation, and this paper simulates it by suing a 3D single-phase model. According to Fig. 3a, the flow fields of different casting speeds are similar that the velocity of the mainstream becomes weaken gradually along the casting direction, and as the casting speed arising, the impacting depth increases gradually. From Fig. 3b, the molten steel flow will form circuiting zone around inner and outer arcs, and obviously, the range and strength of outer arcs are bigger than inner arcs, and it is because of the arc structure of billet mold so that the stream is closer to inner arc and the flow room gets smaller. To the liquid surface velocity field, its uniformity, symmetry, velocity gradient, and maximum velocity value need mainly considering. In Fig. 4, the maximum velocity of four casting speeds is all about 0.04–0.05 m/s, the uniformity of distribution and symmetry are good, as well as the velocity gradient is reasonable. And these can let mold liquid surface keep appropriately active, contributing to flux slag melting.

Table 2 Impact depth comparison of two simulations at four casting speeds

Casting speed (m/min)	3.5	4.5	5.5	6.5
Numerical simulation (mm)	635	665	720	725
Water modeling (mm)	563	579	597	617
Difference (mm)	72	86	123	108

The Verification of Mathematical Model

In order to verify the reasonability of mathematical model, the water modeling at four casting speeds was conducted in this paper. In the water modeling, the flow field in the mold was visual by adding blue ink into water. Figure 5a, b indicates that the flow field is similar between two methods.

And in Fig. 5b, this paper thinks that the distance from the stream bottom that it spreads just to touch the mold shell to meniscus defined the impacting depth. The comparison of impacting depths of two simulations is listed in Table 2. The mathematical is slightly lager but it is within the allowable range of error. Therefore, the model established in this paper is reasonable.

The Study of Level Fluctuation at the Top Free Surface

The level fluctuation is related with slag entrapment, the uniform distribution of liquid slag, and the activity of slag layer, and finally, these effects will embody into the quality of the strand. For this reason, this paper investigated the level fluctuation of top free surface by using 2D multiphase model. Due to the level fluctuation pattern of free surface is similar among four casting speeds, this paper only presents a representative level fluctuation contours of different times at casting speed 6.5 m/min. As shown in Figs. 6 and 7, at 5 s, the fluctuation is also small. The reason is that the molten steel has just entered the mold and the circuiting zone has not developed completely. At 10 s, the level fluctuation of the free surface has changed obviously, forming a wave peak near the submerged nozzle and the wall of the mold, and forming a trough in the middle part. And then the level fluctuation of free surface changes periodically.

For analyzing the level fluctuation quantitatively, by choosing the isoline of 0.1 volume fraction of molten steel as interface and setting the initial free surface of molten steel as 0 point, this paper extracted the level fluctuation value at four casting speeds. So that the corresponding wave heights were obtained by using the maximum level fluctuation minus the minimum. Table 3 shows wave heights of different times at four casting speeds.

In general, the wave height ±3–5 mm is a critical value [12], and if exceeding it, there occurs slag entrapment. From Table 3, most level fluctuations are within ±3 mm and the average wave height is less than 3 mm. It is appropriate to protect from slag entrapment and keep the activity of free surface. Moreover, the maximum

Table 3 Wave height of different times at four casting speeds

Casting speed (m/min)	Wave height (mm)							
	Left side				Right side			
	5 s	10 s	15 s	30 s	5 s	10 s	15 s	30 s
3.5	2.514	0.958	2.543	2.406	2.313	1.699	2.408	0.942
4.5	2.369	1.763	1.244	3.975	1.632	0.567	0.413	5.654
5.5	2.090	1.430	2.353	2.481	2.154	1.748	0.975	0.857
6.5	1.996	1.846	2.203	0.978	2.495	2.150	2.109	2.642

range of level fluctuation in free surface is ±5 mm and the maximum wave height is 5.654 mm, but it did not result in slag entrapment during the numerical simulation.

Conclusions

In this paper, 3D single-phase and 2D multiphase numerical simulations were, respectively, conducted to study the flow field in ultrahigh-speed billet continuous casting mold. Conclusions are as follows:

(1) The velocity of the mainstream becomes weakened gradually along the casting direction, and as the casting speed rises, the stream impact depth increases gradually. The molten steel flow forms a circuiting zone around inner and outer arcs, and obviously, the range and strength of outer arcs is bigger than inner arcs.
(2) The maximum velocity of four casting speeds is about 0.04–0.05 m/s, the uniformity of distribution and symmetry is good, and the velocity gradient is reasonable.
(3) The mathematical is slightly larger, and the difference between the both is 70–120 mm, but it is within the allowable range of error.
(4) Most level fluctuations are within ±3 mmm and the average wave height is less than 3 mm. Therefore, no slag entrapment happens.

Acknowledgements The work is financially supported by the Natural Science Foundation of China, Project No. 51874060 and 51874059.

References

1. Thomas BG, Mika LJ, Najjar FM (1990) Simulation of fluid flow inside a continuous slab-casting machine. Metall Trans B 21(2):387–400
2. Lait JE, Brimacombe JK, Weinberg F (1974) Mathematical modeling of heat flow in the continuous casting of steel. Ironmak Steelmak 1(2):90–97

3. Arcos-Gutierrez H, Barrera-Cardiel G, de Jesus Barreto J et al (2014) Numerical study of internal SEN design effects on jet oscillations in a funnel thin slab caster. ISIJ Int 54(6):1304–1313
4. Miranda R, Barron MA, Barreto J et al (2005) Experimental and numerical analysis of the free surface in a water model of a slab continuous casting mold. ISIJ Int 45(11):1626–1635
5. Ren BZ, Zhu MY, Wang H, Chen Y (2008) 3D numerical simulation of electromagnetic field and flow field in bloom continuous casting mold with electromagnetic stirring. Acta Metall Sin 44(4):507–512
6. Zheng S (2006) Water model study on removing inclusion in a ladle with argon injected through nozzle and porous plug. Acta Metall Sin 42(11):1143–1148
7. Jin X, Chen DF, Xie X, Shen J, Long MJ (2013) Investigation on water model for fluid flow in slab continuous casting mold with consideration of solidified process. Steel Res Int 84(1):31–39
8. Wang Y, Zhang LF (2010) Transient fluid flow phenomena during continuous casting: part I—cast start. ISIJ Int 50(12):1777–1782
9. Tan L, Shen H, Liu B (2003) Numerical simulation of surface oscillation in continuous casting mold. Acta Metall Sin 39(4):435
10. Launder BE, Spalding DB (1972) Lectures in mathematical models of turbulence. Academic Press, Cambridge, MA, USA
11. Brackbill JU, Kothe DB, Zemach C (1992) A continuum method for modeling surface tension. J Comput Phys 100(2):335–354
12. Lu QT, Wang XH, Yu HX et al. (2007) Computation of F number and its relation with slag entrapment in slab continuous casting mold. J Univ Sci Technol Beijing 29(8):811–815

Thermodynamic Properties of Layered Tetradymite-like Compounds of the Ag–Ge–Sb–Te System

M. Moroz, F. Tesfaye, P. Demchenko, M. Prokhorenko, D. Lindberg, O. Reshetnyak and L. Hupa

Abstract The phase equilibria of the Ag–Ge–Sb–Te system in the part Ag_8GeTe_6–Ge–GeTe–Sb_2Te_3 were investigated by the electromotive force (EMF) method. The determined phase relations were used to express the chemical reactions. The potential-forming reactions were performed by applying electrochemical cells (−) C | Ag | Ag_2GeS_3-glass | D | C (+), where C is graphite, Ag_2GeS_3-glass is the fast purely Ag^+ ions conducting electrolyte, and D is an equilibrium mixture of phases. According to the experimental data on the EMF versus temperature of each electrochemical cells, analytical equations for the Gibbs energies of $GeSb_8Te_{13}$, $GeSb_6Te_{10}$, $GeSb_4Te_7$, $GeSb_2Te_4$, $Ge_2Sb_2Te_5$, $Ge_3Sb_2Te_6$, and $Ge_4Sb_2Te_7$ compounds were obtained. The thermodynamic properties of silver-saturated tetradymite-like compounds have been calculated for the first time. A good correspondence between experimental values and structure data reported in the literature has been established.

Keywords Thermoelectric materials · Layered compounds · Phase equilibria · Thermodynamic properties · EMF method

---

M. Moroz (✉)
Department of Chemistry and Physics, National University of Water and Environmental Engineering, Rivne 33028, Ukraine
e-mail: m.v.moroz@nuwm.edu.ua

M. Moroz · O. Reshetnyak
Department of Physical and Colloid Chemistry, Ivan Franko National University of Lviv, Lviv 79005, Ukraine

F. Tesfaye · L. Hupa
Laboratory of Inorganic Chemistry, Johan Gadolin Process Chemistry Centre, Åbo Akademi University, 20500 Turku, Finland

P. Demchenko
Department of Inorganic Chemistry, Ivan Franko National University of Lviv, Lviv 79005, Ukraine

M. Prokhorenko
Department of Cartography and Geospatial Modeling, Lviv Polytechnic National University, Lviv 79013, Ukraine

D. Lindberg
School of Chemical Engineering, Aalto University, 02150 Espoo, Finland

J. Lee et al. (eds.), *Materials Processing Fundamentals 2020*, The Minerals, Metals & Materials Series, https://doi.org/10.1007/978-3-030-36556-1_23

Introduction

In recent decades, the study of thermoelectric materials (TM) has attracted increasing attention both from an energy and environmental point of view. Thermoelectric materials provide a large variety of applications in heating, cooling, and power generation due to direct conversion of heat into electrical energy [1–3]. The efficiency of TM is defined by the figure-of-merit ZT parameter, which is $ZT = S^2\sigma T/(k_{el} + k_{lat})$, where S, σ, k_{el} and k_{lat} represent the Seebeck coefficient, electrical, electronic, and lattice vibrations conductivity, respectively [4]. Many studies in the past decades are focused on increasing of ZT values in several ways by studying the multicomponent materials. The ternary compounds with mixed layered tetradymite-like structure of the homologous series nGeTe·$m$$Sb_2Te_3$ (n, m = 1–4) in the GeTe–Sb_2Te_3 system were identified as good TM [5–10].

The T–x phase diagram of the GeTe–Sb_2Te_3 system has been studied by Abrikosov et al. [11]. The formation of the following compounds was established $GeSb_4Te_7$, $GeSb_2Te_4$, and $Ge_2Sb_2Te_5$. These ternary compounds melt incongruently at 878, 888 and 903 K, respectively. Skums et al. [12] and Skoropanov et al. [13] reported that the $GeSb_4Te_7$ and $GeSb_2Te_4$ compounds melt congruently at 880 and 888 K, respectively. Later on, four additional compounds $Ge_4Sb_2Te_7$, $Ge_3Sb_2Te_6$, $GeSb_6Te_{10}$, and $GeSb_8Te_{13}$ of the homologous series nGeTe·$m$$Sb_2Te_3$ were found [5–9]. These compounds were identified by high-resolution electron microscopy, electronic diffraction [5, 6, 14, 15] and X-ray diffraction [8, 14] techniques. In the layered ternary compounds, Te atoms form a cubic close packing structure with octahedral voids, occupied by Ge and Sb atoms. The hexagonal unit cells are formed by multilayer packets of different types, ordered in alternating directions in the direction perpendicular to hexagonal c axis. The layered compounds can be divided into two groups, based on the compositional variations of "$GeSb_2Te_4$". The structures of the GeTe-rich compounds are formed by identical slabs. The structures of Sb_2Te_3-rich compounds include five- and seven-layer slabs [8].

For the practical application of compounds in electronic cooling devices and transformation of thermal energy into electrical power, absence of structural transformations in the operating temperature range is important. Shelimova et al. [9] measured the Hall coefficient, electrical resistivity, and thermoelectric power of compounds of the homologous series in the range of 100–600 K. According to Skums et al. [12], some of these compounds undergo a phase transition at $T > 600$ K. Below $T = 600$ K, the energy of thermal vibrations of atoms may not be sufficient to change the lattice crystal structure. Under these conditions, the phase will be in a metastable state. The examples of transition of compounds from the metastable to the thermodynamic equilibrium state were published in several works [16–21]. Such transition was carried out by including compound as one of the components into a positive electrode of the electrochemical cells (ECCs). A key role in achieving the thermodynamic equilibrium state of the compounds belongs to Ag^+ ions migrated from the Ag reference electrode to multi-phase mixture of a positive electrode [22].

Moroz and Prokhorenko [16] reported that continuous solid solution of the PbSe–PbTe system at $T < 570$ K is metastable. Decomposition of the solid solution and formation in the right electrode of ECC the intermediate phase was established.

Formation of the Ag_2GeSe_3 compound in the $Ag_2Se–GeSe_2$ system was found under carefully controlled conditions in ECC [17]. Furthermore, Ag_2GeSe_3 was observed to undergo a phase transition at $T = 535$ K.

The phase diagram of the Bi–Te system in the entire composition range is given in [19]. It is characterized by the presence of Bi_2Te_3, Bi_4Te_5, Bi_8Te_9, BiTe, Bi_6Te_5, Bi_4Te_3, Bi_2Te, and $Bi_{14}Te_6$ compounds. Abrikosov et al. [20] investigated the homogeneity ranges of the Bi_2Te_3, BiTe, Bi_2Te, and $Bi_{14}Te_6$ compounds. It was found that BiTe and Bi_2Te are the phases of variable composition. The phase equilibria of the Ag–Bi–Te system and thermodynamic properties of the compounds have been investigated by the electromotive force (EMF) method in the temperature range of 490–550 K [18]. It is characterized by the presence of the Ag_2Te–Bi, $Ag_2Te–Bi_{14}Te_6$, $Ag_2Te–Bi_2Te$, Ag_2Te–BiTe, and $Ag_2Te–Bi_2Te_3$ quasi-binary sections. A good agreement between calculated [18] and the literature [23] thermodynamic data for BiTe and Bi_2Te_3 compounds confirms the phase equilibria of the Ag–Bi–Te system presented in [18]. It is mean that in thermodynamic equilibrium BiTe and Bi_2Te are stoichiometric compounds due to the decomposition of their solid solutions under potential-forming process.

According to [21], the solid solubility of PbS in Ag_8GeS_6 does not exceed 13 mol%. Metastable state of this solid solution was established by the EMF method in [24]. It was found that the maximum solubility occurs up to 19 mol% PbS. Furthermore, the existence of intermediate phase with the approximated composition $Ag_{6.62}Pb_{0.16}Ge_{0.84}S_{5.20}$ was predicted.

Herein, we present the thermodynamic properties of the saturated solid solutions of the nGeTe·$m$$Sb_2Te_3$ compounds in the phase region Ag_8GeTe_6–Ge–$Ge_4Sb_2Te_7$–Sb_2Te_3 of the Ag–Ge–Sb–Te system. These thermochemical data of the compounds can be used to optimize or complete phase diagrams in the investigated system and for selection of thermodynamically stable materials with optimal values of the *ZT* parameter.

Experimental

The starting materials for synthesis were high-purity elements: Ag, 99.99 wt% (Alfa Aesar, Germany); Ge, 99.999 wt% (Alfa Aesar, Germany); Sb, 99.99 wt% (Alfa Aesar, Germany); S, 99.999 wt% (Alfa Aesar, Germany); Te, 99.999 wt% (Alfa Aesar, Germany).

For the EMF measurements [24–29], the following electrochemical cell was used:

$$(-)\text{C}|\text{Ag}|\text{Ag}_2\text{GeS}_3 - \text{glass}|\text{D}|\text{C}(+),$$

where C is graphite, Ag_2GeS_3-glass is the fast purely Ag^+ ions conducting electrolyte [30, 31], and D is a four-element equilibrium mixture of phases. The vertical lines in ECC indicate phase boundaries or contacts between cell components. The cell polarities and half-cell reactions in the ECCs were established according to rules described in [24, 32].

The positive (right) electrodes D of the EECs were synthesized by melting of chemical elements in thin-walled evacuated quartz glass ampoules at $T = 1070$ K for 5 h. Slowly cooled to room temperature, samples were grounded into a fine powder with the particle size of ≤ 5 μm, evacuated, and then annealed at $T = 550$ K for 250 h. The composition of positive electrodes of ECCs was calculated based on equations of electrochemical reactions for each of 7 four-phase regions of the Ag_8GeTe_6–Ge–$Ge_4Sb_2Te_7$–Sb_2Te_3 system. The Ag_2GeS_3-glass [30, 33] was obtained by melt quenching of the corresponding elements from $T = 1200$ K in ice water. The phase compositions of the right electrode compounds were characterized by differential thermal analysis (Paulik–Paulik–Erdey derivatograph fitted with chromel–alumel thermocouples and an H307-1 XY recorder) and by X-ray powder diffraction (STOE STADI P diffractometer, transmission mode, $CuK\alpha_1$ radiation, a bent Ge (111) monochromator on primary beam, $2\theta/\omega$ scan mode) techniques. X-ray phase analysis was performed using STOE WinXPOW [34] and PowderCell [35] program packages using crystallographic data on the structures of known phases taken from databases [36, 37].

Components of the ECCs in powder form were pressed at 10^8 Pa through a 2-mm-diameter hole arranged in the fluoroplast matrix up to density $\rho = (0.93 \pm 0.02)\cdot\rho_0$, where ρ_0 is the experimentally determined density of cast samples. Fivefold thermal cycling of ECCs in the temperature range of 450–510 K was performed to eliminate possible defects due to plastic deformation during sample pressing. The heating and cooling rates were of 2 K min^{-1}. Experiments were performed in a horizontal resistance furnace, similar to that described in [38]. As a protective atmosphere, we used a continuously flowing highly purified (0.9999 volume fraction) Ar(g) at $P = 1.2\cdot10^5$ Pa, with a flow rate of 2×10^{-3} m^3 h^{-1} from the right to left electrode of the ECCs. The temperature was maintained with an accuracy of $\pm$ 0.5 K. The EMF of the cells was measured by high-resistance (input impedance of $>10^{12}$ Ω) universal U7-9 digital voltmeter. The equilibrium in ECCs at each temperature was achieved within 2 h. After equilibrium has been attained, the EMF values were constant or their variation did not exceed $\pm$0.2 mV.

In our previous works [39, 40] we have described in details the scheme of ECCs and procedure of the EMF measurements.

Results and Discussion

As can be seen in Table 1 and Fig. 1, the concentration space of the Ag–Ge–Sb–Te system in the Ag_8GeTe_6–Ge–$Ge_4Sb_2Te_7$–Sb_2Te_3 part consists of 7 four-phase regions. In Fig. 1, the phase regions' borders are marked by two-phase equilibrium

Table 1 Four-phase regions of the Ag–Ge–Sb–Te system in the Ag_8GeTe_6–Ge–$Ge_4Sb_2Te_7$–Sb_2Te_3 part and the EMF values of ECCs at $T = 480$ K in corresponding phase areas

No.	Phase region	E/mV
1	$Ge_4Sb_2Te_7$–Ge–Ag_8GeTe_6–$Ge_3Sb_2Te_6$	275.09
2	$Ge_3Sb_2Te_6$–Ge–Ag_8GeTe_6–$Ge_2Sb_2Te_5$	272.44
3	$Ge_2Sb_2Te_5$–Ge–Ag_8GeTe_6–$GeSb_2Te_4$	270.17
4	$GeSb_2Te_4$–Ge–Ag_8GeTe_6–$GeSb_4Te_7$	265.98
5	$GeSb_4Te_7$–Ge–Ag_8GeTe_6–$GeSb_6Te_{10}$	258.38
6	$GeSb_6Te_{10}$–Ge–Ag_8GeTe_6–$GeSb_8Te_{13}$	254.52
7	$GeSb_8Te_{13}$–Ge–Ag_8GeTe_6–Sb_2Te_3	247.50

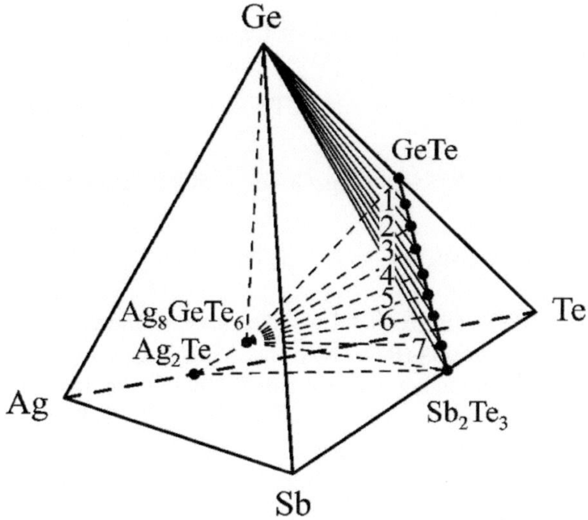

Fig. 1 Phase equilibria of the Ag–Ge–Sb–Te system in the Ag_8GeTe_6–Ge–GeTe–Sb_2Te_3 part, below $T = 520$ K. 1 is $Ge_4Sb_2Te_7$, 2 is $Ge_3Sb_2Te_6$, 3 is $Ge_2Sb_2Te_5$, 4 is $GeSb_2Te_4$, 5 is $GeSb_4Te_7$, 6 is $GeSb_6Te_{10}$, and 7 is $GeSb_8Te_{13}$

lines. The lines of two-phase equilibria were determined in [9, 41, 42] as well as by our investigation by the EMF method. The spatial location of the established phase regions relative to composition of Ag was used to determine the thermodynamic properties of ternary compounds.

In the phase region (1), listed in Table 1, the electrochemical process of the formation of Ag_8GeTe_6, Ge, and $Ge_3Sb_2Te_6$ from the pure Ag and $Ge_4Sb_2Te_7$ compound follows:

$$8Ag = 8Ag^+ + 8e^- - \text{left electrode (reference system)}$$

$$8Ag^+ + 8e^- + 6Ge_4Sb_2Te_7 = Ag_8GeTe_6 + 5Ge + 6Ge_3Sb_2Te_6$$

– rightelectrode (samplesystem),

$$8Ag + 6Ge_4Sb_2Te_7 = Ag_8GeTe_6 + 5Ge + 6Ge_3Sb_2Te_6 \text{ – overall cell reaction.} \quad (R1)$$

For the phase regions (2)–(7), listed in Table 1, the overall cell reactions can be expressed as:

$$8Ag + 6Ge_3Sb_2Te_6 = Ag_8GeTe_6 + 5Ge + 6Ge_2Sb_2Te_5, \quad (R2)$$

$$8Ag + 6Ge_2Sb_2Te_5 = Ag_8GeTe_6 + 5Ge + 6GeSb_2Te_4, \quad (R3)$$

$$8Ag + 12GeSb_2Te_4 = Ag_8GeTe_6 + 5Ge + 6GeSb_4Te_7, \quad (R4)$$

$$8Ag + 18GeSb_4Te_7 = Ag_8GeTe_6 + 5Ge + 12GeSb_6Te_{10}, \quad (R5)$$

$$8Ag + 24GeSb_6Te_{10} = Ag_8GeTe_6 + 5Ge + 18GeSb_8Te_{13}, \quad (R6)$$

$$8Ag + 6GeSb_8Te_{13} = Ag_8GeTe_6 + 5Ge + 24Sb_2Te_3. \quad (R7)$$

Reactions (R1)–(R7) were realized to take place in the ECCs at different temperatures. Based on equations (R1)–(R7), the positive electrodes D of ECCs were prepared from pure elements Ag, Ge, Sb, and Te taken in molar ratios: (1) 8 : 3 : 10 : 22, (2) 8 : 4 : 14 : 29, (3) 8 : 4 : 10 : 23, (4) 8 : 4 : 6 : 17, (5) 8 : 5 : 4 : 15, (6) 8 : 7 : 4 : 17, and (7) 8 : 9 : 4 : 19, respectively.

The relationship of EMF versus temperature of reactions (R1)–(R7) in ECCs was approximated by Eqs. (1)–(7) and are shown in Fig. 2. The crystal lattices of tetradymite-like compounds are highly disordered, due to the presence of mixed occupancy of cations in octahedral voids [43]. The bends of the curves at different temperatures in Fig. 2 arise from the individual activation changes of occupancy sites by Ge, Sb, and Ag cations that are characteristic of each of the studied phases.

$$E_{(1)}/\text{mV} = (162.98 \pm 0.21) + (233.56 \pm 0.43) \times 10^{-3}\text{T/K} \quad 462 \le \text{T/K} \le 503, \quad (1)$$

$$E_{(2)}/\text{mV} = (171.25 \pm 0.21) + (210.82 \pm 0.44) \times 10^{-3}\text{T/K} \quad 458 \le \text{T/K} \le 495, \quad (2)$$

$$E_{(3)}/\text{mV} = (185.46 \pm 0.16) + (176.48 \pm 0.34) \times 10^{-3}\text{T/K} \quad 457 \le \text{T/K} \le 499, \quad (3)$$

$$E_{(4)}/\text{mV} = (139.20 \pm 0.17) + (264.13 \pm 0.35) \times 10^{-3}\text{T/K} \quad 460 \le \text{T/K} \le 499, \quad (4)$$

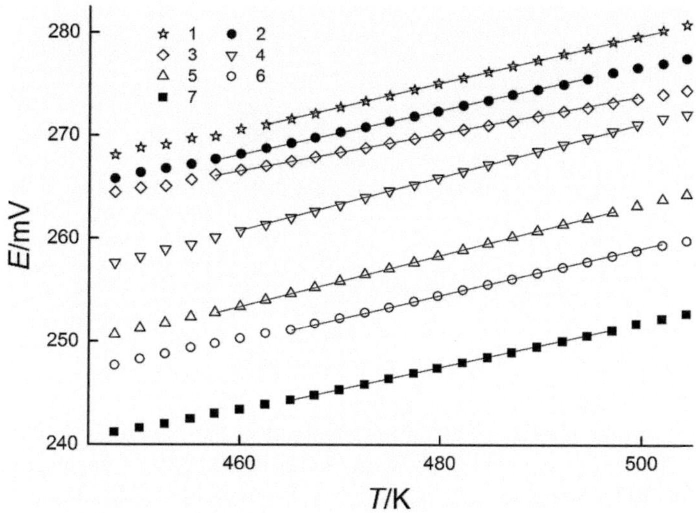

Fig. 2 Temperature dependences of EMF of the cells with the positive electrodes D of the phase regions: 1 is $Ge_4Sb_2Te_7$–Ge–Ag_8GeTe_6–$Ge_3Sb_2Te_6$, 2 is $Ge_3Sb_2Te_6$–Ge–Ag_8GeTe_6–$Ge_2Sb_2Te_5$, 3 is $Ge_2Sb_2Te_5$–Ge–Ag_8GeTe_6–$GeSb_2Te_4$, 4 is $GeSb_2Te_4$–Ge–Ag_8GeTe_6–$GeSb_4Te_7$, 5 is $GeSb_4Te_7$–Ge–Ag_8GeTe_6–$GeSb_6Te_{10}$, 6 is $GeSb_6Te_{10}$–Ge–Ag_8GeTe_6–$GeSb_8Te_{13}$, and 7 is $GeSb_8Te_{13}$–Ge–Ag_8GeTe_6–Sb_2Te_3

$$E_{(5)/\mathrm{mV}} = (139.39 \pm 0.18) + (247.89 \pm 0.38) \times 10^{-3}\mathrm{T/K} \quad 457 \le \mathrm{T/K} \le 496, \tag{5}$$

$$E_{(6)}/\mathrm{mV} = (147.71 \pm 0.22) + (222.53 \pm 0.46) \times 10^{-3}\mathrm{T/K} \quad 465 \le \mathrm{T/K} \le 501, \tag{6}$$

$$E_{(7)}/\mathrm{mV} = (145.75 \pm 0.19) + (211.97 \pm 0.39) \times 10^{-3}\mathrm{T/K} \quad 465 \le \mathrm{T/K} \le 497. \tag{7}$$

In addition, Eqs. (1)–(7) confirm the correctness of the spatial location of the established phase regions. In particular, EMF values of ECCs remain constant independent of the general composition in the phase region but change drastically on their boundaries [24]. Furthermore, as can be seen in Table 1, the phase region further away from the composition of Ag is characterized by higher EMF value at $T =$ const.

The Gibbs energies, entropies, and enthalpies of the reactions (R1)–(R7) can be calculated by combining the measured EMF values of each ECC and the thermodynamic Eqs. (8)–(10):

$$\Delta_r G = -n \cdot F \cdot E, \tag{8}$$

$$\Delta_r H = -n \cdot F \cdot [E - (\mathrm{d}E/\mathrm{d}T)T], \tag{9}$$

Table 2 Standard thermodynamic values of reactions (R1)–(R7) in ECCs at $T = 298$ K

Reaction	$-\Delta_r G^{\circ}$	$-\Delta_r H^{\circ}$	$\Delta_r S^{\circ}$
	kJ mol^{-1}		J mol^{-1} K^{-1}
(R1)	179.53 ± 0.62	125.80 ± 0.16	180.28 ± 0.33
(R2)	180.68 ± 0.66	132.18 ± 0.16	162.73 ± 0.34
(R3)	183.75 ± 0.57	143.15 ± 0.12	136.22 ± 0.26
(R4)	168.20 ± 0.48	107.45 ± 0.13	203.88 ± 0.27
(R5)	164.61 ± 0.52	107.59 ± 0.14	191.34 ± 0.29
(R6)	165.20 ± 0.64	114.01 ± 0.17	171.77 ± 0.36
(R7)	161.26 ± 0.56	112.50 ± 0.15	163.62 ± 0.30

$$\Delta_r S = n \cdot F \cdot (\mathrm{d}E/\mathrm{d}T), \tag{10}$$

where $n = 8$ is the number of electrons involved in the reactions (R1)–(R7), $F = 96485.33289$ C·mol^{-1} is Faraday constant, and E is the EMF of the electrochemical cells.

By assuming $\left(\frac{\partial \Delta_r H}{\partial T}\right)_p = 0$ and $\left(\frac{\partial \Delta_r S}{\partial T}\right) = 0$, which implies that $\Delta_r C_p = 0$ [32], the thermodynamic function of reactions (R1)–(R7) were calculated through extrapolation of the linear temperature dependences of the EMF of ECCs to $T = 298$ K and using Eqs. (8)–(10). The results of the calculations are listed in Table 2.

Standard Gibbs energy and entropy of reactions (R1)–(R7) are related to the Gibbs energy of formation and entropy of compounds and pure elements by the following equations:

$$\Delta_{r(1)} G^{\circ} = \Delta_f G^{\circ}_{\mathrm{Ag_8GeTe_6}} + 6\Delta_f G^{\circ}_{\mathrm{Ge_3Sb_2Te_6}} - 6\Delta_f G^{\circ}_{\mathrm{Ge_4Sb_2Te_7}}, \tag{11}$$

$$\Delta_{r(1)} S^{\circ} = S^{\circ}_{\mathrm{Ag_8GeTe_6}} + 5S^{\circ}_{\mathrm{Ge}} + 6S^{\circ}_{\mathrm{Ge_3Sb_2Te_6}} - 8S^{\circ}_{\mathrm{Ag}} - 6S^{\circ}_{\mathrm{Ge_4Sb_2Te_7}}, \tag{12}$$

$$\Delta_{r(2)} G^{\circ} = \Delta_f G^{\circ}_{\mathrm{Ag_8GeTe_6}} + 6\Delta_f G^{\circ}_{\mathrm{Ge_2Sb_2Te_5}} - 6\Delta_f G^{\circ}_{\mathrm{Ge_3Sb_2Te_6}}, \tag{13}$$

$$\Delta_{r(2)} S^{\circ} = S^{\circ}_{\mathrm{Ag_8GeTe_6}} + 5S^{\circ}_{\mathrm{Ge}} + 6S^{\circ}_{\mathrm{Ge_2Sb_2Te_5}} - 8S^{\circ}_{\mathrm{Ag}} - 6S^{\circ}_{\mathrm{Ge_3Sb_2Te_6}}, \tag{14}$$

$$\Delta_{r(3)} G^{\circ} = \Delta_f G^{\circ}_{\mathrm{Ag_8GeTe_6}} + 6\Delta_f G^{\circ}_{\mathrm{GeSb_2Te_4}} - 6\Delta_f G^{\circ}_{\mathrm{Ge_2Sb_2Te_5}}, \tag{15}$$

$$\Delta_{r(3)} S^{\circ} = S^{\circ}_{\mathrm{Ag_8GeTe_6}} + 5S^{\circ}_{\mathrm{Ge}} + 6S^{\circ}_{\mathrm{GeSb_2Te_4}} - 8S^{\circ}_{\mathrm{Ag}} - 6S^{\circ}_{\mathrm{Ge_2Sb_2Te_5}}, \tag{16}$$

$$\Delta_{r(4)} G^{\circ} = \Delta_f G^{\circ}_{\mathrm{Ag_8GeTe_6}} + 6\Delta_f G^{\circ}_{\mathrm{GeSb_4Te_7}} - 12\Delta_f G^{\circ}_{\mathrm{GeSb_2Te_4}}, \tag{17}$$

$$\Delta_{r(4)} S^{\circ} = S^{\circ}_{\mathrm{Ag_8GeTe_6}} + 5S^{\circ}_{\mathrm{Ge}} + 6S^{\circ}_{\mathrm{GeSb_4Te_7}} - 8S^{\circ}_{\mathrm{Ag}} - 12S^{\circ}_{\mathrm{GeSb_2Te_4}}, \tag{18}$$

$$\Delta_{r(5)} G^{\circ} = \Delta_f G^{\circ}_{\mathrm{Ag_8GeTe_6}} + 12\Delta_f G^{\circ}_{\mathrm{GeSb_6Te_{10}}} - 18\Delta_f G^{\circ}_{\mathrm{GeSb_4Te_7}}, \tag{19}$$

$$\Delta_{r(5)}S^{\circ} = S^{\circ}_{Ag_8GeTe_6} + 5S^{\circ}_{Ge} + 12S^{\circ}_{GeSb_6Te_{10}} - 8S^{\circ}_{Ag} - 18S^{\circ}_{GeSb_4Te_7}, \quad (20)$$

$$\Delta_{r(6)}G^{\circ} = \Delta_f G^{\circ}_{Ag_8GeTe_6} + 18\Delta_f G^{\circ}_{GeSb_8Te_{13}} - 24\Delta_f G^{\circ}_{GeSb_6Te_{10}}, \quad (21)$$

$$\Delta_{r(6)}S^{\circ} = S^{\circ}_{Ag_8GeTe_6} + 5S^{\circ}_{Ge} + 18S^{\circ}_{GeSb_8Te_{13}} - 8S^{\circ}_{Ag} - 24S^{\circ}_{GeSb_6Te_{10}}, \quad (22)$$

$$\Delta_{r(7)}G^{\circ} = \Delta_f G^{\circ}_{Ag_8GeTe_6} + 24\Delta_f G^{\circ}_{Sb_2Te_3} - 6\Delta_f G^{\circ}_{GeSb_8Te_{13}}, \quad (23)$$

$$\Delta_{r(7)}S^{\circ} = S^{\circ}_{Ag_8GeTe_6} + 5S^{\circ}_{Ge} + 24S^{\circ}_{Sb_2Te_3} - 8S^{\circ}_{Ag} - 6S^{\circ}_{GeSb_8Te_{13}}. \quad (24)$$

The entropy of formations of the $GeSb_8Te_{13}$ compound can be calculated as:

$$\Delta_f S^{\circ}_{GeSb_8Te_{13}} = S^{\circ}_{GeSb_8Te_{13}} - S^{\circ}_{Ge} - 8S^{\circ}_{Sb} - 13S^{\circ}_{Te}. \quad (25)$$

For $Ge_3Sb_2Te_6$, $Ge_4Sb_2Te_7$, $Ge_2Sb_2Te_5$, $GeSb_2Te_4$, $GeSb_4Te_7$, and $GeSb_6Te_{10}$ compounds, the corresponding reactions to determine $\Delta_f S^{\circ}$ can be written similar to Eq. (25) with their respective moles.

By combining Eqs. (8)–(25) and using data of the pure elements [23], Ag_8GeTe_6, and Sb_2Te_3 [23, 24], the standard Gibbs energy of formations of the layered tetradymite-like compounds of the homologous series nGeTe·mSb$_2$Te$_3$ were calculated to be:

$$\Delta_f G^{\circ}_{GeSb_8Te_{13}}/(\mathrm{kJ\,mol^{-1}}) = -(248.06 \pm 2.17) - (3.90 \pm 0.03) \times 10^{-3}\mathrm{T/K}, \quad (26)$$

$$\Delta_f G^{\circ}_{GeSb_6Te_{10}}/(\mathrm{kJ\,mol^{-1}}) = -(191.51 \pm 1.56) - (1.74 \pm 0.02) \times 10^{-3}\mathrm{T/K}, \quad (27)$$

$$\Delta_f G^{\circ}_{GeSb_4Te_7}/(\mathrm{kJ\,mol^{-1}}) = -(135.32 \pm 1.01) + (0.75 \pm 0.01) \times 10^{-3}\mathrm{T/K}, \quad (28)$$

$$\Delta_f G^{\circ}_{GeSb_2Te_4}/(\mathrm{kJ\,mol^{-1}}) = -(79.14 \pm 0.53) + (3.54 \pm 0.02) \times 10^{-3}\mathrm{T/K}, \quad (29)$$

$$\Delta_f G^{\circ}_{Ge_2Sb_2Te_5}/(\mathrm{kJ\,mol^{-1}}) = -(96.16 \pm 1.21) + (6.50 \pm 0.06) \times 10^{-3}\mathrm{T/K}, \quad (30)$$

$$\Delta_f G^{\circ}_{Ge_3Sb_2Te_6}/(\mathrm{kJ\,mol^{-1}}) = -(114.99 \pm 2.13) + (10.79 \pm 0.16) \times 10^{-3}\mathrm{T/K}, \quad (31)$$

$$\Delta_f G^{\circ}_{Ge_4Sb_2Te_7}/(\mathrm{kJ\,mol^{-1}}) = -(134.90 \pm 3.29) + (15.96 \pm 0.31) \times 10^{-3}\mathrm{T/K}. \quad (32)$$

The thermodynamic values of the saturated solid solutions of compounds in the Ag_8GeTe_6–Ge–$Ge_4Sb_2Te_7$–Sb_2Te_3 part are given in Table 3 and Fig. 3.

Table 3 Summary of the standard thermodynamic properties of layered tetradymite-like compounds of the GeTe–Sb_2Te_3 system at $T = 298$ K determined in this work

Phase	$-\Delta_f G^\circ$	$-\Delta_f H^\circ$	$T\Delta_f S^\circ$	S°	$\Delta_f S^\circ$
	kJ mol^{-1}			J mol^{-1} K^{-1}	
$GeSb_8Te_{13}$	251.96 ± 1.38	248.06 ± 2.17	3.90 ± 0.03	1051.80 ± 14.73	13.09 ± 0.09
$GeSb_6Te_{10}$	193.25 ± 1.28	191.51 ± 1.56	1.74 ± 0.02	805.02 ± 13.61	5.84 ± 0.04
$GeSb_4Te_7$	134.57 ± 1.18	135.32 ± 1.01	−0.75 ± 0.01	557.16 ± 6.48	−2.50 ± 0.02
$GeSb_2Te_4$	75.60 ± 1.09	79.14 ± 0.53	−3.54 ± 0.02	308.25 ± 3.35	−11.88 ± 0.06
$Ge_2Sb_2Te_5$	89.66 ± 1.21	96.16 ± 1.21	−6.50 ± 0.06	378.86 ± 3.34	−21.84 ± 0.21
$Ge_3Sb_2Te_6$	104.20 ± 1.38	114.99 ± 2.13	−10.79 ± 0.16	445.06 ± 5.91	−36.23 ± 0.53
$Ge_4Sb_2Te_7$	118.94 ± 1.57	134.90 ± 3.29	−15.96 ± 0.31	508.32 ± 8.99	−53.54 ± 1.03

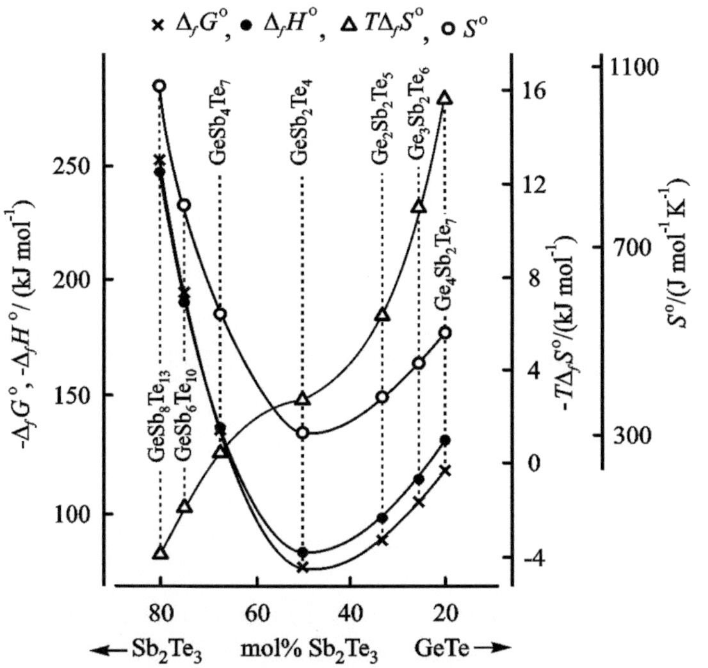

Fig. 3 Concentration changes of thermodynamic functions of layered tetradymite-like compounds of the GeTe–Sb_2Te_3 system

As can been seen in Fig. 3, at 50 mol% Sb_2Te_3, a composition that corresponds to $GeSb_2Te_4$ compound, the $\Delta_f G^\circ$, $\Delta_f H^\circ$, S° values is minimum, and the sign of the second derivative of $T\Delta_f S^\circ$ function changes. Gibbs energy, enthalpy, and entropy of the compounds are observed to increase with the addition of GeTe or Sb_2Te_3 to $GeSb_2Te_4$ composition. Results shown in Table 3 and Fig. 3 are in a good agreement with those data reported in [7–9] regarding the difference in the crystal structures of the compounds of the GeTe–Sb_2Te_3 system on either side of the $GeSb_2Te_4$. The unit cell of the $GeSb_2Te_4$ compound contains 3 seven-layer slabs across the c axis. According to [7], in GeTe-rich compounds' thermodynamic values increasing is accounted with more quantities of atomic layers in the packet and the number of the same type packets that determine the unit cell parameter c. For the Sb_2Te_3-rich compounds, increases are due to the number and alternation of 5- and 7-layer packets along the c axis.

Conclusions

1. The reproducibility of the values $E_{(1)\text{–}(7)}(T)$ of the ECCs during the heating and cooling cycles characterizes saturated solid solutions of the compounds of the GeTe–Sb_2Te_3 system as thermodynamically stable phases in the temperature range of 455–505 K.
2. The measured EMF values were applied to calculate the Gibbs energies, enthalpies, and entropies of silver-saturated solid solutions of the nGeTe·$m$$Sb_2Te_3$ (n, m = 1–4) compounds.
3. Concentration dependences of the thermodynamic values of compounds confirm the previously established difference between the crystalline structures of GeTe-rich and Sb_2Te_3-rich phases of the GeTe–Sb_2Te_3 system.

Acknowledgements This work was partially supported by the Ministry of Education and Science of Ukraine (grant No. 0119U002208). This work was also financially supported by the Academy of Finland project "Thermodynamic investigation of complex inorganic material systems for improved renewable energy and metals production processes" (Decision number 311537), as part of the activities of the Johan Gadolin Process Chemistry Center at Åbo Akademi University.

Conflict of Interest The authors declare that they have no conflict of interest.

References

1. Rogacheva EI, Nashchekina ON, Tavrina TV, YeO Vekhov, Sipatov AYu, Dresselhaus MS (2003) Non-stoichiometry in SnTe thin films and temperature instabilities of thermoelectric properties. Mater Sci Semicond Process 6(5–6):497–501
2. Rogacheva EI, Nashchekina ON, Sipatov AYu, Fedorov AG, Grigorov SN (2009) Growth mechanism and thermoelectric properties of PbSe/EuS superlattices. Phys Status Solidi C 6(5):1149–1153
3. Chu S, Majumdar A (2012) Opportunities and challenges for a sustainable energy future. Nature 488(7411):294–303
4. Tesfaye F, Moroz M (2018) An overview of advanced chalcogenide thermo-electric materials and their applications. J Electron Res Appl 2(2):28–41
5. Kuypers S, van Tendeloo G, van Landuyt J, Amelinckx S (1988) Electron microscopic study of the homologous series of mixed layer compounds $R_2Te_3(GeTe)_n$ (R = Sb, Bi). J Solid State Chem 76(1):102–108

6. Frangis N, Kuypers S, Manolikas C, Van Landuyt J, Amelinckx S (1989) Continuous series of one-dimensional structures in the compounds $Bi_{2+x}Se_3$, $Bi_{2+x}Te_3$, $Sb_{2+x}Te_3$, $(Bi_2Te_3)_nGeTe$ and $(Sb_2Te_3)_nGeTe$. Solid State Commun 69(8):817–819
7. Shelimova LE, Karpinskii OG, Zemskov VS, Konstantinov PP (2000) Structural and electrical properties of layered tetradymite-like compounds in the GeTe—Bi_2Te_3 and GeTe—Sb_2Te_3 systems. Inorg Mater 36(3):235–242
8. Shelimova LE, Karpinskii OG, Kretova MA, Kosyakov VI, Shestakov VA (2000) Homologous series of layered tetradymite-like compounds in the Sb-Te and GeTe-Sb_2Te_3 systems. Inorg Mater 36(8):768–775
9. Shelimova LE, Karpinskii OG, Konstantinov PP, Kretova MA, Avilov ES, Zemskov VS (2001) Composition and properties of layered compounds in the GeTe–Sb_2Te_3 system. Inorg Mater 37(4):342–348
10. Horichok I, Ahiska R, Freik D, Nykyruy L, Mudry S (2016) Phase content and thermoelectric properties of optimized thermoelectric structures based on the Ag-Pb-Sb-Te system. J Electron Mater 45(3):1576–1583
11. Abrikosov NK, Danilova-Dobryakova GT (1965) Sb_2Te_3-GeTe phase diagram. Izv Akad Nauk SSSR Neorg Mater 1(2):204–209 (in Russian)
12. Skums VF, Valevskii BL, Pashko VA (1985) Quantitative differential thermal analysis of phase relations in the GeTe-Sb_2Te_3 system. Zh Fiz Khim 59(9):2159–2162
13. Skoropanov AS, Valevsky BL, Skums VF, Samal GI, Vecher AA (1985) Physico-chemical study of Ge(Pb)Te-$Bi_2(Sb_2)Te_3$ system ternary compounds. Thermochim Acta 90:331–334
14. Karpinsky OG, Shelimova LE, Kretova MA, Fleurial J-P (1998) An X-ray study of the mixed-layered compounds of $(GeTe)_n$ $(Sb_2Te_3)_m$ homologous series. J. Alloys Compd 268(1–2):112–117
15. Karpinskii OG, Shelimova LE, Kretova MA, Fleurial JP (1998) Structural study of ternary layered compounds in the $(GeTe)_n{\cdot}(Bi_2Te_3)_m$ and $(GeTe)_n{\cdot}(Sb_2Te_3)_m$ homologous series. Inorg Mater 34(3):225–232
16. Moroz MV, Prokhorenko MV (2015) Phase relations in PbSe-PbTe alloys of the Ag-Pb-Se-Te system studied by EMF measurements. Inorg Mater 51(4):302–306
17. Moroz MV, Prokhorenko MV (2015) Determination of thermodynamic properties of saturated solid solutions of the Ag–Ge–Se system using EMF technique. Russ J Electrochem 51(7):697–702
18. Prokhorenko MV, Moroz MV, Demchenko PYu (2015) Measuring the thermodynamic properties of saturated solid solutions in the Ag_2Te-Bi-Bi_2Te_3 system by the electromotive force method. Russ J Phys Chem A 89(8):1330–1334
19. Shelimova LE, Karpinsky OG, Kretova MA, Avilov ES (1996) Phase equilibria in the Ge-Bi-Te ternary system at 570–770 K temperature range. J Alloys Compd 243(1–2):194–201
20. Abrikosov NKh, Stasova MM (1985) Solid solutions based on bismuth and antimony tellurides and bismuth selenides. Izv Akad Nauk SSSR Neorg Mater 21(12):2011–2015 (in Russian)
21. Moroz MV, Demchenko PYu, Prokhorenko MV, Reshetnyak OV (2017) Thermodynamic properties of saturated solid solutions of the phases Ag_2PbGeS_4, $Ag_{0.5}Pb_{1.75}GeS_4$ and $Ag_{6.72}Pb_{0.16}Ge_{0.84}S_{5.20}$ of the Ag-Pb-Ge-S system determined by EMF method. J Phase Equilibria Diffus 38(4):426–433
22. Karapetyans MKh (1953) Chemical thermodynamics. Goskhimizdat, Moscow (in Russian)
23. Barin I (1995) Thermochemical data of pure substance. VCH, Weinheim
24. Babanly, MB, Yusibov, YA, Babanly, NB (2011) The EMF method with solid-state electrolyte in the thermodynamic investigation of ternary copper and silver chalcogenides. In: Kara S (ed) Electromotive force and measurement in several systems. InTech, pp 57–78
25. Voronin MV, Osadchii EG (2013) Thermodynamic properties of silver and bismuth sulfosalt minerals, pavonite ($AgBi_3S_5$) and matildite ($AgBiS_2$) and implications for ore deposits. Econ Geol 108(5):1203–1210
26. Aliev ZS, Musayeva SS, Imamaliyeva SZ, Babanly MB (2017) Thermodynamic study of antimony chalcoiodides by EMF method with an ionic liquid. J Therm Anal Calorim 133:1115–1120

27. Voronin MV, Osadchii EG, Brichkina EA (2017) Thermochemical properties of silver tellurides including empressite (AgTe) and phase diagrams for Ag–Te and Ag–Te–O. Phys Chem Miner 44(9):639–653
28. Babanly NB, Orujlu EN, Imamaliyeva SZ, Yusibov YA, Babanly MB (2019) Thermodynamic investigation of silver-thallium tellurides by EMF method with solid electrolyte Ag_4RbI_5. J Chem Thermodyn 128:78–86
29. Imamaliyeva SZ, Musayeva SS, Babanly DM, Jafarov YI, Taghiyev DB, Babanly MB (2019) Determination of the thermodynamic functions of bismuth chalcoiodides by EMF method with morpholinium formate as electrolyte. Thermochim Acta 679:178319
30. Robinel E, Carette B, Ribes M (1983) Silver sulfide based glasses (I): glass forming regions, structure and ionic conduction of glasses in GeS_2–Ag_2S and GeS_2–Ag_2S–AgI systems. J Non-Cryst Solids 57(1):49–58
31. Moroz MV, Demchenko PYu, Prokhorenko SV, Moroz VM (2013) Physical properties of glasses in the Ag_2GeS_3-AgBr system. Phys Solid State 55(8):1613–1618
32. Osadchii EG, Rappo OA (2004) Determination of standard thermodynamic properties of sulfides in the Ag-Au-S system by means of a solid-state galvanic cell. Am Mineral 89(10):1405–1410
33. Moroz M, Tesfaye F, Demchenko P, Prokhorenko M, Lindberg D, Reshetnyak O, Hupa L (2018) Determination of the thermodynamic properties of the $Ag_2CdSn_3S_8$ and Ag_2CdSnS_4 phases in the Ag–Cd–Sn–S system by the solid-state electrochemical cell method. J Chem Thermodyn 118:255–262
34. Diffractometer. Stoe WinXPOW, Version 3.03 (2010) Stoe Cie GmbH Darmstadt
35. Kraus W, Nolze G (1996) POWDER CELL—a program for the representation and manipulation of crystal structures and calculation of the resulting X-ray powder patterns. J Appl Crystallogr 29:301–303
36. Downs RT, Hall-Wallace M (2003) The American Mineralogist crystal structure database. Am Mineral 88(1):247–250
37. Villars P, Cenzual K (eds) (2014) Pearson's crystal data: crystal structure database for inorganic compounds, release 2014/15. ASM International, Ohio, USA
38. Tesfaye F, Taskinen P (2014) Electrochemical study of the thermodynamic properties of matildite (β-$AgBiS_2$) in different temperature and compositional ranges. J Solid State Electrochem 18(6):1683–1694
39. Moroz M, Tesfaye F, Demchenko P, Prokhorenko M, Lindberg D, Reshetnyak O, Hupa L (2018) Thermodynamic properties of magnetic semiconductors $Ag_2FeSn_3S_8$ and Ag_2FeSnS_4 determined by the EMF method. In: Lambotte G, Lee J, Allanore A, Wagstaff S (eds) Materials processing fundamentals 2018. Springer, New York, pp 87–98
40. Moroz M, Tesfaye F, Demchenko P, Prokhorenko M, Lindberg D, Reshetnyak O, Hupa L (2019) Thermal stability and thermodynamics of the Ag_2ZnGeS_4 compound. In: Lambotte G, Lee J, Allanore A, Wagstaff S (eds) Materials processing fundamentals 2019. Springer, New York, pp 215–226
41. Blachnik R, Gather B (1978) Mischungen von GeTe, SnTe und PbTe MIT Ag_2Te lin beitrag zur klärung der konstitution der ternären Ag-IVb-Te systeme (IVb = Ge, Sn, Pb). J. Common Met. 60(1):25–32
42. Wu H-J, Chen S-W (2011) Phase equilibria of Ag–Sb–Te thermoelectric materials. Acta Mater 59(16):6463–6472
43. Shelimova LE, Karpinskii OG, Svechnikova TE, Nikhezina IYu, Avilov ES, Kretova MA, Zemskov VS (2008) Effect of cadmium, silver, and tellurium doping on the properties of single crystals of the layered compounds $PbBi_4Te_7$ and $PbSb_2Te_4$. Inorg Mater 44(4):371–376

Author Index

J. Lee et al. (eds.), *Materials Processing Fundamentals 2020*, The Minerals, Metals & Materials Series, https://doi.org/10.1007/978-3-030-36556-1

Subject Index

J. Lee et al. (eds.), *Materials Processing Fundamentals 2020*, The Minerals, Metals & Materials Series, https://doi.org/10.1007/978-3-030-36556-1

MIX
Papier aus verantwortungsvollen Quellen
Paper from responsible sources
FSC® C105338

If you have any concerns about our products,
you can contact us on
ProductSafety@springernature.com

In case Publisher is established outside the EU,
the EU authorized representative is:
Springer Nature Customer Service Center GmbH
Europaplatz 3, 69115 Heidelberg, Germany

Printed by Libri Plureos GmbH
in Hamburg, Germany